Advanced Ceramic Materials

Edited by Mohsen Mhadhbi

Published in London, United Kingdom

IntechOpen

Supporting open minds since 2005

Advanced Ceramic Materials
http://dx.doi.org/10.5772/intechopen.87703
Edited by Mohsen Mhadhbi

Contributors
Mohsen Mhadhbi, Faïçal Khlissa, Chaker Bouzidi, Abdulkarim Ziedan Khalf, Mohamed Guendouz, Alexandra Bourdot, Djamila Boukhelkhal, Oussama Babachikh, Amine Hamadouche, Alexey Vereschaka, Sergey Grigoriev, Marina Volosova, Nikolay Sitnikov, Filipp Milovich, Nikolay Andreev, Catherine Sotova, Hengchang Nie, Genshui Wang, Xianlin Dong, Touseef Para, Shaibal Kanti Sarkar, Dongmian Zang, Xiaowei Xun, Le Dai Vuong, Vo Thanh Tung, George Kordas, Upsorn Boonyang, Namon Hirun, Reedwan Bin-Zafar Auniq, Ravinder Dachepalli, N. Maramu, S. Katlakunta, Vladimir Solomonov, Vladimir Osipov, Vecheslav Platonov, Vladislav Shitov, Phan Dinh Gio, Rapolu Sridhar, J. Laxman Naik, K. Vijaya Kumar

First published in London, United Kingdom, 2021 by IntechOpen
IntechOpen is the global imprint of INTECHOPEN LIMITED, registered in England and Wales, registration number: 11086078, 5 Princes Gate Court, London, SW7 2QJ, United Kingdom
Printed in Croatia

British Library Cataloguing-in-Publication Data
A catalogue record for this book is available from the British Library

Additional hard and PDF copies can be obtained from orders@intechopen.com

Advanced Ceramic Materials
Edited by Mohsen Mhadhbi
p. cm.
Print ISBN 978-1-83881-204-1
Online ISBN 978-1-83881-212-6
eBook (PDF) ISBN 978-1-83881-213-3

For EU product safety concerns:
IN TECH d.o.o., Prolaz Marije Krucifikse Kozulić 3, 51000 Rijeka, Croatia, info@intoohopen.com or visit our website at intechopen.com.

Meet the editor

Dr. Mohsen Mhadhbi is an assistant professor at the National Institute of Research and Physicochemical Analysis, Ariana, Tunisia. He obtained an MSc in Physics and a Ph.D. from the Faculty of Sciences of Sfax, Tunisia. His research areas include nanomaterials, nanotechnology, ceramics, composites, powder technology, and modeling. Dr. Mhadhbi has supervised several researchers and authored several research papers in national and international journals and books. He has also presented numerous papers at national and international conferences and seminars. He is a member of various associations and journal editorial boards.

Contents

Preface

This book provides an overview of theory, methods of synthesis, and applications of ceramic materials based on recent and innovative works. This book also provides a detailed description of synthesis methods, properties, and advanced applications of ceramics. Ceramic materials have proved to be advantageous for researchers and industries.

Advanced Ceramic Materials, which covers a wide area, helps students, academics, engineers, scientists, and other interested readers to understand the value of ceramics.

The book is divided into six sections. Section 1 presents an introduction to the latest developments in the field of ceramic materials. Section 2 focuses on the various types of ceramics materials and Section 3 presents different fabrication processes for ceramic materials. Section 4 presents the proprieties of ceramics. Section 5 discusses the various applications of ceramic materials, and Section 6 describes the challenges in Rietveld refinement and structure visualization.

This book is a valuable reference for all those working in the field of materials science, especially practicing engineers in materials processing and manufacturing technologies.

The editor would like to warmly thank all the authors for their contributions to this book. Thanks also to the whole team at IntechOpen for their support and help in putting the book together.

All suggestions for the improvement of this book will be gratefully acknowledged.

Mohsen Mhadhbi
National Institute of Research and Physical-Chemical Analysis,
Technopole Sidi Thabet 2020 Ariana,
Tunsia

Section 1

Introduction

Chapter 1

Recent Advances in Ceramic Materials for Dentistry

Mohsen Mhadhbi, Faïçal Khlissa and Chaker Bouzidi

Abstract

Dental ceramics constitute a heterogeneous group of materials with desirable optical and mechanical proprieties combined with chemical stability. They are inorganic non-metallic materials used in several applications. These materials are biocompatible to tissue, highly esthetic, with satisfying resistance to tensile and shear stress. Over the past years, several developments in new ceramic materials in dental restoration were achieved, including processing techniques and high mechanical properties. Thus, concepts on the structure and strengthening mechanisms of dental ceramic materials are also discussed. The dental practitioner requires best knowledge concerning indications, limitations, and correct use of started materials. The purpose of this book chapter is to overview advances in new ceramic materials and processes, which are used in dentistry. The properties of these materials are also discussed.

Keywords: ceramics, dentistry, oxide ceramics, glass ceramics, zirconia, alumina, silicate, composites

1. Introduction

Nowadays, ceramics are applied in a wide range of industrial applications such as cutting tools, nuclear reactors, automotive, biomedical, etc. Therefore, ceramics have been the subject of considerable researches because their chemical stability and good mechanical properties compared to many other materials. Physical and mechanical tests, combined with fundamentals of engineering science, were made to structure designs and restorative materials. However, although these materials are esthetically attractive and resistant to wear, they are prone to brittle fracture at stress concentrations.

Dental ceramics are mainly consisting of glass ceramics enriched with lithium disilicate, leucite, polycrystals of zirconia and alumina, etc. [1–4].

Additionally, dental ceramics, available for dental castings, were first developed in early 1960s, increasing demands for ceramic restorations materials with respect to both esthetic and mechanical properties.

In 1800s, the long process of moving ceramics from a craft to a science based technology was underway under the direction of engineers [5].

In 1830s, a new composition made from 78% feldspar, 15% kaolin, dehydtared borax, and potash silicate was developed [6]. Thus, the surface of this ceramic was glazed using a glossary glass forming materials.

In 1889s, the utilization of ceramics to build crowns was patented as the all ceramic "jacket crown" [7]. These crowns were often failed because of internal micro-cracking, but, in 1950s, they are utilized considerably until the invention of porcelain fused to metal (PFM) crowns.

In the early of 1980s, the first major breakthrough in full ceramic restoration with the application of computer-aided technologies [8].

Therefore, several works have been made in the field of dentistry by using of computer-aided digitizing (CAD), computer-aided design (CAD), and computer-aided manufacturing (CAM) [9, 10].

The purpose of this chapter is to provide an overview of the recent progress of new materials and processing methods for dentistry.

2. Classification of ceramic materials

Ceramic materials are utilized for several dental applications and are distinguished by their good mechanical properties, high electrical resistance, high thermal conductivity, and excellent biocompatibility. Thus, the oxides, particularly alumina, zirconia, and silica are the most commonly used ceramic materials in the area of dentistry. These materials are classified based on their chemical compositions or based on processing methods.

2.1 Based on chemical compositions

In the past decades, the ceramic materials have attired much attention due to their excellent properties depend to their chemical composition. However, several researches have been studied to develop nanoceramic materials, will be further expanded in future.

2.1.1 Silicon oxide ceramics

Silicon oxides ceramics have been widely employed in biomedical applications because its mechanical stability, biocompatibility, and high specific surface, which can be modified [11–15]. The silanol group on the support of the silicon atom can be activated to make a chemical bond with organosilane, which can also lead to providing various functional groups that can mediate a vast selection of particular bioconjugation strategies [16]. When stable silanes layers are formed on the silicon surface conventional bioconjugation process are used to physisorb or chemisorb a broad bioactive nanoparticles and molecules on the silicon surface.

Zhang et al. [15] observed a reduction of 90% in albumin adsorption on silicon surfaces by 0.05% Tween 20 over 4 h. The self-assembly of polyethylene glycol (PEG) with monomethoxypoly(ethylene glycol) (MPEG) have been used with great success for functionalization of silicon surfaces and for suppressing the adsorption of platelets, fibroblasts, and Hela cells. The water contact angles of the different silicon surfaces are showed in **Table 1**. The maximum value was found for the case of the methylated silicon surface with 2% of dichlorodimethylsilane (in the range of 99–102 °).

Porous silicon (pSi) is a biocompatible and biodegradable material due to its high surface area, which induces a fast oxidation of silicon in aqueous solution [17–19]. Hence, it is shown that particles synthesized out of pSi are biodegradable in plasma, blood, and tissue and then stable [20].

Additionally, the internalization of pSi particles by endothelial cells and macrophages in vivo and in vitro with no adverse effects associated to particles partitioning and cell proliferation [21]. The controlled release of cytokine is near to that of controls, showing that pSi particles are also non-immunogenic. Hence, no toxicity has been revealed in healthy receiving several injections of these pSi particles [22].

Surface	Dominating surface groups	Wetability (°)
Hydrophilic silicon	O^- and Si-OH	<10
Intermediate silicon (with HF)	Si-H and Si-OH	55–60
Intermediate methylated silicon (with 0.008% DDS)	Si-CH_3 and Si-OH	55–60
Methylated silicon (with 2% DDS)	Si-CH_3	99–102

Table 1.
The water contact angles of different surfaces [15].

Indeed, polymers coatings have been employed to coat pSi particles to protect them from cellular degradation although the conjugation of antibodies has promoted the efficient delivery of payloads [23].

2.1.2 Aluminum oxide ceramics

Bioceramics (like alumina, zirconia, etc.) are mainly employed in orthopedic and dental reparation. Alumina (aluminum oxide) is the only solid oxide form of aluminum (Al_2O_3). Thus, corundum is the crystalline form of alumina.

Alumina was first used since the 1970s and its clinical results revealed a fracture rate greater than 13% [24]. However, the disadvantage of these materials was related to the fact that they could not be processed to full final density. In the late of 1980s, a second generation ceramic materials, with a smaller grain sizes and a higher density, was developed. The fracture rate of these materials was less than 5% [25]. Today, a third generation ceramic materials, characterized by high purity, full density, and finer microstructure was appeared. The properties of biomedical grade alumina are illustrated in **Table 2**.

Additionally, it is shown that the degree of tensile bending strength of ultrafine Al_2O_3 particles is remarkably over that of all other ceramics [27]. The ceramics for substructures of "jacket crowns" enriched by alumina (up to 60% of weight) of different grain size (10–30 μm) have been used to increase the stability. Hence, intense refraction of light takes place at the alumina (in the feldspar) due to the difference in the refraction index between feldspar and corundum.

2.1.3 Aluminum oxide ceramics reinforced with zirconium oxides

Mechanical properties of alumina were improved by addition of ceramic composites, as reinforcing agents, like zirconia. Generally, these ceramic–ceramic composites present a great hardness as compared to the all composites. Although most ceramic second phases improve strength and hardness they modestly improve fracture toughness [28, 29]. Al_2O_3-SiC nanocomposite has been reported to have the

Property	Value
Density (g/cm^3)	3.96–3.98
Fracture toughness ($MPa.m^{1/2}$)	3.4–4
Bending strength (MPa)	550–630
Hardness (GPa)	19–20
Young's modulus (GPa)	380

Table 2.
Mechanical properties of alumina [26].

most improved properties [30]. Thus, it has been shown that SiC increases significantly the wear resistance of aluminum oxide.

Doğan and Hawk [31] revealed that the toughness of alumina with 34 vol%SiC increased from 3.4 to 4.6 MPa.m$^{1/2}$. Similarly, Belmonte et al. [32] showed that the fracture toughness of the sample of alumina with 20 vol%SiC reached a value of 5.9 MPa.m$^{1/2}$.

However, the zirconia system uses a mixture of zirconium oxide and aluminum oxide as a framework to achieve a marked increase in the flexural strength. Alumina constitute approximately two third of the structure and the remaining structure was composed of tetragonal zirconia. In addition, the proportion of glass phase covers 20–25% of the total structure. The increase over alumina is due to the zirconia particles that protect the structure against crack propagation. It has a very high strength of around 700 MPa and very poor translucency.

Tuan et al. [33] incorporated zirconia particles into alumina and reported that the fracture toughness was improved. For zirconia-toughened alumina including 10 vol% zirconia, they recorded that fracture strength and fracture toughness were 943 MPa and 11.8 MPa.m^{½}, respectively. Toughness values of 10 MPa.m^{½} for 10 vol% zirconia [34] and 7.02 MPa.m^{½} for 50 vol% zirconia content have also been reported [35]. Zirconia is a bioinert ceramic and can suffer from low cellular attachment, which could be compensated when mixed with biopolymers [36]. Alumina-zirconia composites have received great attention in dentistry as promoted bioceramics due to their excellent biocompatibility [37].

In the last years, many recent studies were focused on the investigation of the tribological-mechanical behaviors and biosafety of alumina toughened zirconia (ATZ) composites [38–41]. Thus, the benefits of these composites are the combination of the properties of alumina and zirconia.

Daskalova et al. [42] studied the effect of surface modification by femtosecond laser on zirconia based ceramics for screening of cell-surface interaction. The X-ray diffraction analysis demonstrated preservation of the tetragonal phase of Zr ceramic materials for a particular fs-laser treatment conditions (see **Figure 1**).

Moreover, scaffolds design and fabrication are major areas in dentistry for tissue engineering applications that need controlled positioning of cells on solid substrates with predefined orientation. Hence, surface functionalization generated by defined surface structure was strongly depending on the quality and surface texture.

2.1.4 Zirconium oxides ceramics

Zirconia (ZrO_2) is a ceramic material which has been applied in the health field and distinguished by its high mechanical properties, biocompatibility, and chemical stability [43]. The polycrystal tetragonal zirconia, stabilized with yttria (3Y-TPZ) contains 3 mol% yttria oxide (Y_2O_3), was first applied in the field of medical. The 3Y-TPZ has been the most studied and utilized in dentistry [44]. Thus, the 3Y-TPZ was fabricated in small grains (0.2–0.5 mm in diameter), which minimizes the phenomenon of structural deterioration or destabilization in the presence of saliva, decreasing the subcritical crack growth [45]. **Figure 2** shows the SEM micrograph of the powder after sintering. However, for the formation of a great amount of monolclinic zirconia a powerful machining should be used because of the high compression applied by machining, leading to the formation of micro-cracks on the surface of material [3].

Similar to that of stainless steel, zirconia is characterized by good chemical stability, good biocompatibility, mechanical strength, toughness, and Young's modulus [46]. No adverse reactions have been found, when osteoblasts were seeded on zirconia and were able to proliferate and differentiate on it [47]. Zirconia ceramics are becoming a prevalent biomaterial in dentistry and dental implantology [48].

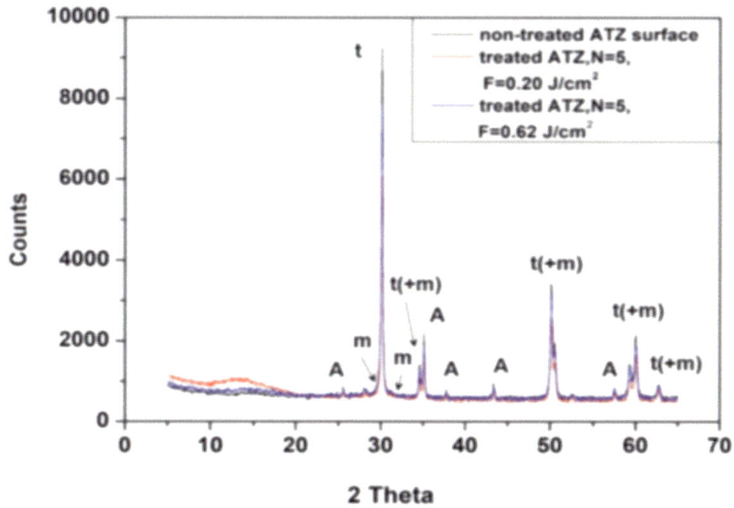

Figure 1.
XRD patterns of the surface of sample [42].

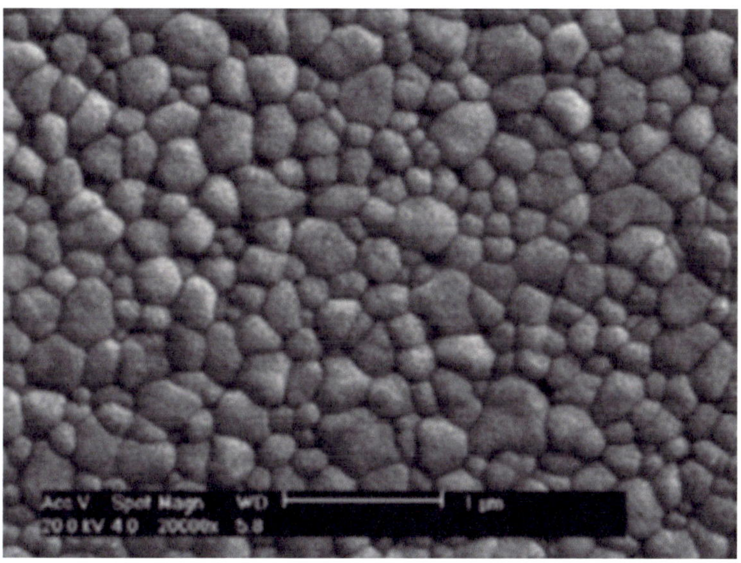

Figure 2.
SEM micrograph of the sample obtained after sintering [45].

2.1.5 Hybrid ceramics

In recent years, all ceramic and composite restorations have been widely used because of their biocompatibility and esthetic features compared with metal-ceramic restorations (MCR) [49, 50]. A range of ceramic systems are commercially available like leucite, alumina, zirconia, and feldspar based ceramics [51]. Thus, many indirect composites categories, with various size of filler particles, are also used [52].

Recently, new ceramic/polymer materials, used in CAD/CAM technology, have been developed [53]. In fact, the use of mixed materials enhances crack propagation and reduces fracture stress. The protocol for glass–ceramic materials requires acid etching followed by silanization and the application of resin cement [54–56]. Silane coupling agent is a molecule having two functional groups, which has the ability to form a durable bond between organic and inorganics materials [56, 57].

The adhesive cementation technique improves the clinical performance of all-ceramic dental restorations due to the ability of resin to penetrate the microporosities that are created by etching process [58, 59]. However, the indirect restorations with composite, such as resins, can be produced by the application of physical conditioning technique using air-particle abrasion and silanization to achieve optimum adhesive bond between different materials [60].

Recent studies [61] showed that hardness of hybrid ceramic materials was given from the ceramic content because the indenter was highly sensitive to making this portion. Moreover, there was a significant interaction in the interface between resin cements and hybrid ceramic materials. The longevity of restorations can be affected by storage because of the high concentration of the water and smaller molecules, which cause a reduction in free spaces between functional groups and polymer chains [62].

New hybrid ceramics filled and un-filled polyamide 12 (PA 12) were developed by a fused deposition modeling framework [63]. The proprieties of hybrid ceramics filled and un-filled polyamide 12 are summarized in **Table 3**. The highest tensile strength was recorded at 40% filled PA 12 as compared to un-filled PA 12. The highest tensile modulus was recorded at 35% filled PA 12 as compared to un-filled PA 12. Indeed, the highest impact strength was recorded at 35% filled PA 12 as compared to un-filled PA 12.

2.2 Based on processing methods

In the last few decades, there have been remarkable advances in the mechanical properties and methods of fabrication of ceramic materials.

2.2.1 Casting

Casting is based on the solidification of a fluid that has been poured or injected into a mold. The final product is also known as a casting. Thus, casting process consists of three steps: melting, casting, and recovery.

The biocompatibility effects of indirect exposure of base-metal dental casting alloys were analyzed [64].

The effects of the rare earth element lanthanum on the metal-ceramic bond strength of Co-Cr alloys prepared by casting were studied [65]. XRD and SEM analysis of the samples revealed the presence of dendritic microstructures with some defects and an island shaped intermetallic compounds rich in Cr and Mo. The increasing of the number of "La" leads to the increasing of the diffusion layer at the interface, the increasing of thickness of the native oxide layer, and to improving the wettability. In addition, the results showed that the debonded surfaces of the specimens exhibit mixed fracture modes (adhesive and cohesive failure).

Samples	Tensile strength (MPa)	Tensile modulus (MPa)	Impact strength (kJ/m^2)
Un-filled PA 12	41.38 ± 2.93	1006.28 ± 101.66	6.02 ± 2.51
30% filled PA 12	36.82 ± 1.61	1087.08 ± 126.89	11.92 ± 1.49
35% filled PA 12	36.71 ± 1.81	1382.34 ± 89.21	16.96 ± 7.01
40% filled PA 12	36.99 ± 1.41	1327.06 ± 157.62	12.42 ± 2.96

Table 3.
Proprieties of hybrid ceramics filled and un-filled polyamide 12 [63].

Similarly, Atwood et al. [66] modeled the surface contamination of dental titanium produced by casting. They showed that the contamination of the wedge sample was established to extended range from 30 to 120 mm. Hence, they concluded that the addition of micro- and nano-models revealed the predictions are shown to be in good agreement for the pattern of contamination. **Figure 3** shows the image of the mold metal interface, which was characterized by three layers: (i) irregular contact surface and with topography in the scale of 20 μm, (ii) globular structure, and (iii) dendritic structure.

2.2.2 Sintering

Sintering is a heat treatment under pressure applied to a powders compact without melting. The final product is a solid or porous mass with excellent properties.

Fan et al. [67] studied the mechanical properties of sintering temperature on the microstructure of dental zirconia-toughened alumina (ZTA). By increasing temperature, they concluded that the mechanical properties of the samples were improved, the crystal structure of ZrO_2 was changed (from tetragonal into monoclinic), and the porosity was decreased. However, the ceramics sintered at 1450 °C showed greatest fracture toughness (5.23 $MPa.m^{1/2}$) and greatest flexural strength (348 MPa). The authors concluded that the properties of ZTA ceramic depend on sintering temperature, and the optimal temperature was about 1200–1250 °C. Ghayebloo et al. [68] revealed that it is possible to fabricate ZLS glass–ceramics by sintering (**Figure 4**). The results showed a highest flexural strength of 255.10 ± 15.44 MPa, a fracture toughness of 3.15 ± 0.62 $MPam^{1/2}$, a Vickers microhardness of 7.96 ± 0.13 GPa, and a bulk density of 2.63 ± 0.02 g/cm^3. Thus, the lowest water absorption was of 0.11 ± 0.12 and the apparent porosity was of 0.25 ± 0.32.

2.2.3 Partial sintering

Partial sintering is considered as the most straightforward processing route for macro-porous scaffolds and involves the partial sintering of initially porous powder compacts.

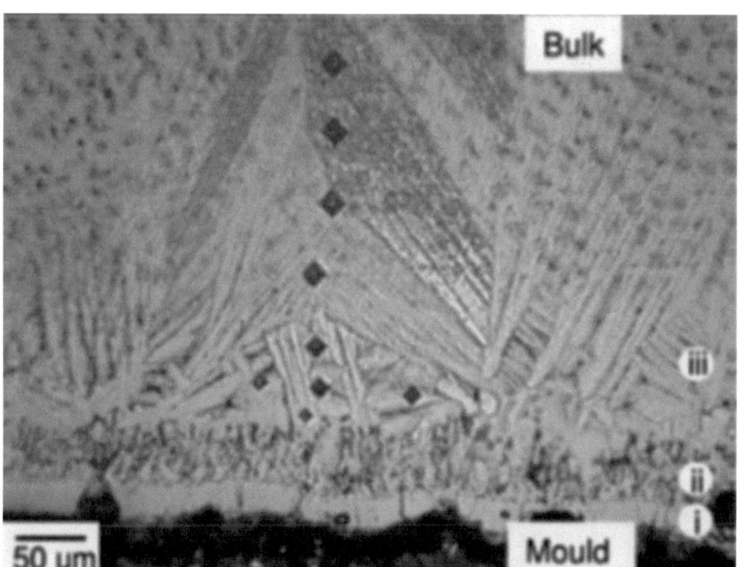

Figure 3.
Light micrograph of the bulk sample [66].

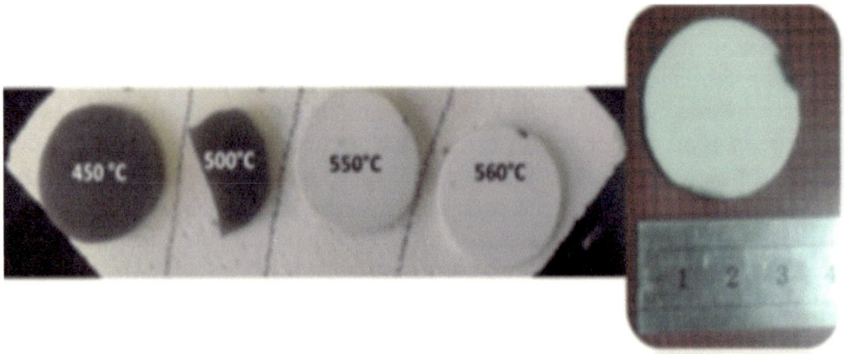

Figure 4.
The final samples obtained by spark plasma sintering [68].

A homogenous although closed pore structure can be produced when sintering is terminated before full densification [69]. The pore size and porosity are controlled by the size of the powder particles and the degree of partial sintering, and the size of the raw powder should generally be 2–5 times larger than the desired pore size. Chen et al. [70] studied the properties of YB_2C_2 ceramics prepared by partial sintering. Thus, a porous YB_2C_2 ceramics were prepared by partial sintering. The results showed a good mechanical behavior: high porosity (57.17–75.26%) and a high compressive strength (9.32–34.78 MPa). In another work [71], alumina powder agglomerates were prepared by partial sintering. The SEM micrographs of the porous ceramic obtained after sintering are given in **Figure 5**. The final ceramic material is characterized by a hierarchical porous network that can contain three levels of interconnected pores: the voids existing between the agglomerates (≥10 μm in size), the porosity remaining inside the agglomerates after partial sintering (≈100–1000 nm in size), and the pores that may exist within the initial ceramic particles (<100 nm in size). Such porous ceramic structures are developed to be applied in the field of dentistry.

2.2.4 Glass infiltration

The glass infiltration processing is a powerful technique for the fabrication of ceramic/glass composite with exceptional mechanical properties and low shrinkage.

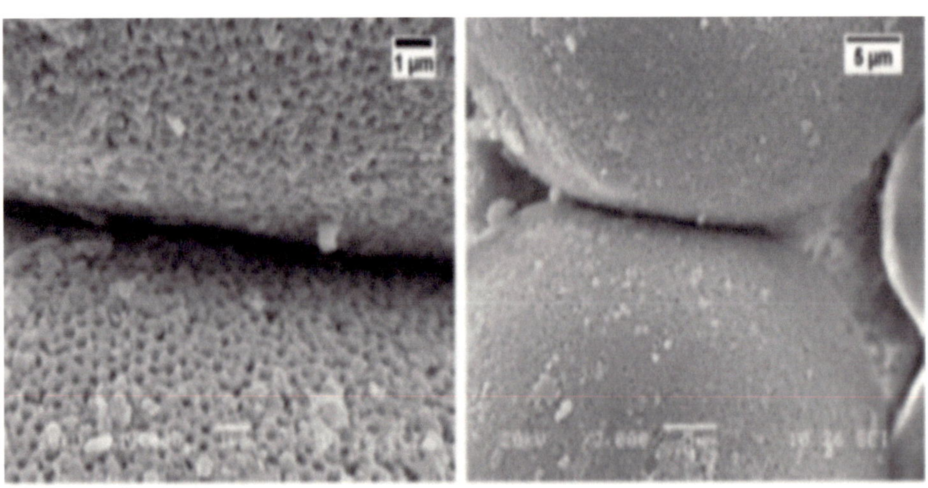

Figure 5.
SEM micrographs of the porous ceramic after sintering [71].

Porous Y-TZP nano-ceramics, with hierarchical heterogeneities, were prepared by partial sintering method from meso-porous powder [72]. The results showed that the products have a crystallite sizes between 34 and 71 nm for relative densities between 54 and 81.7%. They also revealed a surface area of 18 m^2/g, a thermal conductivity of 0.63–1.88 $W.m^{-1}.K^{-1}$, an elastic modulus of 32–156 GPa, and a strength in the range of 70 and 540 MPa.

Yang et al. [73] investigated the effects of process parameters and material characteristics in glass infiltration of gel cast zirconia-toughened alumina (ZTA) ceramic for dental applications. They showed that the strength of the obtained ceramic was 291 MPa and the shrinkage was 1.8548%.

In another work [74], biocomposites were obtained by infiltrating porous alumina-titania (Al_2O_3-TiO_2) substrates with a lanthania-rich (La_2O_3) glass. Al_2O_3-TiO_2 substrates were fabricated using high energy milled powder mixtures of two different compositions. The sintered substrates presented α-Al_2O_3 and β-Al_2TiO_5 as crystal phases and relative densities ranging between 65.5 ± 2 and 69.4 ± 1.2%. These products were then infiltrated by lanthania containing glass at a higher temperature (1140 °C) for 2 hours. These ceramics showed a fracture toughness up to 2.6 $MPa.m^{1/2}$, a fracture strength in the order of 218–254 MPa, a high density of 94–99% (**Figure 6**), and a Vickers hardness in the order of 895–1036 HV. However, phase identification of the samples by XRD indicated the decomposition of aluminum titanate into alumina and titania besides the formation of lanthanum borosilicate ($LaBSiO_5$). In addition, all studied compositions presented non-cytotoxic behavior and low chemical solubility (inferior to 75 $\mu g/cm^2$).

2.2.5 Slip casting and sintering

In-Ceram zirconia bulk composites were synthesized via slip casting of alumina or zirconia. Slip was a dispersion of particles of ceramic powders in a liquid (such as water). Thus, the pH of water was then regulated to the desired value to charged particles.

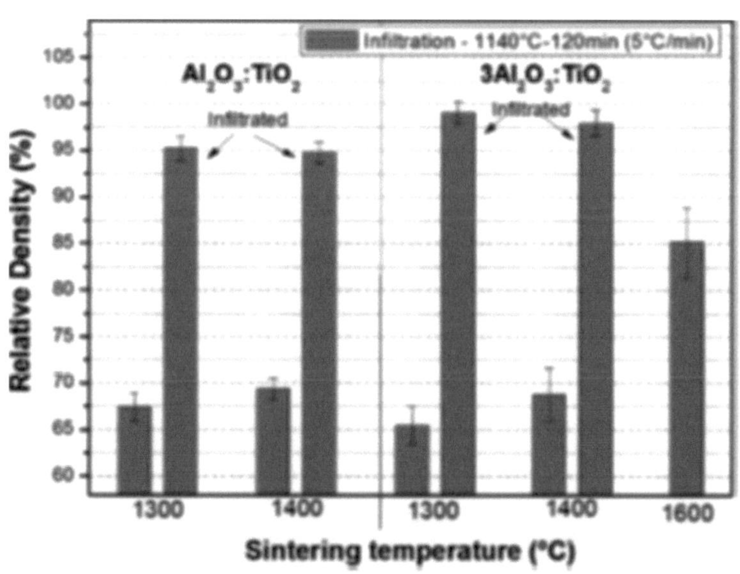

Figure 6.
Variation of relative density with sintering temperature of the Al_2O_3-TiO_2 and $3Al_2O_3$-TiO_2 composites after glass infiltration [74].

Kim et al. [75] fabricated dense zirconia compacts by slip casting and sintering from zirconia nanopowders. Thus, the green compacts obtained from slip casting were cold isostatic pressed to enhance the close packing and densified by sintering at 1450 °C for 2 h. Highly dense zirconia compacts with a relative density of 99.5% and grain size of 350 nm were obtained based on the powder type and solid loading in the slurry. The microstructure and mechanical hardness of the sintered specimen after slip casting were dependent on the yttria content in the 3 mol% yttria-stabilized tetragonal zirconia polycrystal powder and the solid loading within the slurry.

Additionally, Kim et al. [76] prepared dental zirconia implants by sintering. They showed that the zirconia blocks have many surface cracks that lead to the deterioration of mechanical strength and the failure of the implant in the body. Thus, highly dense 3Y-TZP samples with a relative density of 99% and grain size of 200–400 nm were obtained at a solid loading of 50–65 wt%. Recently, removable partial dentures (RPD) cobalt-chromium (Co-Cr) alloys are fabricated using a casting technique [77]. New additive manufacturing processes based on laser-sintering has been developed for quick fabrication of RPD metal frameworks at low cost. **Figure 7** illustrates the SEM micrograph of the fractured surface of Co-Cr alloy after casting. As can be seen, the Co-Cr alloy exhibited smaller grain size, higher microstructural homogeneity, and low porosity (2.1–3.3%). It has been shown that laser sintered alloys are more precise and present better mechanical and fatigue properties than cast alloys for RPD.

2.2.6 Hot isostatic pressing

For a decade, hot isostatic pressing (HIP) has been used successfully by manufacturers around the world to increase productivity. HIP was used to eliminate pores and remove casting defects (such as oxides and carbides) to dramatically increase the material properties.

Gionea et al. [78] synthesized zirconia powders by HIP at 500 °C for 2 h. The results showed that a pure cubic phase, with average particle dimension

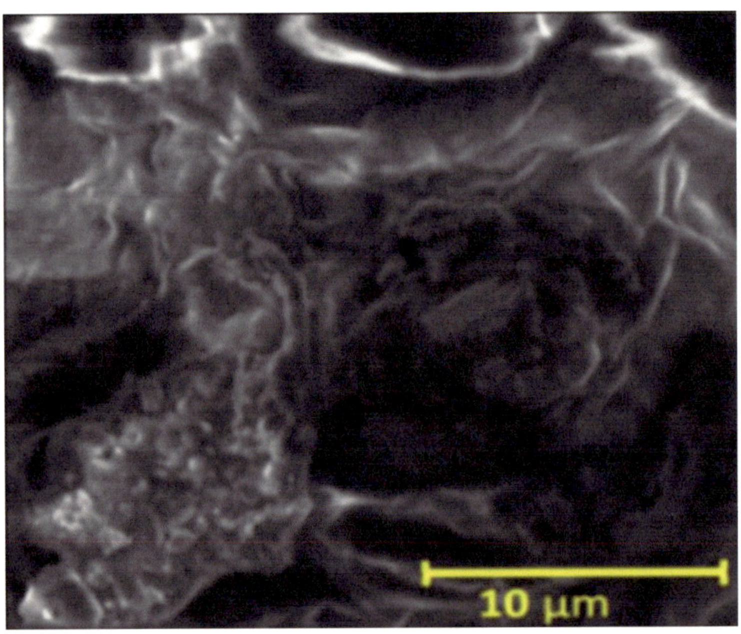

Figure 7.
SEM image of the fractured surface of Co-Cr alloy after casting [77].

about 70 nm, was obtained. Thus, the obtained samples presented a mixture of monoclinic-tetragonal or monoclinic-cubic phases. Final dense ceramic materials (relative density of 94%) were achieved. However, ZrO_2-CaO ceramics have high biocompatibility and excellent mechanical properties characterized by strength of 500–708 MPa and Young's modulus of 1739–4372 MPa. Hu et al. [79] synthesized tetragonal zirconia polycrystalline (3Y-TZP) ceramics by HIP. The grain size of the final products reached about 138 nm. This fine grain size leads to an increase in Vickers hardness to achieve 13.79 MPa. These materials also revealed an elevated transmittance (in the range of 76–78%). The result showed that HIP was an effective process to prepare infrared-transparent 3Y-TZP ceramics with small grain size and with good optical and mechanical properties. Similarly, Klimke et al. [80] fabricated ZrO_2 ceramics by HIP. They demonstrated that the particle size, determined by TEM, was less than 50 nm (**Figure 8**) and the maximum in-line transmission was about 77%, which observed at IR wavelengths in the range of 3–5 μm.

2.2.7 CAD/CAM milling and copy milling

CAD/CAM milling and copy milling is an important field of dentistry and prosthodontics using CAD/CAM (computer-aided design and computer-aided manufacturing) to improve the design and creation of dental restorations [81, 82]. Thus, leucite-reinforced glass–ceramics involve Authentic and Empress CAD. Both have the identical microstructure and containing feldspathic glass with about 45 wt%. These blocks may characterize utilizing external strains and containing finer leucite crystals (about 5–10 μm in size). Additionally, the strength behavior of Empress CAD was comparable with Vitablocs. During the last two decades, dental CAD/CAM technology has been used to replace the laborious and time consuming, conventional lost wax technique for efficient fabrication of restorations [83]. Hence, this technology enables dentists to produce complex shapes of ceramics.

Typically, CAD/CAM dental restorations are milled from solid blocks of ceramic or composite resin that closely match the basic shade of the restored tooth. Metal alloys including zirconia can also be milled. The software sends this data to a milling machine where the prosthesis was milled [84]. CAD/CAM allows easy production of precise, esthetic, and durable prostheses [85]. CAD/CAM complements earlier technologies employed for these goals by enhancing

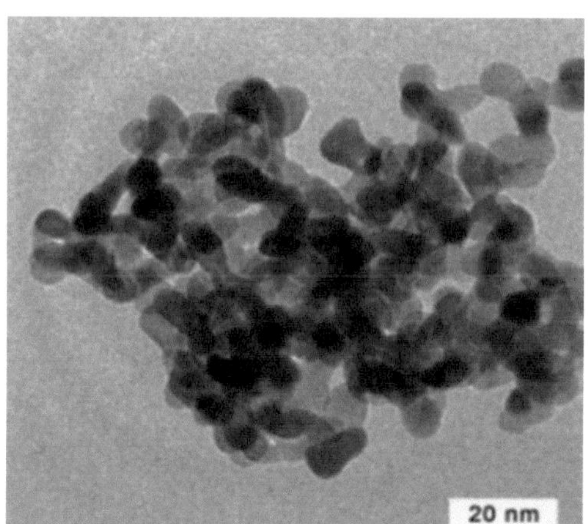

Figure 8.
TEM image of the ZrO_2 ceramic powder [80].

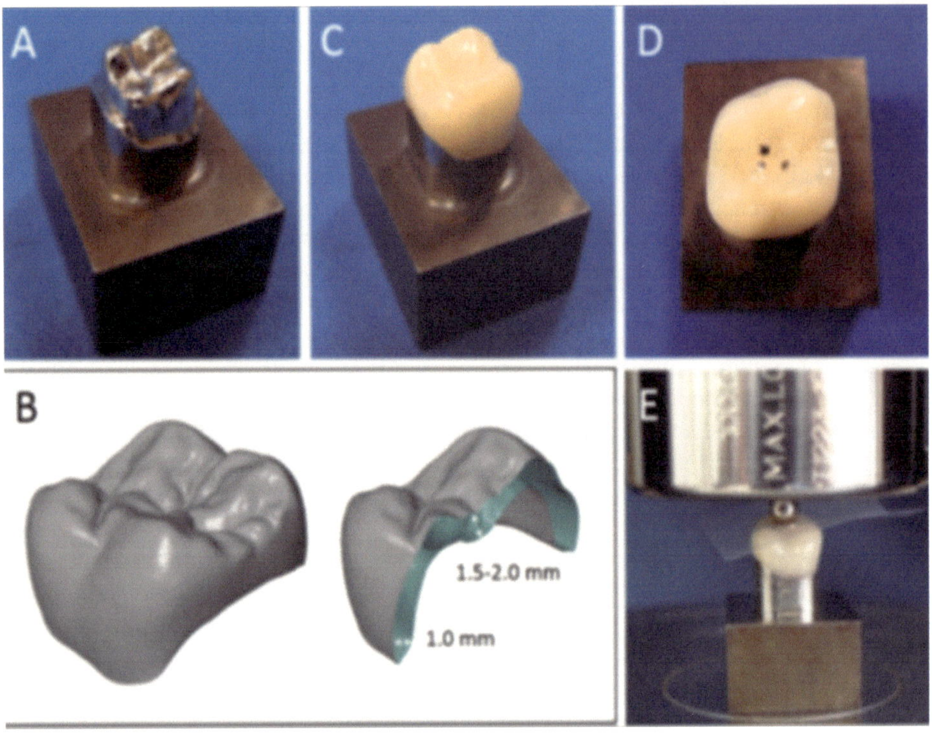

Figure 9.
Photographs of (A) stainless steel abutment, (B) CAD design for a crown specimen, (C) CAD/CAM composite resin crown bonded to the abutment, (D) loading points on the occlusal surface, and (E) the used machine [87].

the speed of design and creation, making affordable restorations, reducing unit coast, etc. Nevertheless, chair-side CAD/CAM equipment requires more time on the part of the dentists, and the fee was much higher than conventional restorative treatments. During the years 2015–2018, the data of 21 patients undergoing fibula free flap reconstructive surgery with CAD/CAM patient-specific reconstruction plates were analyzed, including the applicability of the virtual plan, flap survival, duration of surgery, ischemia time, simultaneous dental implantation, implant exposure, and postoperative complications [86]. At the time of primary reconstruction, a number of 76 dental implants were inserted in the 21 patients. The results showed that, in the secondary surgery, the implant can be uncovered 38.1% of the total patients in 7.6 months. Thus, the overall success rate for implants was 97.4%. Virtual surgical planning with CAD/CAM plates allows early and functional dental rehabilitation. Okada et al. [87] fabricated composite crowns using four computer aided design/computer aided manufacturing (CAD/CAM) blanks composed of a resins (sample 1) and a lithium disilicate (sample 2), which exhibited distinct tendencies (see **Figure 9**). The results revealed that the flexural strength was in the range of 175 to 247 MPa for sample 1 and 360 MPa for sample 2 while the fracture strength was in the range of 3.3 to 3.9 kN for the sample 1 and 3.3 kN for the sample 2.

3. Dental implants

Ceramic materials are good biomaterials widely used in dental implant because of their excellent biocompatibility and mechanical properties.

3.1 Bioglass implants

For the first time, the bioglass implants were introduced by Hench's team at the late 1960s [88]. Because of the great bioactivity of bioglasses, they are one of the best biomaterials for renovation and bone repair. Moreover, to combine great mechanical strength and excellent bioactivity of bioglasses, they can be successfully used as coatings on inert substrates [89]. A bioactive surface can be considered as important agent to avoid many simultaneous reactions, which take place between the implant and the targeted tissue at the implant surface. **Figure 10** shows the interfacial reactions involved in forming a bond between bioactive glass and bone. As can be seen, the first five stages take place at the periphery of bioactive glass and involve release of alkali ions (bacterial growth is inhibited as a result of pH increase) along with the formation of crystallized hydoxycarbonate apatite (HCA). Hence, the implanted material fastens down with the tissue from step 6 to 11 in consequence of osteostimulation and bone growth.

The bioactive glass with its composite coatings can be classified as following categorization according to the coating structure:

3.1.1 45S5 Bioglass

Ceramic 45S5 bioglass with 24.5% sodium oxide, 45% silicon dioxides, 24.5% calcium dioxide and 6% phosphorus pentoxide, shown on **Figure 11**, have attracted the attentions of researchers as a biomaterial substance because of its osseointegration capability, bioactive surface and the ability of healing bone damages [90–92]. It can be prepared by melt-cast method with various crystallinity including amorphous and crystalline [93]. But, to guarantee the perfect amorphous sample crystallization, the heat-treatment should be continued for 1 hour at 1000 °C.

3.1.2 58S bioglass

Ceramic 58S bioactive glass is a great bioactive, biodegradable glass with the capability of bone bonding. This bioactive glass, having 33% calcium oxide, 58% silicon dioxide 9% phosphorus pentoxide, is able to receive specific attention as scaffold substance [94–96]. The reaction of 58S bioactive glass with physiological fluids occurs after implantation quickly and makes bond to the tissue of bone. This

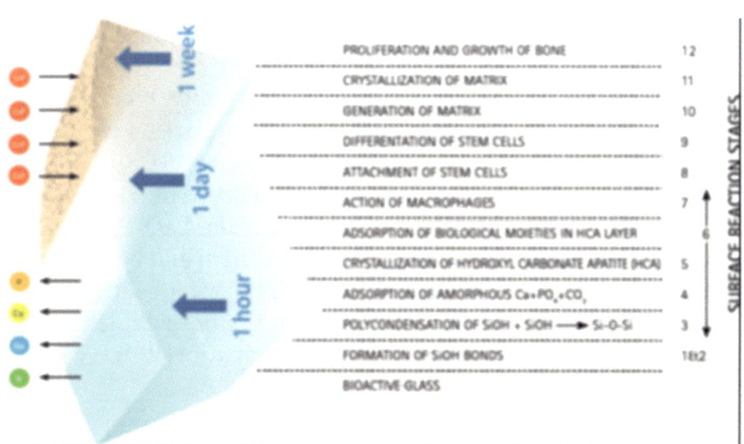

Figure 10.
Sequence of interfacial reactions involved in forming a bond between bioactive glass and bone [89].

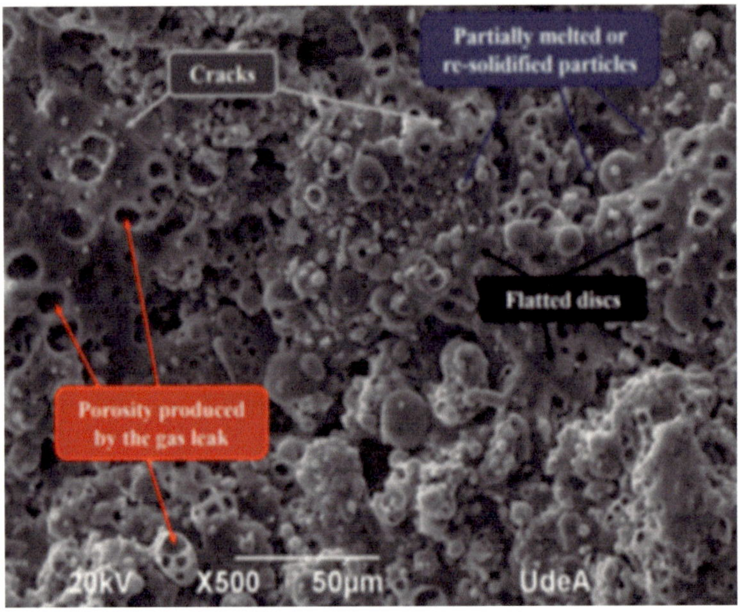

Figure 11.
The typical surface structure of 45S5 bioglass coatings sprayed [90].

happens without inflammatory, toxicity and foreign-body reaction. The fast ionic dissolution as well as the hydroxyl-carbonated apatite layer formation was seen after the rapid in-vivo surface reactions. With the release of calcium, silicon and phosphorous ions, the gene expression and the proliferation of osteoblast happens to form the bone quickly. This produced hydroxyapatite layer was as the bioactivity evidence of prepared bioglass, which both X-ray diffraction and scanning electron microscope images can confirm.

3.2 β-Tricalcium phosphate implants

β-tricalcium phosphate (β-$Ca_3(PO_4)_2$) and hydroxyapatite ($Ca_{10}(PO_4)_6(OH)_2$) belong to the family of calcium ceramics, which are widely and successfully used bioceramics for bone regeneration, drug delivery, biological cement, and tissue engineering scaffolds [97]. They exhibit pronounced resemblance to bone tissue minerals, excellent biocompatibility, good cell attachment properties for ensuring natural biodegradability, and bioresorption [98–100]. However, β-tricalcium phosphate is a bone substitute that has high biocompatibility, favorable resorption properties, and osteoconductivity [101–103]. In comparison with other bone substitutes, tricalcium phosphate, α-tricalcium phosphate (α-$Ca_3(PO_4)_2$) and β-tricalcium phosphate (β-$Ca_3(PO_4)_2$), are two polymorphs of tricalcium phosphate (TCP). The study of the phase equilibrium diagram of the CaO-P_2O_5 system shows that β-TCP, which crystallizes in the rhombohedral system and belongs to the space group *R3C*, transforms into α-TCP (monoclinic, *P21/a*), after heating at 1125 °C. The α-TCP has a less densely packed structure but it is more soluble than β-TCP. Its rapid hydrolyses produces calcium deficient hydroxyapatite [104].

3.3 Alumina implants

Alumina is very inert and resistant to corrosion in an in vivo environment [105]. It elicits minimal response from the tissues, and remains stable for many years of service. Few minutes after the implantation of alumina device, proteins and other biomolecules adsorb on its surface, to form a fibrous capsule around the implant that protects it from immune system. The fact that alumina is biocompatible does not mean that tiny particles formed by the implant wear cannot generate a significant foreign body reaction [106]. Hence, α-alumina is dense (with a specific gravity of 3.97), nonporous, and nearly inert material. It is extremely hard and scratch-resistant (9 on the Mohs scale, next only to diamond). It has excellent corrosion resistance in vivo environments (**Figure 12**). Dense alumina implants were used as dental implants, since the 70's, because of their excellent wettability, allowing them to easily adsorb water and biomolecules, resulting in a low coefficient of friction. However, the most disabling property of alumina is its brittleness (high elastic modulus), hence the need to optimize the composition, the porosity and the grain size to improve the mechanical properties of alumina, such as strength, fatigue resistance and fracture resistance. Because of the better resistance to fracture and the higher bending strength (13.000 kg/cm^2) of single crystal alumina, compared to that of polycrystalline alumina (3500 kg/cm^2), single crystal alumina is used for dental implants. Thus, a typical alumina implant is made of single crystal alumina cylindrical core around which polycrystalline alumina is fused. Currently, alumina dental implants are declining in popularity and being replaced by other material having better properties [107].

3.4 Zirconia implants

The demand for zirconia dental implants are increasing recently. In comparison with the Ti dental implants, their increased esthetic, due to similarity to the human tooth color, is the main benefit of these implants [88, 108].

Zirconia with better optical, esthetic, mechanical and biological qualifications, is a hopeful substitute to traditional Ti implant system for oral recovery [109], and is produced by the oxidation of zirconium [110]. Zirconium, which is a transition metal [111], with gray white color [112], can be used to make zirconia implant. Segments of the metal implant can be uncovered by recession of gingiva and the loss of apical bone, which this can disclose a discolored overlying gingiva [113]. These concerns make an opportunity to use the zirconia ceramics because they enjoy great esthetic, biological and mechanical characteristics and they also lack electrically corrosion. Polyethylene and Ti show more inflammatory reactions than zirconia. Less inflammatory response along with the lack of mutagenicity and toxicity in zirconia, can be considered as the most attractive zirconia properties [114]. Zirconia-based ceramics are attractive materials because they exhibit satisfying strength (more than 1000 MPa) and toughness (about 6–10 MPa $m^{1/2}$), allowing them to contribute

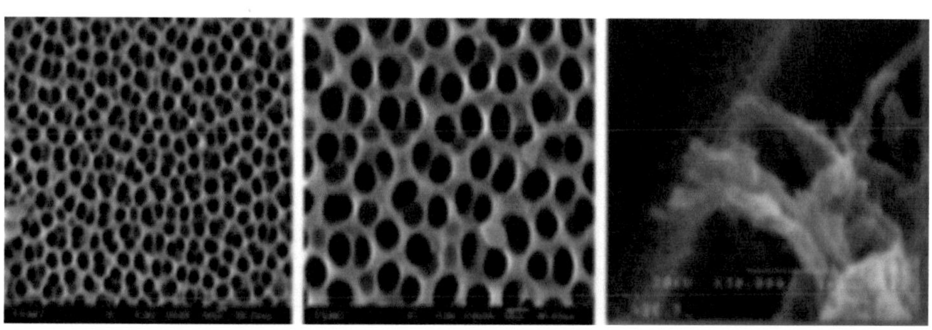

Figure 12.
Nanoporous alumina fabricated using the anodization process (left and center). Osteoblast interaction with the nanoporous architecture (right) [106].

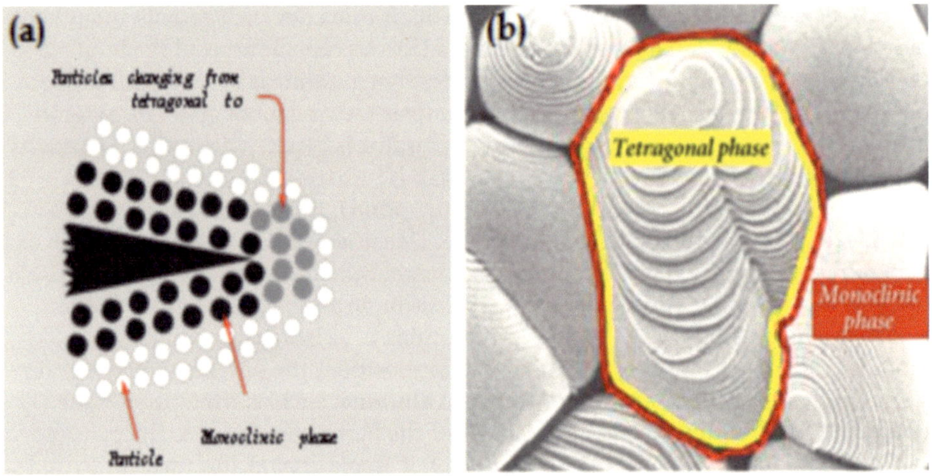

Figure 13.
(a) Zirconia has the ability to phase change from a tetragonal phase to a monoclinic phase to stop ensuing cracks, which is referred to as "transformation toughening" and (b) during phase change, there is approximately 3 to 5% volume increase from tetragonal to monoclinic phase [7].

to solve the problem of the fragility encountered when using alumina, as implant. Indeed, zirconia owes its importance to the stress-induced transformation of the metastable tetragonal crystallites phase into monoclinic phase, when it is localized around a propagating crack. As shown on **Figure 13**, this transformation contributes to increase the fracture toughness and the resistance to crack propagation by expanding the volume (3–6%) and absorbing a part of the energy required for crack propagation, which leads to crack shielding. The negative aspect of this transformation is "aging", which happens when it is localized on the surface of the implant, in presence of acidic liquids. Hence, the surface of the implant presents some signs of "aging", such as the roughness and the appearance of micro-cracks [115].

3.5 Hydroxyapatites implants

Hydroxyapatite is a bioceramic of great clinical interest due to its nontoxicity, bioactivity, good biocompatibility, osteoconductivity, and its non-inflammatory nature. However, since it has a high elasticity modulus (brittle), hydroxyapatite is usually associated to other materials to form an implant in load-bearing applications. In fact, implants for load bearing, such as titanium screw, can be coated with

Property	Value
Thickness	—
Crystallinity (%)	62
Phase purity (%)	95
Density (g/cm^3)	24
Tensile strength (MPa)	>50.8
Shear strength (MPa)	>22
Ca/P ratio	1.67–1.76
Heavy metals (ppm)	<50

Table 4.
Properties of hydroxyapatites coat [117].

hydroxyapatite. The application of hydroxyapatite coatings is an interesting surface amendment on dental implants [116]. As its coatings apply on implanted material, it provides enough calcium and phosphate ions at initial implantation stage and makes the implant material biocompatible [117]. The properties of hydroxyapatites are given in **Table 4**.

However, hydroxyapatite has osteogenic nature and is able to form strong bond with host tissues, so it is widely used in biomedical field for osteointegration, bone replacement and regeneration, coating metallic implants, and to fill the defects generated in bones [118].

4. Challenges

Despites their benefits, all ceramic dental materials and their applications shows challenges which still need to be tacked.

The challenges in dentistry remain in understanding and improving the clinical performance of the biocompatible restorative materials by improving definition of failures, laboratory testing, and clinical studies. In fact, material factors, including differences in thermal conductivity and coefficient of thermal expansion between core and veneer, likely create residual stresses that redispose a restoration to chipping. Only requirements of patients further complicate the challenge of understanding factors that contribute to long term success of restoration. In this context, some works include report patient or provider factors or patient control groups.

Few recent works [119, 120] have been reported on clinical trials. Several improvements have been recently made in structural reliability via damage tolerance and flaw control [121, 122]. Predictive laboratory tests can reduce the need for expensive and time-consuming clinical tests, which sometimes exceed the commercial lifetime of the materials being evaluated. In addition, laboratory tests, likely over estimate clinical lifetimes, can replicate clinical failure modes.

Several parameters like dimensional accuracy, surface, and mechanical properties of ceramic dental materials should be improved to obtain high quality final products [123]. Another challenge is bacteriological safety of the final products which are in contact with human organs and tissues. However, it is necessary to make sterilized protocols while keeping intrinsic properties [123]. In addition, Lee et al. [124] showed that the accuracy in Z-direction is harder to enhance than in the directions X and Y because the presence of uncontrolled parameters like evaporation of material during machining, shrinkage, and spreading densification of the powder. Hence, the porosity of the ceramics is another challenge in dentistry. It has been showed that the porosity was reduced by adding dopants or viscous liquid forming phase, choosing the corresponding powder granulometry, and applying HIP to the green body [125]. Several studies revealed that the surface quality of ceramic materials depends strongly on the technique, raw material characteristics, and processing conditions [126, 127].

Moreover, there maining challenges for future advances are present abundant arenas for future innovations. Moreover, it will be important to determine where and how informed simplifications in testing conditions can be made.

Machining techniques and design methods should to be improved and innovated to achieve good ceramic restorations with subsurface damage and little surface.

CAD and CAM and fabrication processes creating veneers and cores separately will further evolve [128]. Thus, these approaches will be complemented by additive approaches, laying down materials only in places where it is needed to create a restoration [129]. However, these approaches have shown significant substantial hurdles.

5. Conclusion

For over 200 years, ceramic materials have been utilized in dentistry. This chapter shows that dental ceramics can be fabricated by different techniques. The CAD/CAM technology is the most widely used in dentistry.

Dentistry as an art of oral health is one of the major affiliates of dental science. Operative dentistry continues to evolve toward bright future with the innovations and development of new materials, techniques, and equipments. Several numbers of dental ceramic materials have been developed with respect to strength, survival, applications, and esthetics. The success of dental ceramic materials depends on various factors like design, type of material, cementation media, clinical data, etc. These factors help the dentist to enhances the relation between laboratory studies and clinical data and to choice the appropriate ceramic material.

Although the remarkable evolution of ceramics in dentistry not all the challenges have been solved.

Acknowledgements

We would like to thank the Author Service Manager Mrs. Mia Vulovic for his time and support. We also thank our institutions for give us support.

Conflict of interest

The authors declare no conflict of interest.

Thanks

We thank IntechOpen Publisher for accepting this work to participate in the book project under the working title "Ceramic Materials", ISBN 978–1–83881-212-6.

Author details

Mohsen Mhadhbi[1*], Faïçal Khlissa[2] and Chaker Bouzidi[2]

1 Laboratory of Useful Materials, National Institute of Research and Physical-Chemical Analysis, Technopole, Ariana, Tunisia

2 Laboratory of Physical Chemistry of Mineral Materials and Their Applications, National Center of Material Science Researches, Technopole, Borj-Cedria, Tunisia

*Address all correspondence to: mhadhbi_mohsen@yahoo.fr

References

[1] Denry I, Holloway JA, Ceramics for dental applications: A review. Materials. 2010;3:351-368. DOI:10.3390/ma3010351

[2] Kelly JR, Dental ceramics: what is this stuff anyway? The Journal of the American Association. 2008;139:4S–7S. DOI:10.14219/jada.archive.2008.0359

[3] Kelly JR, Denry I, Stabilized zirconia as a structural ceramics: an overview. Dental Materials. 2008;24:289-298. DOI:10.1016/j.dental.2007.05.005

[4] Rekow ED, Silva NRFA, Coelho PG, Zhang Y, Guess P, Thompson VP. Performance of dental ceramics: challenges for improvements. Journal of Dental Research. 2011;90:937-952. DOI:10.1177/0022034510391795

[5] Cahn RW, editor. Handbook of The Coming of Materials Science. 1st ed. Oxford: Pergamon; 2001. 590 p. DOI:9780080529424

[6] Al-Wahadni A, The roots of dental porcelain: A brief historical perspective. Dental News. 1999;2:43-44. DOI: not available

[7] Helvey G, A history of dental ceramics. Compendium. 2010;31:309-311. www.aegisdentalnetwork.com

[8] Mörmann WH, Innovationen bei ästhetischen Restaurationen im Seitenzahngebiet (Keramik): Computer- unterstütze Systeme. Dtsch Zahnärztl Z. 1988;43:900-903. DOI: not availablc

[9] Reich SM, Wichmann M, Rinne H, Shortall A, Clinical performance of large, all-ceramic CAD/CAM-generated restorations after three years: A Pilot study. The Journal of the American Dental Association. 2004;135:605-612. DOI:10.14219/jada.archive.2004.0248

[10] Wittneben JG, Wright RF, Weber HP, Gallucci GO, A systematic review of the clinical performance of CAD/CAM single-tooth restorations. The International Journal of Periodontics and Restorative Dentistry. 2009;22:466-471. DOI:10:11607/prd.3519

[11] Voskerician G, Shive MS, Shawgo RS, Von Recum H, Anderson JM, Cima MJ, Langer R, Biocompatibility and biofouling of MEMS drug delivery devices. Biomaterials. 2003;24:1959-1967. DOI:10.1016/S0142-9612(02)00565-3

[12] Wang YC, Ferrari M, Surface modification of micromachined silicon filters. Journal of Materials Science. 2000;35:4923-4930. DOI:10.1023/A:1004830408749

[13] La Flamme KE, Popat KC, Leoni L, Markiewicz E, La Tempa TJ, Roman BB, Grimes CA, Desai TA, Biocompatibility of nanoporous alumina membranes for immunoisolation. Biomaterials. 2007;28:2638-2645. DOI:10.1016/j.biomaterials.2007.02.010

[14] Zhang MQ, Desai T, Ferrari M, Proteins and cells on PEG immobilized silicon surfaces. Biomaterials. 1998;19:953-960. DOI:10.1016/s0142-9612(98)00026-x

[15] Zhang MQ, Ferrari M, Reduction of Albumin Adsorption onto Silicon Surfaces by Tween 20. Biotechnology and Bioengineering. 1997;56:618-625. DOI:10.1002/(SICI)1097-0290(19971220)56:6<618::AID-BIT4>3.0.CO;2-Q

[16] Rother D, Sen T, East D, Bruce IJ, Silicon, silica and its surface patterning/activation with alkoxy-and amino-silanes for nanomedical applications. Nanomedicine. 2011;6:281-300. DOI:10.2217/nnm.10.159

[17] Canham LT, Bioactive silicon structure fabrication through nanoetching techniques. Advanced Materials. 1995;7:1033-1037. DOI:10.1002/adma.19950071215

[18] Anderson SHC, Elliott H, Wallis DJ, Canham LT, Powell JJ, Dissolution of different forms of partially porous silicon wafers under simulated physiological conditions. Physica Status Solidi (a). 2003;197:331-335. DOI:10.1002/pssa.200306519

[19] Anglin EJ, Cheng L, Freeman WR, Sailor MJ, Porous silicon in drug delivery devices and materials. Advanced Drug Delivery Reviews. 2008;60:1266-1277. DOI:10.1016/j.addr.2008.03.017

[20] Tasciotti E, Godin B, Martinez JO, Chiappini C, Bhavane R, Liu X, Ferrari M, Near-infrared imaging method for the in vivo assessment of the biodistribution of nanoporous silicon particles. Molecular Imaging. 2011;10:56-58. DOI:10.2310/7290.2011.00011

[21] Serda RE, Ferrati S, Godin B, Tasciotti E, Liu X, Ferrari M, Mitotic trafficking of silicon microparticles. Nanoscale. 2009;1:250-259. DOI:10.1039/b9nr00138g

[22] Tanaka T, Mangala LS, Vivas-Mejia PE, Nieves-Alicea R, Mann AP, Mora E, Han HD, Shahzad MM, Liu X, Bhavane R, Gu J, Fakhoury JR, Chiappini C, Lu C, Matsuo K, Godin B, Stone RL, Nick AM, Lopez-Berestein G, Sood AK, Ferrari M, Sustained small interfering RNA delivery by mesoporous silicon particles. Cancer Research. 2010;70:3687-3696. DOI:10.1158/0008-5472.CAN-09-3931

[23] De Rosa E, Chiappini C, Fan DM, Liu XW, Ferrari M, Tasciotti E, Agarose surface coating influences intracellular accumulation and enhances payload stability of a nano-delivery system. Pharmaceutical Research. 2011;28:1520-1530. DOI:10.1007/s11095-011-0453-2

[24] Willmann G, Ceramic femoral heads retrievals data. Clinical Orthopaedics and Related Research. 2000;379:173-177. DOI:10.1097/00003086-200010000-00004

[25] Willmann G, Von Chamier W, Bioceramics in orthopedics: New applications. Stuttgart, Germany, Enke Verlag, 1998. DOI: not exist

[26] Solarino G, Piconi C, De Santis V, Piazzolla A, Moretti B, Ceramic Total Knee Arthroplasty Ready to Go? Georg Thieme Verlag KG Stuttgart, New York. 2017. ISSN 2282-4324. DOI:10.1055/s-0037-1607428

[27] Rosenblum MA, Schulman A, A review of all-ceramic restorations. The Journal of the American Dental Association. 1997;128:297-307. DOI:10.14219/jada.archive.1997.0193

[28] Pillai SKC, Baron B, Pomeroy MJ, Hampshire S, Effect of oxide dopants on densification, microstructure and mechanical properties of alumina-silicon carbide nanocomposite ceramics prepared by pressureless sintering. Journal of the European Ceramic Society. 2004;24:3317-3326. DOI:10.1016/j.jeurceramsoc.2003.10.024

[29] Lu HX, Sun HW, Li GX, Chen CP, Yang DI, Hu X, Microstructure and mechanical properties of Al_2O_3-MgB_2 composites. Ceramics International. 2005;31:105-108. DOI:10.1016/j.ceramint.2004.03.040

[30] Carroll L, Sternitzke M, Derby B, Silicon carbide particle size effects in alumina-based nanocomposites. Acta Materialia. 1996;44:4543-4552. DOI:10.1016/1359-6454(96)00074-2

[31] Doğan CP, Hawk JA, Influence of whisker reinforcement on the abrasive wear behavior of silicon nitride and

alumina-based composites. Wear. 1997;203-204:267-277. DOI:10.1016/S0043-1648(96)07348-6

[32] Belmonte M, Nieto MI, Osendi MI, Miranzo P, Influence of the SiC grain size on the wear behaviour of Al_2O_3/SiC composites. Journal of the European Ceramic Society. 2006;26:1273-1279. DOI:10.1016/j.jeurceramsoc.2005.01.049

[33] Tuan WH, Chen RZ, Wang TC, Cheng CH, Kuo PS. Mechanical properties of Al_2O_3/ZrO_2 composites. Journal of the European Ceramic Society. 2002;22:2827-2833. DOI:10.1016/S0955-2219(02)00043-2

[34] Wang YS, He C, Hockey BJ, Lacey PI, Hsu SM, Wear transitions in monolithic alumina and zirconia-alumina composites. Wear. 1995;181-183:156-164. DOI:10.1016/0043-1648(95)90020-9

[35] Huang XW, Wang SW, Huang XX, Microstructure and mechanical properties of ZTA fabricated by liquid phase sintering. Ceramics International. 2003;29:765-769. DOI:10.1016/S0272-8842(02)00228-6

[36] Heimann RB, Structure, properties, and biomedical performance of osteoconductivebioceramic coatings. Surface and Coatings Technology. 2013;233:27-38. DOI:10.1016/j.surfcoat.2012.11.013

[37] Fabbri P, Piconi C, Burresi E, Magnani G, Mazzanti F, Mingazzini C, Lifetime estimation of a zirconia-alumina composite for biomedical applications. Dental Materials. 2014;20:138-142. DOI:10.1016/j.dental.2013.10.006

[38] De Aza AH, Chevalier J, Fantozzi G, Schehl M, Torrecillas R, Crack growth resistance of alumina, zirconia and zirconia toughened Alumina ceramics for joint prostheses. Biomaterials. 2002;23:937-945. DOI:10.1016/S0142-9612(01)00206-X

[39] Moraes MCCS, Elias CN, Duailibi Filho J, Oliveira LG, Mechanical properties of alumina-zirconia composites for ceramic abutments. Materials Research. 2004;7:643-649. DOI:10.1590/S1516-14392004000400021

[40] Nevarez-Rascon A, Aguilar-Elguezabal A, Orrantia E, Bocanegra-Bernal MH, On the wide range of mechanical properties of ZTA and ATZ based dental ceramic composites by varying the Al_2O_3 and ZrO_2 content. International Journal of Refractory Metals and Hard Materials. 2009;27:962-970. DOI:10.1016/j.ijrmhm.2009.06.001

[41] Nevarez-Rascon A, Aguilar-Elguezabal A, Orrantia E, Bocanegra-Bernal MH, Al_2O_3(w)-Al_2O_3(n)-ZrO_2 (TZ-3Y)n multi-scale nanocomposite: an alternative for different dental applications. Acta Biomaterialia. 2010;6:563-570. DOI:10.1016/j.actbio.2009.06.029

[42] Daskalova A, Angelova L, Carvalho A, Trifonov A, Nathala C, Monteiro F, Buchvarov I, Effect of surface modification by femtosecond laser on zirconia based ceramics for screening of cell-surface interaction. Applied Surface Science. 2020;513:145914. DOI:10.1016/j.apsusc.2020.145914

[43] Chevalier J, What future for zirconia as a biomaterial? Biomaterials. 2006;27:535-543. DOI:10.1016/j.biomaterials.2005.07.034.

[44] Vagkopoulou, T, Koutayas, SO, Koidis, P, Strub, JR, Zirconia in Dentistry: Part 1. Discovering the nature of an upcoming bioceramic. European Journal of Esthetic Dentistry. 2009;4:130-151. DOI: not available

[45] Denry I, Kelly JR, State of the art of zirconia for dental applications. Dental Materials. 2008;24:299-307. DOI:10.1016/j.dental.2007.05.007.

[46] Jayakumar R, Ramachandran R, Sudheesh Kumara P, Divyarani VV, Srinivasan S, Chennazhi KP, Tamura H, Nair SV, Fabrication of chitin-chitosan/nano ZrO_2 composite scaffolds for tissue engineering applications. International Journal of Biological Macromolecules. 2011;49:274-280. DOI:10.1016/j.ijbiomac.2011.04.020

[47] Josset Y, Oum'Hamed Z, Zarrinpour A, Lorenzato M, Adnet JJ, Laurent Maquin D, In vitro reactions of human osteoblasts in culture with zirconia and alumina ceramics. Journal of Biomedical Materials Research. 1999;47:481-493. DOI:10.1002/(sici)1097-4636(19991215)47:4<481::aid-jbm4>3.0.co;2-y

[48] Michael H, Sophie V, Philippe V, Zirconia: established facts and perspectives for a biomaterial in dental implantology. Journal of Biomedical Materials Research Part B. 2008;88:519-529. DOI:10.1002/jbm.b.31147

[49] Nagai T, Kawamoto Y, Kakehashi Y, Matsumura H, Adhesive bonding of a lithium disilicate ceramic material with resin-based luting agents. Journal of Oral Rehabilitation. 2005;32:598-605. DOI:10.1111/j.1365-2842.2005.01464.x

[50] Roeters JJM, Shortall ACC, Opdam NJM, Can a single composite resin serve all purposes? British Dental Journal. 2005;199:73-79. DOI:10.1038/sj.bdj.4812520

[51] Brentel AS, Ozcan M, Valandro LF, Alarca LG, Amaral R, Bottino MA, Microtensile bond strength of a resin cement to feldspathic ceramic after different etching and silanization regimens in dry and aged conditions. Dental Materials. 2007;23:1323-1331. DOI:10.1016/j.dental.2006.11.011

[52] Quinn JB, Quinn GD, Material properties and fractography of an indirect dental resin composite. Dental Materials. 2010;26:589-599. DOI:10.1016/j.dental.2010.02.008

[53] Bottino M, Campos F, Ramos N, Rippe R, Valandro L, Melo RM, Inlays made from a hybrid material: Adaptation and bond strengths. Operative Dentistry. 2015;40:E83-E91. DOI:10.2341/13-343-L

[54] Wiedhahn K, From blue to white: New high-strength material for Cerec-IPS e.max CAD LT. International Journal of Computerized Dentistry. 2007;10:79-91. DOI: not available

[55] Borges GA, Sophr AM, de Goes MF, Sobrinho LC, Chan DC, Effect of etching and airborne particle abrasion on the microstructure of different dental ceramics. Journal of Prosthetic Dentistry. 2003;89:479-488. DOI:10.1016/S0022-3913(02)52704-9

[56] Ozcan M, Vallittu PK, Effect of surface conditioning methods on the bond strength of luting cement to ceramics. Dental Materials. 2003;19:725-731. DOI:10.1016/S0109-5641(03)00019-8

[57] Amaral R, Ozcan M, Valandro LF, Balducci I, Bottino MA, Effect of conditioning methods on the microtensile bond strength of phosphate monomer-based cement on zirconia ceramic in dry and aged conditions. Journal of Biomedical Materials Research Part B. 2008;85:1-9. DOI:10.1002/jbim.b.30908

[58] Fradeani M, Redemagni M, An 11-year clinical evaluation of leucite-reinforced glass-ceramic crowns: A retrospective study. Quintessence International. 2002;33:503-510. DOI: not available

[59] Passos SP, Valandro LF, Amaral R, Ozcan M, Bottino MA, Kimpara ET, Does adhesive resin application contribute to resin bond durability on etched and silanized feldspathic

ceramic? Journal of Adhesive Dentistry. 2008;10:455-460. DOI: not available

[60] Blank JT, Scientifically based rationale and protocol for use of modern indirect resin inlays and onlays. Journal of Esthetic Dentistry. 2000;12:195-208. DOI:10.1111/j.1708-8240.2000.tb00222.x

[61] Coldea A, Swain MV, Thiel N, Mechanical properties of polymer-infiltrated-ceramic-network materials. Dental Materials. 2013;29:419-426. DOI:10.1016/j.dental.2013.01.002

[62] Soles CL, Yee AF, A discussion of the molecular mechanisms of moisture transport in epoxy resins. Journal of Polymer Science Part B-Polymer Physics. 2000;38:792-802. DOI:0.1002/(SICI)1099-0488(20000301)38:5<792::AID-POLB16>3.0.CO;2-H

[63] Abdullah AM, Tuan Rahim TNA, Wan Hamad WN, Mohamad D, Md Akil H, Rajion ZA, Mechanical and cytotoxicity properties of hybrid ceramics filled polyamide 12 filament feedstock for craniofacial bone reconstruction via fused deposition modelling. Dental Materials. 2018;34: e309-e316. DOI:10.1016/j.dental.2018.09.006

[64] McGinley EL, Moran GP, Fleming GJP, Biocompatibility effects of indirect exposure of base-metal dental casting alloys to a human-derived three-dimensional oral mucosal model. Journal of Dentistry. 2013;41:1091-1100. DOI:10.1016/j.jdent.2013.08.010

[65] Zhou Y, Li N, Wang H, Yan J, Liu W, Xu S. Effects of the rare earth element lanthanum on the metal-ceramic bond strength of dental casting Co-Cr alloys. Journal of Prosthetic Dentistry. 2019;121:848-857. DOI:10.1016/j.prosdent.2018.08.017

[66] Atwood RC, Lee PD, Curtis RV, Modeling the surface contamination of dental titanium investment castings. Dental Materials. 2005.21:178-186. DOI:10.1016/j.dental.2004.02.010

[67] Fan J, Lin T, Hu F, Yu Y, Ibrahim M, Zheng R, Huang S, Ma J, Effect of sintering temperature on microstructure and mechanical properties of zirconia-toughened alumina machinable dental ceramics. Ceramics International 2017;43:3647-3653. DOI:10.1016/j.ceramint.2016.11.204

[68] Ghayebloo M, Alizadeh P, Melo RM, Fabrication of ZrO_2-Bearing lithium-silicate glass-ceramics by pressureless sintering and spark plasma sintering. Journal of the Mechanical Behavior of Biomedical Materials 2020;105:103709. DOI:10.1016/j.jmbbm.2020.103709

[69] Ohji T, Fukushima M, Macro-porous ceramics: processing and properties. International Materials Reviews. 2012;57:115-131. DOI:10.1179/1743280411Y.0000000006

[70] Chen H, Xiang H, Dai F, Liu J, Zhou Y, High strength and high porosity YB_2C_2 ceramics prepared by a new high temperature reaction/partial sintering process. Journal of Materials Science and Technology. 2019;35:2883-2891. DOI: 10.1016/j.jmst.2018.09.071

[71] Jeana G, Sciamanna V, Demuynck M, Cambier F, Gonon M, Macroporous ceramics: Novel route using partial sintering of alumina-powder agglomerates obtained by spray-drying. Ceramics International. 2014;40:10197-10203. DOI:10.1016/j.ceramint.2014.02.089

[72] Kocjan A, Shen Z, Colloidal processing and partial sintering of high-performance porous zirconia nanoceramics with hierarchical heterogeneities. Journal of the European Ceramic Society. 2013;33:3165-3176. DOI:10.1016/j.jeurceramsoc.2013.06.004

[73] Yang Z, Jin Q, Ma J, Tong Y, Wang X, Du R, Wang S, Glass

infiltration of gelcast zirconia-toughened alumina ceramics for dental restoration. Ceramics International. 2012;38:4653-4659. DOI:10.1016/j.ceramint.2012.02.047

[74] Manuel FRPA, Claudinei DS, Caio MFC, Paulo AS, Alfeu SR, Erika CTR, Bruno GS, Kurt S, Development of dense Al_2O_3-TiO_2 ceramic composites by the glass-infiltration of porous substrates prepared from mechanical alloyed powders. Ceramics International. 46(2020)2344-2354. DOI:10.1016/j.ceramint.2019.09.225

[75] Kim WC, Lee JK, Effect of Powder Characteristics on Slip Casting Fabrication of Dental Zirconia Implants. Journal of Nanoscience and Nanotechnology. 2020;20:5385-5389. DOI:10.1166/jnn.2020.17610

[76] Kim DS, Kim WC, Lee JK, Effect of Solid Loading on the Sintered Properties of 3 mol% Yttria-Stabilized Tetragonal Zirconia Polycrystals (3Y-TZP) Ceramics via Slip Casting. Journal of Nanoscience and Nanotechnology. 2019;19:6383-6386. DOI:10.1166/jnn.2019.17045

[77] Alageel O, Abdallah MN, Alsheghri A, Song J, Caron E, Tamimi F, Removable partial denture alloys processed by laser-sintering technique. Journal of Biomedical Materials Research Part B: Applied Biomaterials. 2018;106:1174-1185. DOI:10.1002/jbm.b.33929

[78] Gionea A, Andronescu E, Voicu G, Bleotu C, Surdu VA, Influence of hot isostatic pressing on ZrO2-CaO dental ceramics properties. International Journal of Pharmaceutics. 2016;510:439-448. DOI:10.1016/j.ijpharm.2015.10.044

[79] Hu X, Jiang X, Chen S, Zhu Q, Feng M, Zhang P, Fan J, Jiang B, Mao X, Zhang L, Fabrication of infrared-transparent 3Y-TZP ceramics with small grain size by pre-sintering in an oxygen atmosphere and hot isostatic pressing. Ceramics International. 2018;44:2093-2097. DOI:10.1016/j.ceramint.2017.10.158

[80] Klimke J, Trunec M, Krell A, Transparent tetragonal yttria stabilized zirconia ceramics: influence of scattering caused by birefringence. Journal of the American Ceramic Society. 2011;94:1850-1858. DOI:10.1111/j.1551-2916.2010.04322.x

[81] Davidowitz G, Kotick PG, The use of CAD/CAM in dentistry. Dental Clinics. 2011;55:559-570. DOI:10.1016/j.cden.2011.02.011

[82] Rekow D, Computer-aided design and manufacturing in dentistry: a review of the state of the art. Journal of Prosthetic Dentistry. 1987;58:512-516. DOI:10.1016/0022-3913(87)90285-X

[83] Rekow ED, Erdman AG, Riley DR, Klamecki B, CAD/CAM for dental restorations-some of the curious challenges. IEEE Transactions on Biomedical Engineering. 1991;38:314-318. DOI:10.1109/10.133223

[84] Beuer F, Schweiger J, Edelhoff D, Digital dentistry: An overview of recent developments for CAD/CAM generated restorations. British Dental Journal. 2008;204:505-511. DOI:10.1038/sj.bdj.2008.350

[85] Oen KT, Veitz-Keenan A, Spivakovsky S, Wong YJ, Bakarman E, Yip J, CAD/CAM versus traditional indirect methods in the fabrication of inlays, onlays, and crowns. Cochrane Database of Systematic Reviews. 2014;2014:CD011063. DOI:10.1002/14651858.CD011063

[86] Seier T, Hingsammer L, Schumann P, Gander T, Rücker M, Lanzer M, Virtual planning, simultaneous dental implantation and CAD/CAM plate fixation: A paradigm change in maxillofacial reconstruction. International Journal of Oral and

Maxillofacial Surgery. 2020;49:854-861. DOI:10.1016/j.ijom.2019.11.010

[87] Okada R, Asakura M, Ando A, Kumano H, Ban S, Kawai T, Takebe J. Fracture strength testing of crowns made of CAD/CAM composite resins. Journal of Prosthodontic Research. 2018;62:287-292. DOI:10.1016/j.jpor.2017.10.003

[88] Gul H, Khan M, Khan AS. Bioceramics: types and clinical applications. Handbook of Ionic Substituted Hydroxyapatites: Elsevier; 2020. p. 53-83. DOI:10.1016/B978-0-08-102834-6.00003-3

[89] Mesquita-Guimarães J, Henriques B, Silva F. Bioactive glass coatings. Bioactive Glasses: Elsevier; 2018. p. 103-118. DOI:10.1016/B978-0-08-100936-9.00007-1

[90] Rojas O, Prudent M, López M, Vargas F, Ageorges H. Influence of Atmospheric Plasma Spraying Parameters on Porosity Formation in Coatings Manufactured from 45S5 Bioglass powder. Journal of Thermal Spray Technology. 2020;29:185-198. DOI:10.1007/s11666-019-00952-3

[91] Dimitriadis K, Moschovas D, Tulyaganov DU, Agathopoulos S. Development of novel bioactive glass-ceramics in the $Na_2O/K_2O-CaO-MgO-SiO_2-P_2O_5-CaF_2$ system. Journal of Non-Crystalline Solids. 2020;533:119936. DOI:10.1016/j.jnoncrysol.2020.119936

[92] Eshghinejad P, Farnoush H, Bahrami M, Bakhsheshi-Rad H, Karamian E, Chen X. Electrophoretic deposition of bioglass/graphene oxide composite on Ti-alloy implants for improved antibacterial and cytocompatible properties. Materials Technology. 2020;35:69-74. DOI:10.1080/10667857.2019.1648740

[93] Gao C, Liu T, Shuai C, Peng S. Enhancement mechanisms of graphene in nano-58S bioactive glass scaffold: mechanical and biological performance. Scientific reports. 2014;4:4712. DOI:10.1038/srep04712

[94] Arango-Ospina M, Nawaz Q, Boccaccini AR. Silicate-based nanoceramics in regenerative medicine. Nanostructured Biomaterials for Regenerative Medicine: Elsevier; 2020. p. 255-273. DOI:10.1016/B978-0-08-102594-9.00009-7

[95] Karimi N, Kharaziha M, Raeissi K. Electrophoretic deposition of chitosan reinforced graphene oxide-hydroxyapatite on the anodized titanium to improve biological and electrochemical characteristics. Materials Science and Engineering: C. 2019;98:140-152. DOI:10.1016/j.msec.2018.12.136

[96] Singh AK, Pramanik K, Biswas A. MgO enables enhanced bioactivity and antimicrobial activity of nano bioglass for bone tissue engineering application. Materials Technology. 2019;34:818-826. DOI:10.1080/10667857.2019.1638636

[97] Christel P. Biocompatibility of surgical-grade dense polycrystalline alumina. Clinical Orthopaedics and Related Research. 1992;282:10-18. PMID:1516299

[98] Tateiwa T, Clarke IC, Williams PA, Garino J, Manaka M, Shishido T, Yamamoto K, Imakiire A, Ceramic total hip arthroplasty in the United States: safety and risk issues revisited. The American Journal of Orthopedics. 2008;37:E26-E31. PMID:18401491

[99] Della Bona A, Kelly JR. The clinical success of all-ceramic restorations. The Journal of the American Dental Association. 2008;139:S8-S13. DOI:10.14219/jada.archive.2008.0361

[100] He X, Reichl FX, Milz S, Michalke B, Wu X, Sprecher CM, Yang Y, Gahlert M, Röhling S, Kniha H,

Titanium and zirconium release from titanium-and zirconia implants in mini pig maxillae and their toxicity in vitro. Dental Materials. 2020;36:402-412. DOI:10.1016/j.dental.2020.01.013

[101] Afrashtehfar KI, Del Fabbro M. Clinical performance of zirconia implants: A meta-review. The Journal of Prosthetic Dentistry. 2020;123:419-426. DOI:10.1016/j.prosdent.2019.05.017

[102] Sadowsky SJ. Has zirconia made a material difference in implant prosthodontics? A review. Dental Materials. 2020;36:1-8. DOI:10.1016/j.dental.2019.08.100

[103] Pobloth A-M, Mersiowsky MJ, Kliemt L, Schell H, Dienelt A, Pfitzner BM, Burgkart R, Detsch R, Wulsten D, Boccaccini AR, Bioactive coating of zirconia toughened alumina ceramic implants improves cancellous osseointegration. Scientific Reports. 2019;9:1-16. DOI:10.1038/s41598-019-53094-5

[104] Cai H, Chen J, Li C, Wang J, Wan Q, Liang X. Quantitative discoloration assessment of peri-implant soft tissue around zirconia and other abutments with different colours: A systematic review and meta-analysis. Journal of Dentistry. 2018;70:110-117. DOI:10.1016/j.jdent.2018.01.003

[105] Rivera DS-H, Pozos-Guillen A, Aragón-Piña A, Cerda-Cristerna BI, Masuoka-Ito D, Sánchez-Vargas LO. Glass coatings to enhance the interfacial bond strength between veneering ceramic and zirconia. Odontology. 2020:1-9. DOI:10.1007/s10266-020-00497-w

[106] Roy M, Pompella A, Kubacki J, Piosik A, Psiuk B, Klimontko J, Szade J, Roy RA, Hedzelek W, Photofunctionalization of dental zirconia oxide: Surface modification to improve bio-integration preserving crystal stability. Colloids and Surfaces B: Biointerfaces. 2017;156:194-202. DOI:10.1016/j.colsurfb.2017.05.031

[107] Chevalier J, Gremillard L, Virkar AV, Clarke DR. The tetragonal-monoclinic transformation in zirconia: lessons learned and future trends. Journal of the American Ceramic Society. 2009;92:1901-1920. DOI:10.1111/j.1551-2916.2009.03278.x

[108] Ducheyne P, Qiu Q. Bioactive ceramics: the effect of surface reactivity on bone formation and bone cell function. Biomaterials. 1999;20:2287-2303. DOI:10.1016/S0142-9612(99)00181-7

[109] Yoganand C, Selvarajan V, Cannillo V, Sola A, Roumeli E, Goudouri O, Paraskevopoulos KM, Rouabhia M, Characterization and in vitro-bioactivity of natural hydroxyapatite based bioglass ceramics synthesized by thermal plasma processing. Ceramics International. 2010;36:1757-1766. DOI:10.1016/j.ceramint.2010.02.048

[110] Zhu X, Zhang H, Fan H, Li W, Zhang X. Effect of phase composition and microstructure of calcium phosphate ceramic particles on protein adsorption. Acta Biomaterialia. 2010;6:1536-1541. DOI:10.1016/j.actbio.2009.10.032

[111] Ni S, Chang J, Chou L, Zhai W. Comparison of osteoblast like cell responses to calcium silicate and tricalcium phosphate ceramics in vitro. Journal of Biomedical Materials Research Part B: Applied Biomaterials: An Official Journal of the Society for Biomaterials, The Japanese Society for Biomaterials, and the Australian Society for Biomaterials and the Korean Society for Biomaterials. 2007;80:174-183. DOI:10.1002/jbm.b.30582

[112] Neo M, Herbst H, Voigt C, Gross U. Temporal and spatial patterns of osteoblast activation following

implantation of β-TCP particles into bone. Journal of Biomedical Materials Research: An Official Journal of the Society for Biomaterials, The Japanese Society for Biomaterials, and the Australian Society for Biomaterials. 1998;39:71-76. DOI:10.1002/(SICI)1097-4636(199801)39:1<71::AID-JBM9>3.0.CO;2-F

[113] Xin R, Leng Y, Chen J, Zhang Q. A comparative study of calcium phosphate formation on bioceramics in vitro and in vivo. Biomaterials. 2005;26:6477-6486. DOI:10.1016/j.biomaterials.2005.04.028

[114] Jung Y-S, Lee S-H, Park H-S. Decompression of large odontogenic keratocysts of the mandible. Journal of Oral and Maxillofacial Surgery. 2005;63:267-271. DOI:10.1016/j.joms.2004.07.014

[115] Dorozhkin SV, Epple M. Biological and medical significance of calcium phosphates. Angewandte Chemie International Edition. 2002;41:3130-3146. DOI:10.1002/1521-3773(20020902)41:17<3130::AID-ANIE3130>3.0.CO;2-1

[116] Klein C, Driessen A, De Groot K, Van Den Hooff A. Biodegradation behavior of various calcium phosphate materials in bone tissue. Journal of Biomedical Materials Research. 1983;17:769-784. DOI:10.1002/jbm.820170505

[117] Abdel-Aal E, Dietrich D, Steinhaeuser S, Wielage B. Electrocrystallization of nanocrystallite calcium phosphate coatings on titanium substrate at different current densities. Surface and Coatings Technology. 2008;202:5895-5900. DOI:10.1016/j.surfcoat.2008.06.139

[118] Schenk RK, Buser D. Osseointegration: a reality. Periodontology 2000. 1998;17:22-35. DOI:10.1111/j.1600-0757.1998.tb00120.x

[119] Sailer I, Gottnerb J, Kanelb S, Hämmerle CH. Randomized controlled clinical trial of zirconia-ceramic and metal-ceramic posterior fixed dental prostheses: a 3-year follow-up. The International Journal of Prosthodontics. 2009; 22:553-560. DOI:10.5167/uzh-25315

[120] Etman MK, Woolford MJ. Three-year clinical evaluation of two ceramic crown systems: a preliminary study. Journal of Prosthetic Dentistry. 2010;103:80-90. DOI:10.1016/S0022-3913(10)60010-8

[121] Denry I, Kelly JR. State of the art of zirconia for dental applications. Dental Materials. 2008;24:299-307. DOI:10.1016/j.dental.2007.05.007

[122] Kelly JR, Denry I. Stabilized zirconia as a structural ceramic: An overview. Dental Materials. 2008;24:289-298. DOI:10.1016/j.dental.2007.05.005

[123] Galante R, Figueiredo-Pina CG, Serro AP, Additive manufacturing of ceramics for dental applications: A review. Dental Materials. 2019; 35:825-846. DOI:10.1016/j.dental.2019.02.026

[124] Lee PH, Chung H, Lee SW, Yoo J, Ko J. Review: Dimensional accuracy in additive manufacturing processes. Proceedings of the ASME 2014. International Manufacturing Science and Engineering Conference MSEC2014, 2014, Detroit, Michigan, USA 2014(45806):V001T004A045. DOI:10.1115/MSEC2014-4037

[125] Zocca A, Colombo P, Gomes CM, Günster J. Additive manufacturing of ceramics: issues, potentialities, and opportunities. Journal of the American Ceramics Society. 2015;98(7):1983-2001. DOI:10.1111/jace.13700

[126] Gao W, Zhang Y, Ramanujan D, Ramani K, Chen Y, Williams CB, Wang CCL, Shin YC, Zhang S,

Zavattieri PD. The status, challenges, and future of additive manufacturing in engineering. Computer Aided Design. 2015;69:65-89. DOI:10.1016/j.cad.2015.04.001

[127] Gonzalez-Gutierrez J, Cano S, Schuschnigg S, Kukla C, Sapkota J, Holzer C. Additive manufacturing of metallic and ceramic components by the material extrusion of highly-filled polymers: A review and future perspectives. Materials. 2018;11:1-36. DOI:10.3390/ma11050840

[128] Beuer F, Schweiger J, Eichberger M, Kappert HF, Gernet W, Edelhoff D. High-strength CAD/CAM-fabricated veneering material sintered to zirconia copings-a new fabrication mode for all-ceramic restorations. Dental Materials. 2009;25:121-128. DOI:10.1016/j.dental.2008.04.019

[129] Silva NR, Witek L, Coehlo P, Thompson V, Rekow ED, Smay J. Additive CAD/CAM process for dental prosthesis. Journal of Prosthodontics. 2011;20:93-96. DOI:10.1111/j.1532-849X.2010.00623.x

Section 2

Types of Ceramic Materials

Chapter 2

Ferroelectric Glass-Ceramic Systems for Energy Storage Applications

Abdulkarim Ziedan Khalf

Abstract

An overview of ferroelectric glass ceramics, some literature review and some of the important previous studies were focused in this chapter. Nanocrystalline glass–ceramics containing ferroelectric perovskite-structured phases have been included. All modified glasses having ferroelectric ceramics which prepared by different methods are discussed, that producing nanocrystalline glass–ceramics. Then particular tested to their use as dielectric energy storage materials. These materials exhibit promising dielectric properties, indicating good potential for high energy density capacitors as a result of their nanocrystalline microstructures. The results of the analysis are summarised in this chapter to provide an overview of the energy storage characteristics of the different materials produced during the study.

Keywords: ferroelectric, glass-ceramics, BaTiO3, KNN, energy storage, crystallisation

1. Introduction

Nanocomposite materials can be obtained through the crystallisation of the grain-boundary glass phase in a ceramic matrix; the electrical and structural properties are improved with glass additives [1]. Over the last few decades, the field of electronic ceramics applications has been progressing. Some newer applications, such as in low temperature co-fired ceramics (LTCCs) and dynamic random access memories (DRAMs), utilise the material's dielectric properties. LTCC applications require the sintering temperature to be below the melting point of the electrode materials [2]. The chemical processes of adding glass and using starting materials with ultra-fine particle sizes improve the characteristics of ceramics at low sintering temperatures [3]. Glass additives can have useful effects on the dielectric constant due to their effect in broadening the diffusive phase transition at the Curie temperature, something that is desirable in the application of multilayer ceramic capacitors [4].

Glass–ceramics are ceramic materials that are produced through the controlled nucleation and crystallisation of glass through thermal treatment. Depending upon the chemical composition and microstructure of glass–ceramics, they can exhibit useful thermal, optical, chemical, mechanical, electrical, and magnetic properties. Useful composite materials can be produced by combining glass–ceramics and other materials, such as metals [5]. Low sintering temperatures and high relative permittivity are of primary importance in the manufacture of ferroelectric ceramics. Typical dielectric ceramic materials, such as barium titanate (BaTiO3) and lead titanate zirconate (PZT), have found many applications in the electronics industry. Certain additives for BaTiO3 and PZT, such as LiF and PbO-B2O3-SiO2, can reduce their sintering temperatures to around 900°C and improve their ferroelectric properties, making them suitable for a range of different electronic applications [6, 7].

It is known that the functional properties of ceramic materials are strongly dependent on microstructure, which provides opportunities to develop new or improved ceramic materials through microstructural engineering. One of these approaches involves combining ferroelectric perovskite with glass-forming oxides in order to form ferroelectric glass–ceramics [5, 8]. The microstructure of such materials comprises ferroelectric nanocrystals dispersed within a glass matrix [9], giving rise to novel materials having pore-free, fine-grained microstructures, low thermal expansion coefficients, high mechanical strength, high chemical stability and good dielectric properties [10]. Such materials have potential applications in high energy density capacitors [11], as well as piezoelectric [12] and electro-optic devices [10, 13]. The glass–ceramic processing route can provide well-controlled microstructure, formed by the crystallisation of chemically and microstructurally homogeneous glasses, at relatively low cost [5, 8].

2. Fundamentals of Ferroelectrics

Ferroelectrics are insulating solids that have spontaneous polarisation. This means that they contain a permanent polarisation at the unit cell level, even in the absence of external electric fields. Additionally, ferroelectric materials exhibit the ability to alter the orientation of their polarisation between two or more directions when under the influence of external electric fields. In order to exhibit spontaneous electric polarisation, there must be a noncentrosymmetric arrangement of the ions and their electrons in these materials. Many ferroelectric materials have perovskite structures with a general chemical formula of ABO3 ABO3-type oxides are known to stabilise with a wide range of A (Pb, Ba, Ca, Sr) and B (Ti, Zr, Sn) ions, with A ions having larger ionic radii than B ions.

Ferroelectrics have typical properties which are essential for their use in electronic devices. High relative permittivity and low-loss dielectric characteristics are most important in multilayer ceramic capacitors (MLCC), which are widely used in electronic devices. There have been progressive developments in the manufacture of MLCCs to increase both the relative permittivity and the number of layers, as well as decreasing the layer thickness, t, according to the equation below [14, 15].

$$C = \frac{\varepsilon_o \varepsilon_r A}{t} \tag{1}$$

where ε_r is the relative permittivity or relative dielectric constant. The capacitance itself is dependent upon εr, the area of the parallel plates, A, and the thickness of the dielectric material, t.

Ferroelectrics are polar crystals with the ability to alter their polarisation direction upon the application of an external electric field. They exhibit spontaneous polarisation, even in the absence of external electric fields. In the unit cell, net permanent dipole moments are present in ferroelectric materials. In polycrystalline ceramics, the orientation of the dipole moments are random and therefore a net polarisation is not normally present after cooling through TC in the absence of an external electric field. The overall orientation of the dipole moments in polycrystalline and single crystal ferroelectrics are not completely random at the scale of the unit cell, since they form ordered groups, referred to as domains. Within the domains, there is a uniform alignment of dipoles, with neighboring domains being separated by boundaries known as domain walls.

The direction of spontaneous polarisation in ferroelectrics can be altered through an applied electric field, as shown in **Figure 1**. With the increase of the electric field, the domains begin to align, giving rise to an increase and saturation in the polarisation at high field. In the absence of an external electric field, some of the domains remain aligned. Thus, the crystal displays remnant polarisation. If the field is reversed, the domains change direction. The direction of polarisation flips and produces a hysteresis loop when the external electric field alternates between negative and positive [16, 17].

2.1 Energy storage in capacitors

Significant improvements over the last couple of decades in both the energy storage density and reliability of capacitors have been achieved through a combination of novel materials, diagnostic methods, and manufacturing techniques. Capacitors, inductors, and batteries are means through which electrical energy is stored. **Figure 2** depicts a graph of the specific energy for different energy conversion and storage devices plotted against their specific powers [18].

The characteristics of energy-storage in four types of the most highly studied dielectric materials, namely, relaxor ferroelectrics, polymer-based ferroelectrics, antiferroelectric, and dielectric glass–ceramics were reviewed by Hao [19].

The changes in polarisation upon the application of an electric field are a critical aspect of energy storage dielectrics. This response can be used to estimate the stored energy, which should exclude hysteresis losses. Dielectrics may be grouped into being either linear or non-linear, according to the relationship between the applied electric field and the polarisation. A simple equation (below) may be used to describe their behavior [20].

$$D = \varepsilon_o E + P = \varepsilon E \qquad (2)$$

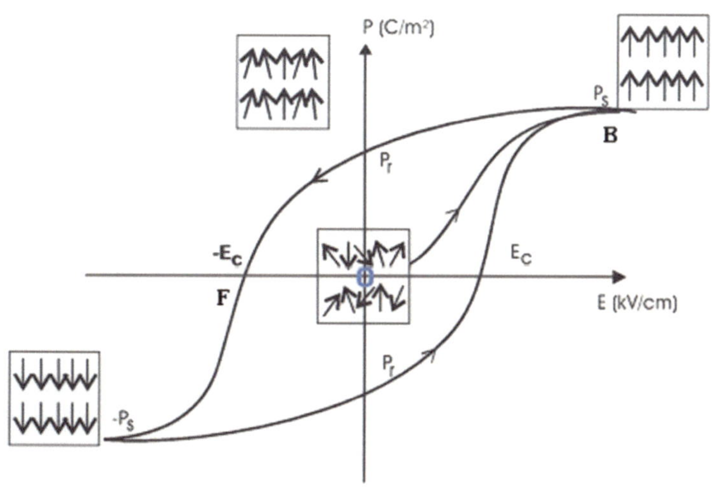

Figure 1.
Illustration of the polarisation-electric field relation, P-E hysteresis loop, for a typical ferroelectric crystal [16].

Therefore:

$$P = \varepsilon_o(\varepsilon_r - 1)E = x_e\varepsilon_o E \qquad (3)$$

where χ is dielectric susceptibility and D is the dielectric displacement.

Energy density, U, is a measure of the energy stored per unit volume. For dielectrics, this can be obtained by the following relationship:

$$U = \int_0^{E_{max}} PdE \qquad (4)$$

Using formula above (Eq. (4)), the U values of the dielectrics can be obtained through the numerical integration of the area between the polarisation and curves for the electric-field polarisation (P-E) loops. **Figure 3**, shows that upon reaching the maximum electric field strength (E_{max}), the polarisation approaches its

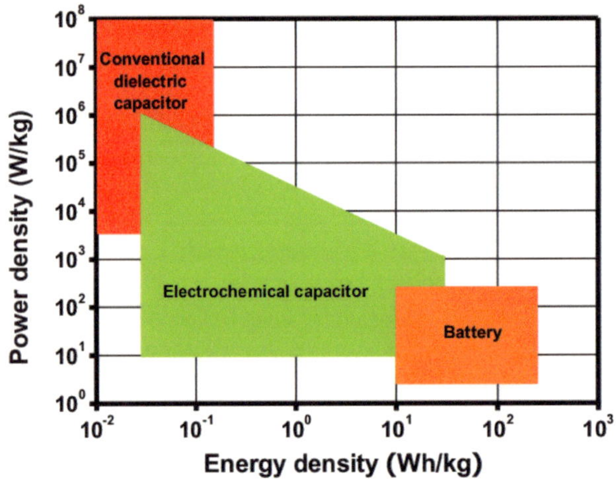

Figure 2.
Diagram of power density as a function of energy density in different energy-storage devices [19].

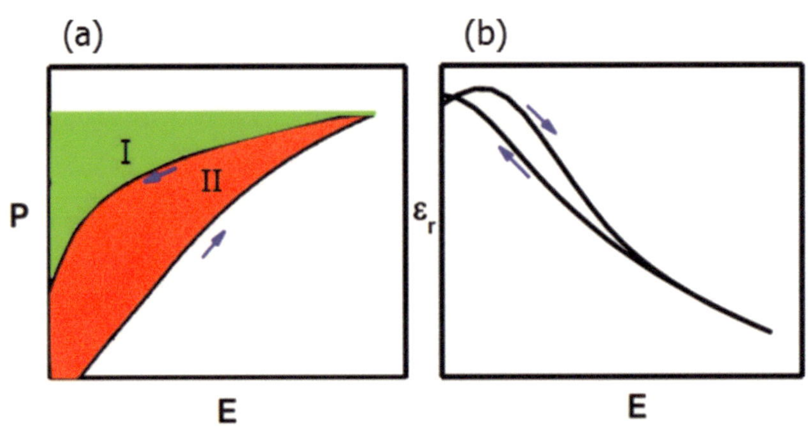

Figure 3.
The typical dependence of (a) polarisation and (b) relative permittivity on the electric field of ferroelectrics in the first quarter shows the charge–discharge cycle. The area I (green shaded area) corresponds to the discharged or recoverable, energy density and area II (red shaded area) correspond to the energy density loss [19].

maximum (P_{max}) and the capacitor holds the electrical energy (U_{store}), as illustrated by the red and green areas.

The recoverable electrical energy density (U_{rec}) is released during the discharge process when the electrical field reduces from E_{max} to zero. This is represented by the green area in **Figure 3**. Therefore, an amount of the stored energy (the red segment surrounded by the loops) is dissipated during the process of depolarisation, denoted the hysteresis loss, U_{loss} [19, 21].

The above analysis indicates that there are three prerequisites to designing an effective dielectric material for practical use with high efficiency and high recoverable energy-storage density. These three requirements need to be satisfied simultaneously and are small remnant polarisation, large saturation polarisation, and a high electric breakdown field [22].

Figure 4(a)-(d) depicts typical P-E loops and an illustration of the energy-storage of four types of dielectrics: (a) linear dielectric with constant permittivity (e.g. Al_2O_3, glass), (b) antiferroelectric with zero net remnant polarisation (e.g. $PbZrO_3$), (c) ferroelectric with spontaneous polarisation (e.g. $PbTiO_3$, $BaTiO_3$), and (d) relaxor ferroelectrics with nanosized domains, e.g. (Pb,La)(Zr,Ti)O_3.

Even though linear dielectrics often have lower energy losses and higher breakdown fields, small polarisation values resulting from the use of low-permittivity dielectrics can reduce their effectiveness for high-energy storage purposes, unless very high breakdown fields can be achieved. Ferroelectrics generally have moderate electric field endurances and larger saturated polarisations, however, due to their larger remnant polarisations, they are often less efficient and have smaller energy-storage densities. **Figure 4** demonstrates that antiferroelectrics and relaxor ferroelectrics are more attractive for high energy storage due to their relatively moderate breakdown fields, smaller remnant polarisations, and larger saturated polarisations.

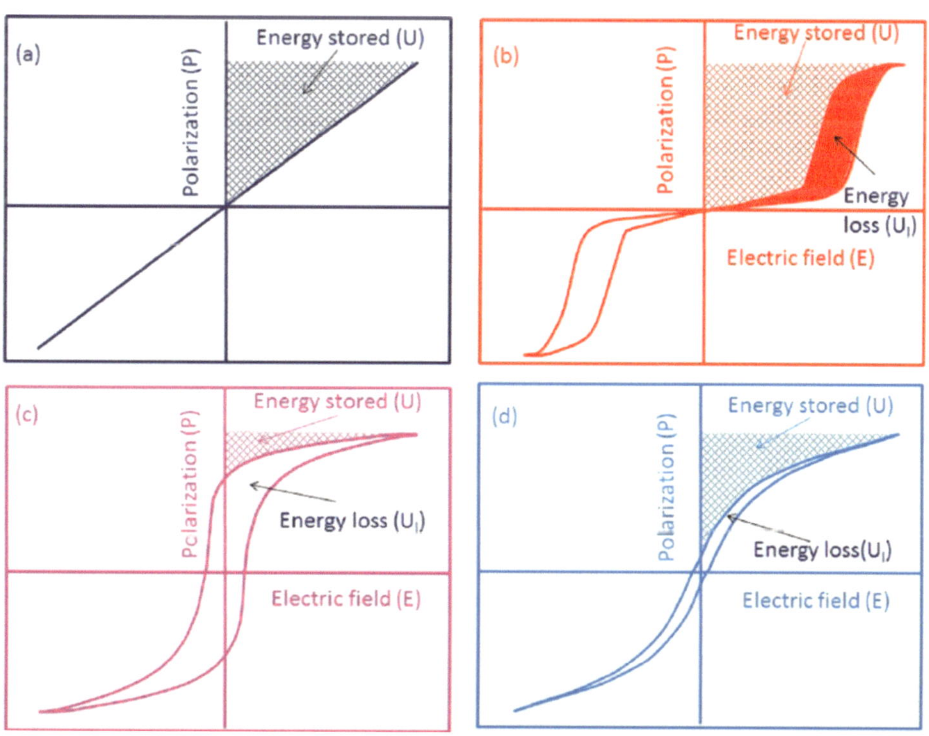

Figure 4.
Schematic description of the energy storage characteristics of (a) linear dielectrics, (b) antiferroelectrics, (c) ferroelectrics, and (d) relaxor ferroelectric ceramics [23].

Novel manufacturing processes, such as the use of composite technology and glass-crystallisation techniques, have allowed for the production of ceramic-polymer composites and glass–ceramics. These materials could potentially combine the larger polarisations of ferroelectrics and the higher breakdown fields of linear dielectrics. Therefore, amongst the aforementioned four groups of dielectrics, namely, relaxor ferroelectrics, ceramic-polymer composites, glass–ceramics, and antiferroelectrics, the former two are generally thought to be the most useful for high energy storage purposes and therefore much research has been conducted on these two types of material [19, 23].

Pb(Zr,Ti)O3 (PZT) based materials have been widely used in energy storage applications because of their high dielectric constant. However, the environmental issues derived from the use of lead have encouraged many searches for more environmentally friendly materials.

The perovskite structure of $BaTiO_3$, capable of high dielectric constant values, spontaneous polarization, low dielectric loss and ferroelectricity offers an alternative for lead-based capacitors. As mentioned earlier on, for energy storage applications a high dielectric breakdown strength is required to allow device miniaturization. It is well known that the energy storage properties of BaTiO3 based ceramics can be improved by reducing the porosity [24], tuning the grain size [25], the addition of glass additives [26], presence of secondary phases, etc. For example, the relative permittivity of BaTiO3 increases as the grain size decreases [27], reaching a maximum of 5000 at grain sizes of about 0.8 to 1.1 μm [28]. This was attributed to domain size and stress effects. Further reductions in the grain size resulted in a rapidly decreased permittivity. Furthermore, the dielectric breakdown strength increases with decreasing grain size [29], being about 8.5 kV mm^{-1} when the grain size is 3.5 μm [30].

The addition of glass additives to induce liquid phase sintering is a widely used technique to improve the energy storage capabilities of BaTiO3 based ceramics. During the liquid phase sintering, a thin layer of the fluxing agent coats the BaTiO3 grains leading to improved relative densities and reduced sintering temperatures. Until now, the use of several glass additives in BaTiO3 ceramics has been proved to show promising results for energy storage applications. For example, Sarkar and Sharma [31] demonstrated that the addition of B2O3 and PbB2O4 to BaTiO3 significantly reduced the sintering temperature to about 800°C, which is suitable for commercial applications as multilayer capacitors. Moreover, they doubled the dielectric breakdown strength of BaTiO3 by the addition of 10 mol% of PbB2O4 [31]. However, this improvement in the dielectric breakdown strength was accompanied by a small decrease in the dielectric constant.

The aliovalent substitution at the Ba2+ and/or Ti4+ sites in the perovskite structure of BaTiO3 has been demonstrated [32] to be an effective approach to tailor the energy storage properties of BaTiO3 to meet industrial application requirements. Recently, Puli et al. [33] investigated the dielectric, ferroelectric and energy density properties of (1-x){BaZr0.2Ti0.8O3}˝-(x){Ba0.7ZCa0.3TiO3} where x = 0.1, 0.15 and 0.20, hereinafter denoted BCZT. They reported a dielectric of the permittivity of 8400 when x = 0.15 and a low loss (tan δ) of 0.014 in samples sintered at 1600°C. **Figure 5** shows the discharge energy density, charge energy density, and energy storage efficiency reported by Puli and co-workers, measured at a maximum electric field of 80 kV cm^{-1} [33]. They achieved an energy storage efficiency of about 70% when x = 0.15.

Wang et al. [34] achieved an energy density of 0.52 J cm^{-3} in a (Ba0.85Ca0.15)(Ti0.9Zr0.1)O3 ceramic prepared by the sol–gel method. They attributed it to the improved microstructure compared to that obtained by the conventional solid-state reaction method. In order to simultaneously attain high dielectric breakdown

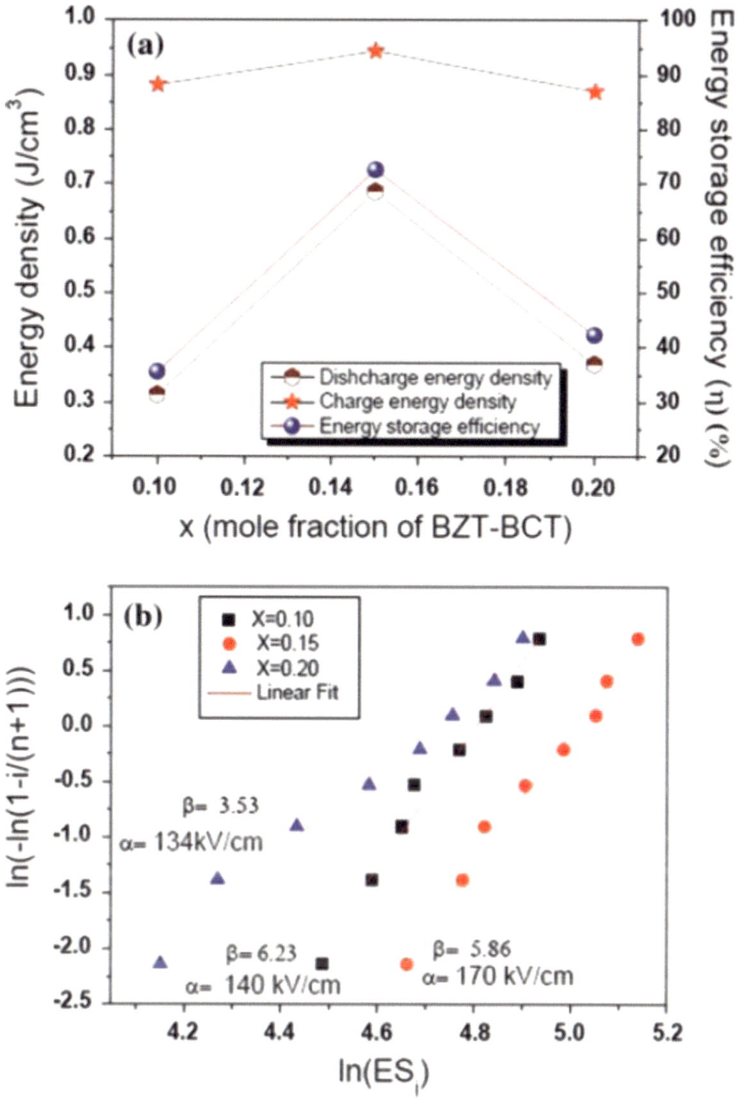

Figure 5.
(a) Composition dependence of recoverable energy density (U_{rec}), stored energy density (U_{stor}), and energy storage efficiency (η %) of (1-x){BaZr0.2Ti0.8O3}-(x){Ba0.7ZCa0.3TiO3} where x = 0.1, 0.15 and 0.20 (b) Weibull plots of the breakdown strength of BCZT ceramics sintered at 1600°C [33].

strength, high energy density and a high dielectric constant in a material, the glass–ceramic concept has been devised. Here, the high dielectric breakdown of the linear dielectric (glass) and the high dielectric constant/large polarization typical of ferroelectric ceramics are combined in a nanostructured composite-type material. Puli et al. [35] followed the glass–ceramic approach to improve the energy storage properties of BCZT ceramics. They added 15 wt% of two different alkali-free glass compositions, namely 0.1BaO + 0.4B2O3 + 0.5ZnO and 0.3BaO + 0.6B2O3 + 0.1ZnO, to BCZT, they reported a slight improvement in the dielectric breakdown field to about 28 kV mm^{-1} but a lower energy density compared to glass-free BCZT. The low energy density values reported were attributed to the low relative permittivity values (≈ 270) for glass–ceramic composition.

Another lead-free perovskite material that exhibits useful ferroelectric properties is the solid solution system potassium-sodium niobate (KNN). The solid solution in the binary system KNbO3-NaNbO3 crystallises as an orthorhombic perovskite, [36], with the composition around K0.5Na0.5NbO_3 being the most popular due to its closeness to the morphotropic phase boundary (MPB) which occurs at about 52.5% Na [37]. The solid solution (K,Na)NnO3 exhibits ferroelectric behaviour which diminishes at high sodium additions until it completely disappears due to the nonpolar, antiferroelectric end-member NaNabO3 [38, 39]. The dielectric constant of K0.5Na0.5NbO_3 at room temperature is about 290 [38] and reaches 990 at 473 K. The use of additives to reduce the grain size and to improve the energy storage abilities of KNN ceramics has shown promising results. Qu et al. [40] achieved an energy storage density of 2.48 J cm^{-3} and a breakdown strength of 29.5 kV mm^{-1} by reducing the grain size of KNN to 0.5 μm through the addition of Sr.(Sc0.5Nb0.5)O_3 (SSN), although they reported the presence of porosity at the grain boundaries. Highly dense KNN-SNN samples were achieved through the addition of 0.5 mol% ZnO, leading to a breakdown strength of 40 kV mm^{-1} and an energy storage density of 2.6 J cm^{-3} [41].

3. History and definition of glass-ceramics

Glass-ceramics are classified as ceramic materials. They are polycrystalline materials that are formed by controlling the crystallisation of an amorphous glass. These materials are an important type of electroceramic and were successfully developed during the 1940 and 1950s. During this period, S. D. Stookey (Corning, USA) successfully used glass–ceramics as electrical insulators in electronics technology [5, 20]. The fundamental patent of Stookey was based on the concept that the TiO2 works as a nucleating agent in a glass system. Additionally, ZrO_2 was used by Tasiro and Wada [5], in 1963 as a nucleating agent. Another discovery was made in the 1950s by Hummel, who discovered the crystal arrangement of the Li_2O-Al_2O_3-2SiO_2 system [42].

Figure 6 shows three types of an atomic structure with different atomic arrangements. A crystalline solid is one which has a long-range order in its atomic structure; an amorphous solid is one in which there is no long-range order in its atomic structure. Crystalline solids have two subdivisions, single crystal and polycrystalline. A single crystal has a periodic atomic arrangement. In this case, there are no grain boundaries. A polycrystalline solid contains many grain boundaries in the structure due to the differences in the orientations of the grains (that have a short-range order) [43].

It is difficult to specifically define a glass since the behaviour of a glass alters with changes in temperature. There are two points at which a glass can be defined; the

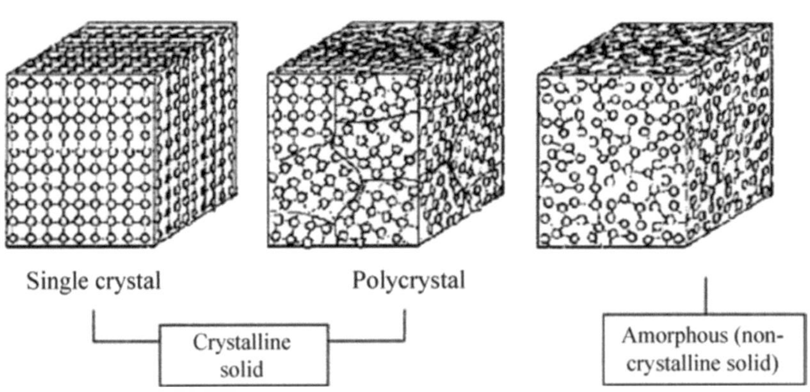

Figure 6.
The structure of ceramics [43].

first is at high temperature, when the glass is a liquid, while the second is at a lower temperature when the glass is considered as a supercooled liquid. Although there are important structural differences between glasses and polycrystalline ceramics, their mechanical and functional properties at room temperature can be similar.

The traditional definition of glass is that it is a supercooled liquid. According to the American Society for Testing and Materials (ASTM), the definition of glass is that it is an inorganic product of fusion which has cooled to a solid state without being crystallized. However, there are alternate definitions for glass, one being that glass is a type of amorphous solid material that lacks long-range order (not a random arrangement) in its atomic structure. Another definition, also put forward by ASTM, it that a glass is a liquid that has lost its ability to flow [5, 16].

3.1 Heat treatment of glass-ceramics

The heat treatment of glass leads to the occurrence of many transitions. Differential scanning calorimetry (DSC) is a form of thermal analysis that depends upon the change in a material's physical properties [42]. In DSC, there is a difference in temperature (ΔT) that is seen between the sample and the reference. Here ΔT represents differences in heat flow as ΔQ. The two quantities, ΔT and ΔQ, are functions of thermal resistance (R), as shown in equation below.

$$\Delta Q = \frac{\Delta T}{R} \tag{5}$$

In the first step of the glass transition, some of the physical properties change for amorphous materials. This change occurs in the heat capacity, which can be measured by DSC as an endothermic change in the sample. The transitions in glass due to the effects of temperature occur in the range of temperature which is known as the glass transition temperature (Tg). Therefore, below Tg, materials display a rigid glassy structure. When the temperature is increased above Tg, these materials display a flexible structure.

Another transition which occurs due to changing temperature is crystallisation. In this case, the amorphous materials are transformed into a crystalline structure. With an increase in temperature, the next conversion is melting. At this point, the crystalline structure converts to a viscous amorphous structure. The melting point is dependent upon the chemical impurity of the materials. After the melting stage, a reaction inside the material causes an increase in the density of the material [5, 42].

3.2 Crystallisation of glass-ceramics

Generally, since 1960, there has been much research undertaken regarding glass systems in the field of glass–ceramics. Glass–ceramics are very important in many fields of application. They have demonstrated many desirable thermal, optical, biological, chemical, and electrical properties. Some of these properties provide advantages to glass–ceramics over more traditional materials. A glass–ceramic is a polycrystalline material formed by controlling the crystallization of glass. Therefore, in order to make glass–ceramics from glass, the main manufacturing process needs to be a thermal one. **Figure 7** shows the steps of glass transforming into glass–ceramic. These steps begin at a low temperature with the formation of nuclei, then at higher temperatures crystallisation occurs by growth of the nuclei; this continues to produce the polycrystalline a glass–ceramic microstructure [5, 43].

Microstructural control is said to be easier when the temperature required for crystallisation lies between but is significantly different from both the glass transition temperature and that of matrix devitrification. In such a case, the desired crystalline phase can be induced to form without devitrification of the glass matrix. The crystallite size generally increases with increasing temperature, as shown by the micrographs in **Figure 8** [44].

In **Figure 9**, the typical thermal preparation of glass–ceramic can be seen. In this case, the raw materials, Li2CO3 and SiO2, are used to create lithium disilicate. There are two main stages in obtaining glass–ceramics: glass formation and glass crystallisation. In each stage, there are many steps which depend upon both temperature and time. The first stage begins by melting the components and then quickly cooling them. The nucleation and crystal growth occurs in the second stage. During this stage, controlled crystallisation of the glass produces nanoscale crystals [5].

3.3 Glass-ceramic properties and applications

Glass-ceramics can be containing a substantial glassy phase if the range is highly crystalline. As a result of that, glass–ceramics can contain grain boundaries as well as it can range from transparent to opaque. Therefore, the properties of glass–ceramics depending on the microstructure and the chemical composition, which can be tuned to meet demanding requirements. In general, the most important of glass–ceramics properties that can be exhibited is about zero thermal expansion with high toughness. as well as, have a high impact resistance with resistant to thermal shock [5, 8].

The potential applications of glass–ceramics in energy storage capacitors was investigated by Du et al. [11]. Here, the Na2O-PbO-Nb2O5-SiO2 glass–ceramics system achieved a highest relative permittivity of >600 after heated the sample at 850°C. The results given by Gorzkowski et al. [46] shown that the interfacial

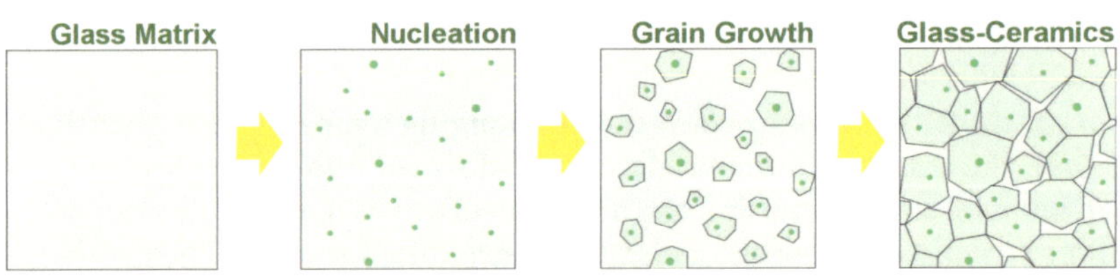

Figure 7.
Schematic representation of glass–ceramic processing [5].

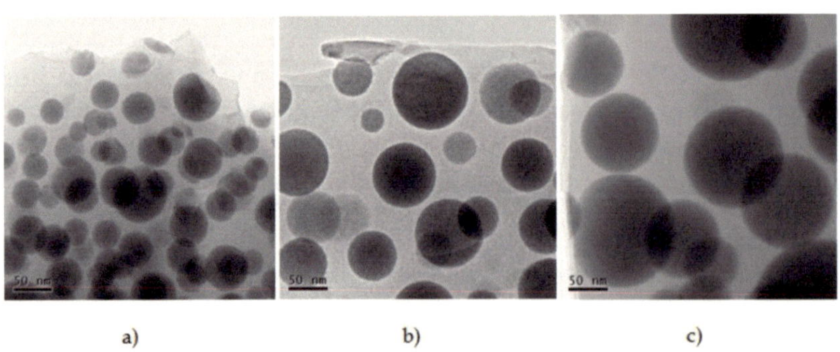

Figure 8.
TEM micrographs showing Nanoscale crystallites precipitated from a glass matrix of SiO2/Lu2Si2O7 heat-treated at 1100°C (a), 1200°C (b) and 1300°C (c) [44].

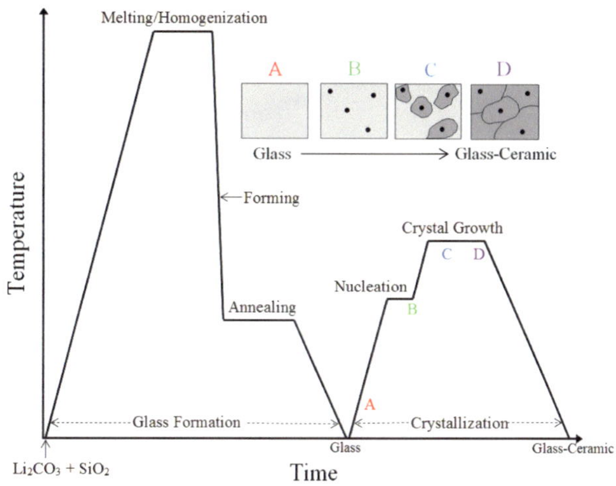

Figure 9.
Schematic of glass processing in the formation of a glass ceramic [45].

polarization due to the difference in the dielectric and conductivity of glass phase and ceramic phase large hysteresis loss, led to low energy density (Max. 0.9 J cm^{-3}) and breakdown strength (Max. 800 kV cm^{-1}) of the BST glass–ceramics that expected much higher energy density ($\approx$ 3.5 J cm^{-3} in case assuming is it ideal linear dielectric behavior) [46].

As the energy crisis intensifies, the search for novel, high performance and environmentally friendly energy storage devices have attracted wide attention. For a material to possess high energy density capabilities, a high relative permittivity and dielectric breakdown strength are needed; this is a challenging task since theoretical and experimental studies on crystalline materials usually show a contrary relationship between these properties [47]. Compromises between relative permittivity and dielectric breakdown strength have encouraged the search for more efficient materials such as polymers and glasses. Recently, a high relative permittivity and a high dielectric breakdown strength were found in fluoropolymers, with energy density values up 25 J cm^{-3} being reported [48]. A high dielectric breakdown of $\approx$ 12 MV cm^{-1} (1.2 MV mm^{-1}) was also reported [49] in a low relative permittivity ($\varepsilon r \approx 6$) alkali-free glass, showing the possibility to reach an energy density of 35 J cm^{-3}. These results are very promising since the relationships between relative permittivity and dielectric breakdown strength in inorganic glasses are not well established.

Xue et al. [38] studied the effect of glass concentrations on the energy storage capabilities of niobate glass ceramics (100-x)(40BaO-10Na2O-50Nb2O5)-x(63SiO2-12BaO-16B2O3-9Al2O3) prepared via controlled-crystallization. They achieved dielectric breakdown strength of about 130 kV mm^{-1} and a charge–discharge efficiency of 92.5%, with a relative permittivity of about 50, see **Figure 10** below.

Glass-ceramics are amongst the most important hosts for transition metal ions [51]. There are several benefits of adding glass to ceramics, such as improving their dielectric properties and reducing both the sintering temperature and the porosity [6]. In glass–ceramics, the ferroelectric and dielectric properties are highly dependent on the characteristics of the glass matrix as well as the crystal volume fraction and crystal size [52]. Glass-ceramics are crucial in the development of more efficiently produced and controlled energy. They are thought to be efficient cathodes or solid electrolyte materials when used in lithium batteries. Dielectric glass ceramics are also of interest due to their uses in high-power microwave systems, distributed power systems, power electronics, and pulsed power.

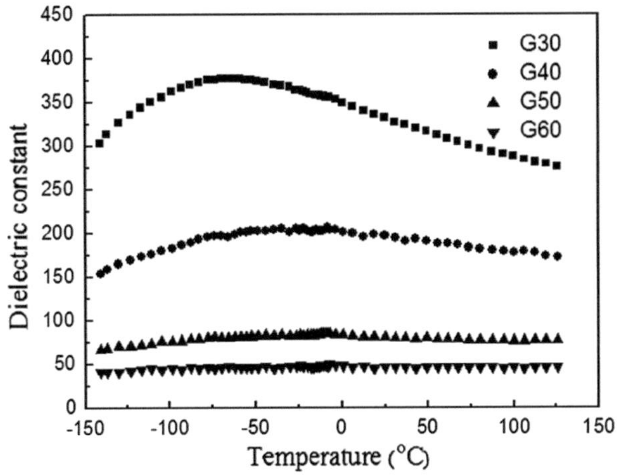

Figure 10.
Relative permittivity versus temperature curves of niobate glass–ceramics prepared via the controlled-crystallisation route, samples were annealed at 1100°C. When the glass content was 30–60%, the samples were labelled G30 to G60, respectively [50].

To decrease the thickness of the dielectric layers in a capacitor and reduce the weight of portable pulsed power systems, it may be necessary to have high dielectric breakdown strength in the materials. In any case, improved control of the porosity, along with enhanced energy storage capabilities, are important aspects of improving the performance of glass–ceramics [53]. The significance of glass–ceramics in many applications lies in the possibility of utilizing key properties such as transparency, strength, resistance to abrasion, and the controlled coefficient of thermal expansion. All these properties can be manipulated by the controlling the composition, extent of crystallisation, crystal morphology, crystal size, and aspect ratio of the materials [54]. Glass-ceramics are used in a range of applications across different fields such as telecommunication, radar, and navigation. With respect to the use of high-density electron device packages, the materials which are used in LTCC applications need to have good dielectric properties and a large coefficient of thermal expansion (CTE). These characteristics are important in order to achieve thermal matching between the components of the package [55, 56].

3.4 Chemical compositions of glass-ceramic systems

3.4.1 Alumina-silica systems

Silicates are salts of silicic acid, H4SiO4. Most of the silicates contain SiO2, while in the case of SiO2 or quartz glass, only SiO2 is present. For all other types of glass, additional oxides, such as: Li2O, BaO, B2O3, Na2O, K2O, MgO, CaO, or Al2O3 are employed as network formers or modifiers. Alkaline and alkaline earth oxides are important in glass due to their effects on the network. Therefore these additives work as effective network modifiers [57]. The alumina-silica system, Al2O3-SiO2, is amongst the most important binary oxides and ceramic systems. **Figure 11** shows the phase diagram of the binary Al2O3-SiO2 system. The low solid solubility for SiO2 in Al2O3, and vice versa, can be seen in this diagram. At a high temperature below 1890°C, part of the Al2O3 component is transformed to a molten state, while the other part remains solid. This also occurs with SiO2 at a temperature below 1600°C. The effect of the SiO2 ratio in reducing the melting point of Al2O3-SiO2, can be seen [58].

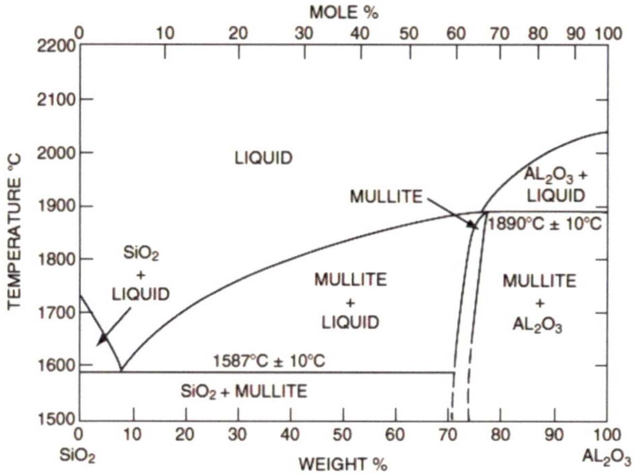

Figure 11.
Phase diagram of the binary alumina-silica system [58].

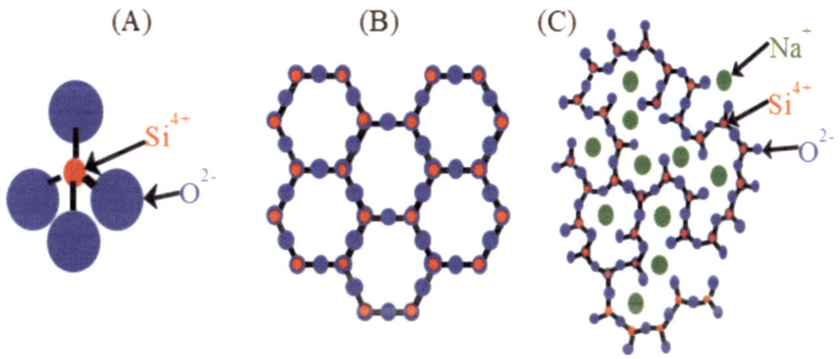

Figure 12.
Silica network structure: (A) silica tetrahedron, (B) pure silica quartz, (C) and a soda lime silicate [57].

Figure 12 shows the silica structure as a network. In **Figure 12-A** the basic units for silica are represented; these are Si4+ with O2-. The network of pure silica, also known as quartz, is presented in **Figure 12-B**, while **Figure 12-C** shows the structure of this 2D network with the addition of some alkaline additives as modifiers [57].

3.4.2 Alkaline and alkaline earth silicates

Glass-ceramics can be modified by a formulation process. Alkaline and alkaline earth oxides are modifiers of glass networks and act to decrease the connectivity of the glass network, effecting changes in the properties of the glass. There are many oxides that are used to modify glass-ceramics, such as: Ba, Li, B, Ca, Na, etc. [57]. These glass-ceramic systems contain binary and ternary systems. Both types of systems contain oxides, such as: BaO, B2O3, MgO, CaO, PbO, ZnO, Al2O3, SiO2, and SrO [5]. Some of the oxides contained within the glass are popular as sintering aids for ceramics. This is because of the high stability of their structures and the low glass transition temperatures, in addition to their thermal and electrical properties. The systems of PbO-containing glasses (e.g. PbO-B2O3-SiO2 and PbO-B2O3-ZnO) are important due to the aforementioned reason [59, 60].

Other glass systems include Bi_2O_3-B_2O_3, BaO-CaO-Al_2O_3-B_2O_3-SiO_2, CdO-Bi_2O-PbO-B_2O_3, and BaO-B_2O_3-SiO_2 which have all been used to reduce the sintering temperature [3, 61]. The structure of PbO is not easy to crystallise, even with a high percentage of it within the glass, because PbO_4 is formed within the glass system. Recently, many glass systems have been developed as alternative lead-free materials. In this research, BaO, ZnO, and Bi_2O_3 are used instead of PbO [61]. Most glasses containing a high proportion of PbO are undesirable due to the toxicity of lead oxide causing environmental problems during or after production. Therefore, most researchers focus on decreasing the PbO content or replacing it with other oxides. This issue was addressed by Bobkova and Khot'ko [62], who were studied the ZnO-SrO-B_2O_3 system. They found that the optimum glass was produced when it contained a high ZnO and low B_2O_3 content. In addition, two ternary systems, B_2O_3-CaO-SiO_2 and B_2O_3-CaO-Al_2O_3, have been studied by Vartanyan et al. [63]. They found that these two systems successfully produced lead-free materials for low-temperature simultaneously fired ceramics (LSCs).

Glass manufacturing processes can be generally complex due to the materials which undergo different changes during the melting and cooling. The ternary systems of B_2O_3-CaO-Al_2O_3, B_2O_3-CaO-SiO_2, and B_2O_3-ZnO-SiO_2–BaO were developed at low melting points below 1000°C then used as sintering aids with ceramics at low sintering temperatures [62, 63]. Consequently, these systems (BaO-B_2O_3-SiO_2) and (BaO-B_2O_3-SiO_2- Al_2O_3) possess their own particular thermal physical properties, where achieved a coefficient of thermal expansion (11–17 ppm $°C^{-1}$) and low permittivity ($\approx$7) [64]. B_2O_3-Bi_2O_3-SiO_2-ZnO (BBSZ) is commonly used as an appropriate additive to reduce the sintering temperature of many dielectric materials and make them suitable for LTCC applications [65, 66]. Therefore in order to improve the electromagnetic properties of Ni–Cu–Zn ferrite (NCZF) and $BaTiO_3$, the BBSZ glass was chosen as a melting agent [67]. Glass-ceramics based on the CaO–MgO–SiO_2 system has been studied with the additives B_2O_3, P_2O_5, Na_2O, and CaF_2. The results showed that these systems required high melting points and the final crystallisation of the glass-ceramic occurred with temperatures of 900°C, producing high levels of density [68].

3.4.3 Barium borate system

The binary system of BaO-B_2O_3 is used in a wide range of glass ceramic production, oxide cathodes and the coloured materials in paints. In addition, these systems are used as sintering aids for low temperature co-fired ceramic applications due to the low softening temperature of the B_2O_3 [69]. **Figure 13** shows the phase diagram for the binary system of BaO-B_2O_3 [70]. The procedures of melting and cooling the BaO-B_2O_3 systems lead to glass formation. In the phase diagram, it can be seen that the lowest melting point with high stability for these systems occurs with approximately 60 to 80% B_2O_3. In practise, the optimum ratio also depends upon the rate of cooling of the glass [70].

The binary system of BaO-B_2O_3 was modified by SiO_2 or Al_2O_3 as a network modifier. Therefore, it was expected that the mineral phase with odd physical characteristic would be obtained. Therefore, glass ceramics are prepared with low sintering temperatures [64]. It should be noted that the melting point of B_2O_3 is around 450°C, which is much lower than SiO_2, which has a melting point of around 1710°C. However, the crystalline glass of B_2O_3 is much more difficult to produce than that of SiO_2 [70].

SiO_2 added to borate glasses generally leads to an improvement in the density of the glass and an increase in its Tg and chemical durability [71]. The effect of the Al_2O_3 on the characteristics of BaO-B_2O_3-SiO_2 was reported by Lim et al. [72].

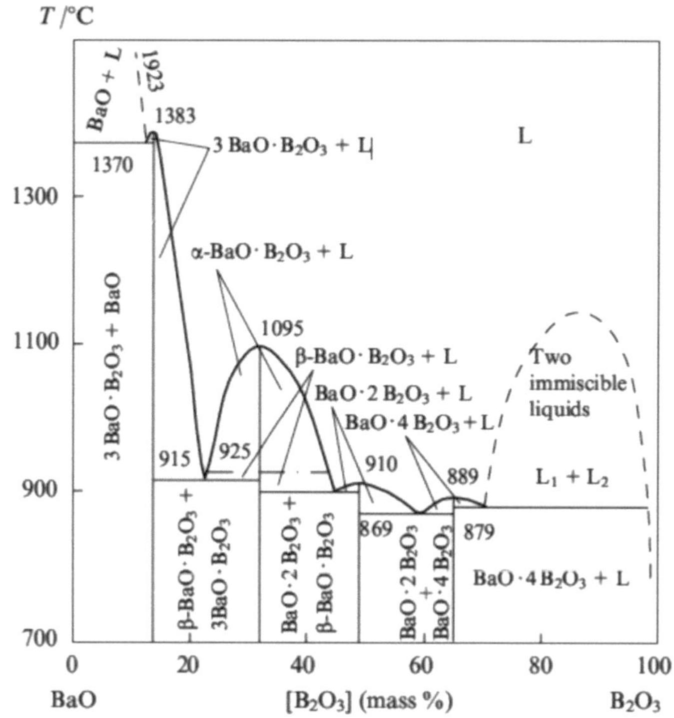

Figure 13.
Phase diagram of the BaO-B2O3 system [59].

It was found that the crystallisation temperature, sintering temperature, and glass transition temperature increased with increased amounts of Al2O3 in the glass systems. The ternary glass system BaO-B2O3-SiO2 was studied by Lim et al. [61, 69]. The SiO2 ratio was fixed at 10% and the effects of the amount of BaO/B2O3 on the thermal and physical properties of the glass system were studied. Increasing B2O3 content gave rise to a clear increase in the dielectric properties. The electrical conductivity of solid silicates shows a complex dependence on glass formation processes, devitrification, and temperature. Here, a slight change in the composition of the glass can lead to marked differences in the electrical conductivity values. In spite of the fact that most studies are about the binary systems, however, there is a scarcity of experimental data on the electrical conductivity of the glasses. Accordingly, the study of the influence of the metal oxide on the electrical conductivity is important for the multicomponent silicate glasses for electrical purposes [73, 74].

It was reported that the electrical conductivity is decreased by the presence of the K+ ions in the alumina-silica glasses while the glass transition temperature increased [73]. Wang et al. noted that the addition of Na2CO3 into Ba0.4Sr0.6TiO3 (BST) ceramics caused a decrease in the room temperature dielectric constant of BST, while the highest Na + and K+ content produced an increase in the energy storage density [75].

4. Ferroelectric glass-ceramic systems

The following sections summarise the results of previous research on the use of glass additives into ferroelectric ceramics, producing ferroelectric glass-ceramic and composites, and on the heat treatment of amorphous glasses to produce glass-ceramics containing nanocrystalline ferroelectric phases. The emphasis of the present research is on lead-free ferroelectric materials and therefore previous work on lead titanate and PZT-based glass-ceramics is not included here.

4.1 Barium titanate systems

BaTiO3-based ceramics have been widely used in a number of electronics and electrical industries, due to their excellent dielectric properties [76]. Glass additives have been used during the sintering process of ceramic materials to form a liquid phase to improve the sintering behaviour of well-known ferroelectric ceramics such as BaTiO3, PZT and KNN. During this process, the reactive liquid formed by the glass additives wets the solid particles and facilitates their rearrangement. This rearrangement of particles gives more effective packing and higher densities, desirable for ferroelectric applications. Furthermore, the capillary pressure developed at each inter-particle space provides an additional driving force for particle rearrangement and re-shaping during sintering, inducing densification.

The pioneering work of Stookey in 1949 [77] on photosensitive glasses led to the discovery of methods for heterogeneous nucleation, where crystal growth takes place from many nuclei dispersed in the glass matrix after the glass has been formed [78]. Later on, Herczog [79] studied the properties and composition of glasses suitable for crystallization of BaTiO3 from glasses. Perovskite BaTiO3 with a minor amount of BaAl2Si2O8 was obtained by heat treating glasses of compositions corresponding to (x)(BaTiO3) + (100-x)(BaAl2Si2O8). The grain size was controlled in the range 0.01 to 1 μm by varying the heating rate and the final heat treatment temperature. When the volume content of BaTiO3 in the glass-ceramic increased from 30 to 60% and the grain size was about 1 μm, the dielectric constant increased from 100 to 1200. For grain sizes less than or equal to 0.2 μm, the relative permittivity was further decreased and found to be independent of temperature; this was attributed to the presence of surface defects in the fine grains. The highest relative permittivity of 1300 at room temperature was achieved at average grain sizes of about 1 μm, as can be observed in **Figure 14**; the breakdown strength and electrical resistivity of this ceramic were also reported to be high.

Nano-sized BaTiO3 (20–80 nm) was produced by McCauley et al. [80] following the same approach as Herczog [79] and using different heat treatments and compositions. They observed intrinsic size effects at crystal sizes lower than 80 nm and predicted a critical size of 17 nm, at which BaTiO3 cannot support a ferroelectric transition. Takahashi et al. [81] reported the glass-ceramic of 0.65(Ba1-xSrx) TiO3–0.27SiO2–0.08Al2O3 (BSTS). The heating temperature of glass-ceramics was 1000°C with a small amount of Sn or Zr; also the relative permittivity also decreased with the addition of both of the Sn and Zr. The εr–T relationship showed a broadening and shift of the Curie point towards room temperature, as shown in **Figure 15**.

The use of solid state reaction for the preparation of BaTiO3 ceramics has advantages where the products have good crystallinity and low cost, as well as a high level of accuracy of stoichiometric control [82]. However, many ferroelectric materials have high sintering temperatures (>1200°C), making them unsuitable for certain applications, such as LTCC. In this case, it is necessary to include some additives to reduce the sintering temperature. There have been many studies aimed at lowering the sintering temperature of BaTiO3. The typical additives used include ZnO, CuO, LiF, and CdO. Amongst these additives, LiF proved to be a very effective sintering aid. The effect of LiF in BaTiO3 was investigated and the results indicate that the sintering point is reduced to approximately 900°C [83, 84].

The use of glass additives in BaTiO3 was also shown to improve densification behaviour and reduce the required sintering temperature. Jeon et al. [85] studied the effects of BaO-B2O3-SiO2 (BBS) glass in BaTiO3 ceramics. It was found that the

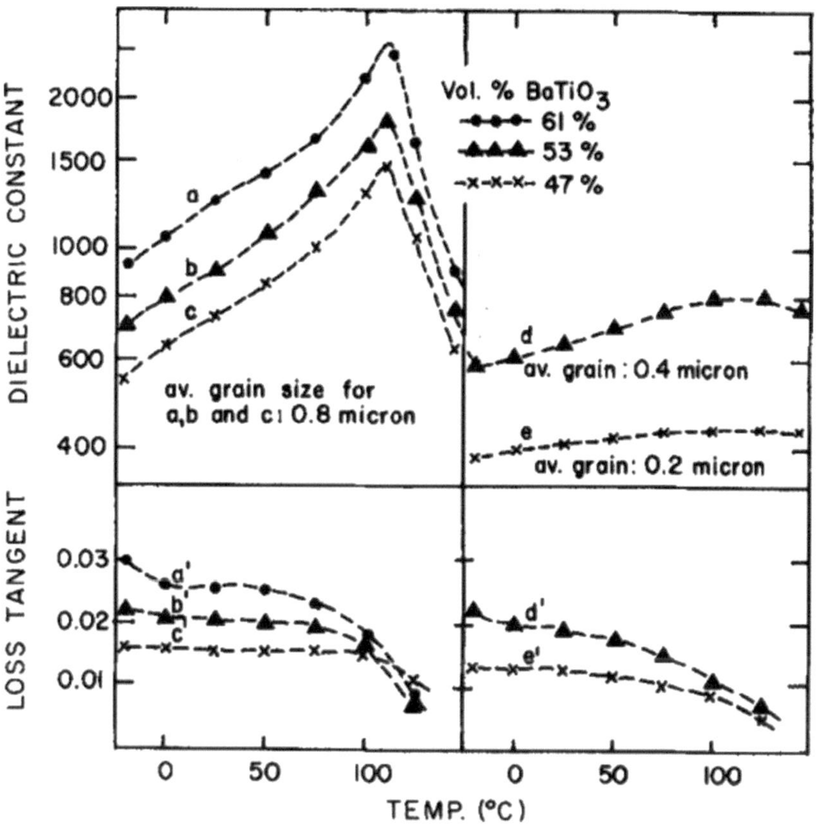

Figure 14.
Variation of dielectric constant and loss tangent with the temperature at 1 kHz [79].

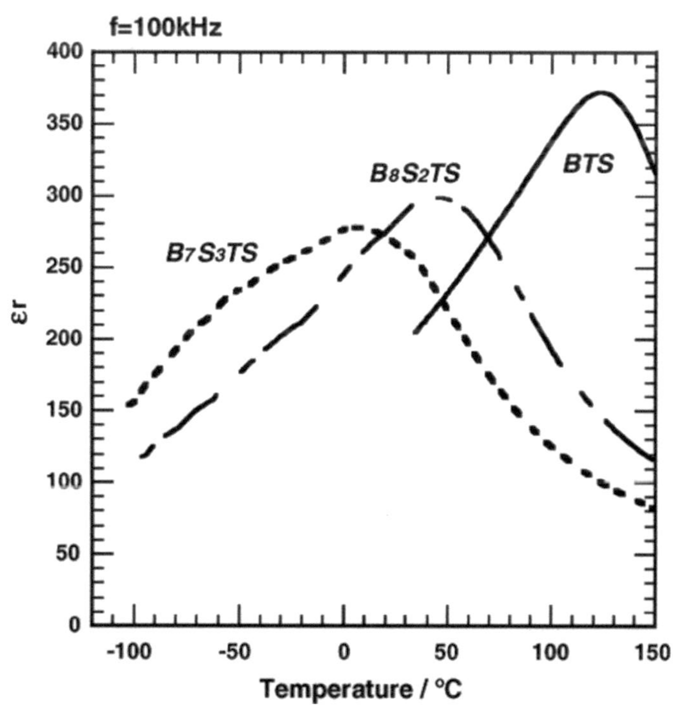

Figure 15.
The εr–T relationship for BSTS bulk glass-ceramic derived from SnO2-containing glasses [81].

sintering temperature required to reach a high density (≈ 93%) could be reduced to ≈1000°C. The influence of glasses on the sintering behavior and properties of ceramics is dependent upon the glass additive content as well as the chemical reaction between the glass and ceramic phases. Three different glass additives for BaTiO3 were studied by Hsianga; these were BaO-B2O3-SiO2 (BBS), PbO-B2O3-SiO2 (PBS), and ZnO-B2O3-SiO2 (ZBS). It was reported that the glasses containing PbO and BaO led to degradation of the dielectric properties and densification of the ceramic. In addition, adding both BBS and PBS caused the formation of a large number of secondary phases [3].

Lin et al. [86] added a manganese oxide-silica glass to pure BaTiO3 and reported the effect of the liquid phase on the dielectric and ferroelectric properties of the material. The addition of the Mn-Si-O glass enabled densification of the nanocrystalline powder at temperatures in the range 1175–1300°C. At high glass concentrations, they observed the formation of Ba2TiSi2O8 and a Mn solid solution in BaTiO3 grains growing at the grain boundaries and inhibiting grain growth. **Figure 16** shows the influence of glass content on the structural and dielectric parameters [86]. It can be observed, that the ceramics with grain sizes in the range of 0.7–1 μm have the higher relative permittivity. Meanwhile, for the glass doped samples, the higher tetragonality of BaTiO3 induced through the liquid phase sintering led to larger dielectric permittivity. However, the dielectric loss was also affected by the Mn solid solution in BaTiO3 and the Ba2TiSi2O8 phases.

More recently, Chen et al. [87] studied the addition of 50–90% of B2O3–Bi2O3–SiO2–ZnO (BBSZ) glass on the dielectric and ferroelectric properties of BaTiO3. By this means, the sintering temperature was reduced to 400–450°C. The second phase of Bi24Si2O40 was observed when samples were sintered at 450°C and the BBSZ concentration was higher than 60 wt%. Once the solubility limit of BBSZ on BaTiO3 was exceeded (BBSZ>60 wt%) and the formation of the Bi24Si2O40 and glass phases took place, some pores of about 1–2 μm appeared due to the capillary pressure, rearranging the particles and affecting the packing; this reduction in density led to a lower relative permittivity for the 60 wt% modified samples. When the amount of liquid increased (up to 70 wt%), the porosity decreased and relative permittivity values of 132 and 207 were achieved at 100 kHz and 100 MHz, respectively. The dielectric loss remained at the same level as that of the pure glass at 100 kHz (0.006). At higher BBSZ concentration (80 and 90 wt%), the overall dielectric properties of the samples decreased due to the lower relative permittivity of the glass phase.

Choi et al. [88] reported that the addition of 1 to 7 wt% of BBS into BaTiO3 (prepared using flame spray pyrolysis) reduced the sintering temperature at 1000°C, yielding an improvement in relative permittivity in comparison with pure BaTiO3 sintered at 1300°C. It was also found that the glass additives improved the density

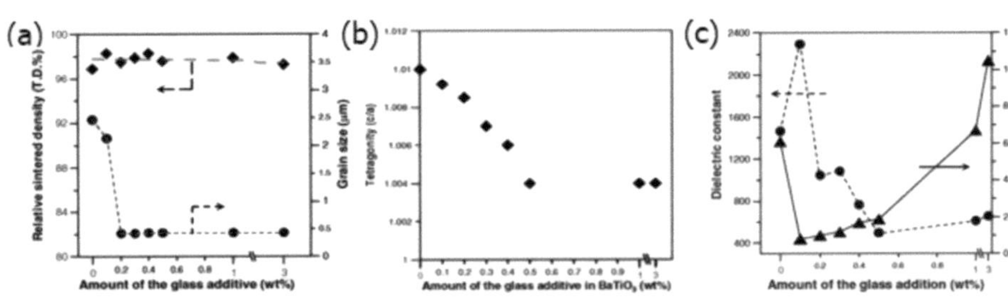

Figure 16.
Effect of glass addition on the (a) grain size (b) tetragonality and (c) dielectric constant and loss of die-pressed BaTiO3 [86].

of the material with an increase in the grain size by several microns. Wang et al. [26] found that the sintering temperature of $BaTiO_3$ ceramics could be reduced to about 1100°C using a glass with the composition 27.68$BaCO_3$–6.92$SrCO_3$-29TiO_2–22SiO_2-12Al_2O_2–2.4BaF_2 (mol %). The effect of this additive was a reduction in the average grain size. It was also found that the energy storage density of the ceramics increased gradually with increasing glass concentration; the highest energy density value of 0.32 J cm^{-3} was obtained for the sample with 7 mol% of the glass. The influence of glass additives on modified barium titanate ceramics was reported by Puli et al. [35], who studied the dielectric and ferroelectric properties of glass-modified BCZT, specifically the composition 0.85($BaZr_{0.2}Ti_{0.8}O_3$)-0.15($Ba_{0.7}ZCa_{0.3}TiO_3$). The addition of 15 wt% (0.1BaO + 0.4B_2O_3 + 0.5ZnO) and (0.3BaO + 0.6B_2O_3 + 0.1ZnO) glass powders yielded dielectric breakdown field strengths of 260 and 280 kV cm^{-1}, with recoverable energy density values of about 1.12 and 0.50 J cm^{-3}, respectively. The resulting samples were composites comprising a mixture of two phases, BCZT and the glass phase. They attributed the high breakdown strength to the presence of alkali free glass composition, and the low loss dielectric to the low dielectric loss of BZT–BCT ceramic composition.

The effect of up to 9 wt% of CaO–B_2O_3–SiO_2 (CBS) glass additions on the microstructure and electrical properties of $Ba_{0.85}Ca_{0.15}Zr_{0.1}Ti_{0.9}O_3$ (BCZT) was studied by Lai et al. [89]. Samples were sintered at 1300°C for 4 hours in air; in the CBS-free sample, they observed the coexistence of orthorhombic and tetragonal BCZT phases with a small amount of $CaTiO_3$ secondary phase by means of XRD. The addition of the CBS glass induced a phase transformation to pseudo-cubic, a shift of the reflections towards lower 2θ values and the coexistence of the orthorhombic and tetragonal phases disappeared. These phase transformations were accompanied by the appearance of another secondary phase, reported to be $Ba_2TiSi_2O_8$. The density of the BCZT samples increased with CBS content, reaching a maximum when the addition of CBS was 2 wt%, and then slowly decreasing at higher CBS concentrations. Regarding the electrical properties, the sample with 2 wt% CBS exhibited the best ferroelectric properties, a dielectric constant of 8874 (at 106 Hz), a Curie temperature of 116°C, Pr of about 3.18 μC cm^{-2}, and d_{33} of 159 pC N^{-1} due to the improved density achieved by the addition of CBS.

4.2 Potassium sodium niobate systems

KNN is one of the most promising candidates for energy storage properties, yet at present much research has focused on the development of, for example, (BaO,SrO)-TiO_2 [90, 91], (BaO,SrO)-Nb_2O_5 [50] and (BaO,Na_2O)-Nb_2O_5 [92] glass-ceramics. Therefore, the study of KNN glass-ceramics still needs more exploration and in-depth research [93]. Many studies have been conducted on sintering of KNN, but there are currently few reports of true glass-ceramics.

Some previous studies have reported the formation of KNN in a glass-ceramic form [56]. Vernacolota et al. [94] reported that glass-ceramics containing KN and KNN phases can be obtained using silicate glasses prepared with alkali metal and niobium additions. The effects of substituting K by Na on thermal and crystallisation behaviour of KNN-SiO_2 glass-ceramics were reported by Aronne et al. [95]. Kioka et al. [96] and Kongputhon et al. [97] studied the control of the crystallisation behaviour in KNN-SiO_2, by varying the K:Na ratio, as a means of modifying the dielectric properties. Alumina-silicate glasses have also been used to fabricate glass-ceramics containing ferroelectric KNN and their dielectric properties reported by Yongsiri et al. [54]. The influence of CeO_2 as a nucleation agent in borosilicate glasses containing KNN have been reported for energy storage capacitor application by Hanyu et al. [93]. On the basis of their results, Yongsiri et al. [54] suggested that

KNN glass-ceramics could be favourable for use in electro-optical applications. The microstructure, crystallite sizes, and crystal quantity of the KNN glass-ceramics were studied as a function of the heat treatment conditions. The heat treated glass-ceramics were found to contain a $KAlSiO2$ phase at heat treatment temperatures between 600 and 675°C, while the perovskite KNN phase was observed at a higher temperature of 700°C, **Figure 17**. The glass-ceramic containing 23.75 mol% of $SiO2$ exhibited a relative permittivity value of ≈ 260 and loss tangent ≈ 0.02 at 10 kHz.

The use of a two-stage incorporation method, which involved the separate preparation of KNN and glass powders prior to melting, was also studied by Yongsiri et al. [98]. The calcined KNN powder was mixed with 25 mol% of $SiO2$ then melted at 1300°C using the conventional melt-quenching technique. The glass was heat treated at temperatures from 525 to 575°C for crystallisation. Increasing heat treatment temperatures were found to improve the crystal size and crystallinity, which in turn plays an important role in controlling the properties of the glass ceramics, including physical, optical, and dielectric properties. It is clear from the SEM results, **Figure 18** that the crystallisation of the KNN phase occurred at temperatures lower than the observed crystallisation temperature of 648°C from the DTA results. Furthermore, the amorphous XRD patterns were observed in the glass-ceramic samples heat treated at temperatures lower than 550°C. The highest relative permittivity value was 474 at a heat treatment temperature of 550°C, while the transparency decreased with increased temperatures.

KNN ceramics prepared by solid state reaction usually require reaction temperatures around 800°C or, sometimes, the double calculation in order to obtain a homogenous powder; sintering temperatures are usually in the range 1100–1200°C [99]. The three starting materials used in KNN are $Nb2O5$, with a high melting point of 1520°C, and two alkali metal carbonates, $K2CO3$ and $Na2CO3$, with melting points of 891°C and 851°C, respectively. Therefore, the alkali metal carbonates/oxides become volatile during the calcination, making it difficult to achieve a chemically homogeneous material. Excess alkali carbonate is used in order to compensate the losses during calcination and to obtain a single phase perovskite product [100]. The studies reported in [96, 100] indicated that a 5 mol% excess of $K2CO3$ and $Na2CO3$ in the raw materials leads to increased stability during calcination at 800°C. The other effect of this excess an increase in particle size, which is probably due to the liquid phase which forms during the calcination [100]. By adding 3% and 1% excess alkali carbonate the density of KNN has improved and, at the same time, this excess leads to reduce the sintering temperature to 1000°C [101, 102].

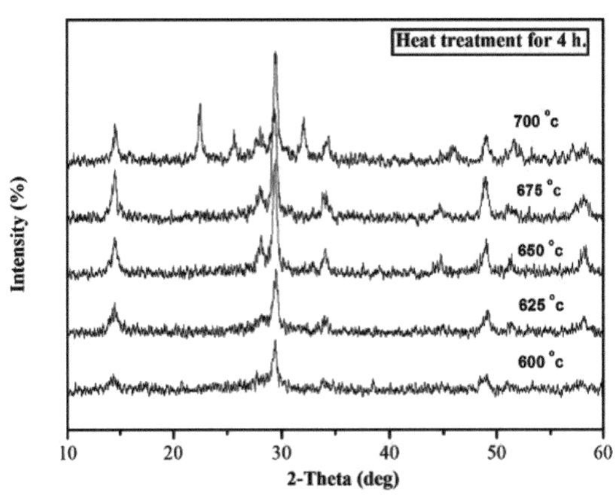

Figure 17.
XRD patterns of the glass-ceramic sample with 5 mol% alumina and 23.75 mol% SiO2 subjected to different heat treatment temperatures [54].

Sintering temperatures for KNN ceramics are usually >1000°C. This needs to be lowered in order to decrease the alkali evaporation during the periods of high temperature, as well as to make it suitable for many desired applications [103]. If lithium additives are added to KNN, the sintering temperature reduces to lower than 1000°C, there is a shift in TC to a higher temperature, and there is an increase in the dielectric constant [104]. Barium borosilicate-based frit (BaO-B2O3-SiO2-Na2O-K2O-CuO-CaO) (abbreviated as BBS) was one glassy additive used previously as a sintering aid to KNN. It has been shown to reduce the sintering temperature to 1000°C. In addition, the additive led to an increase in the mechanical properties and a decrease in the dielectric loss, whilst the TC remained high at 400°C; samples with 1.5 wt% glass frit showed optimal properties as follows: ε_r = 410, tanδ = 0.57 and d33 = 108 pC N^{-1}, **Figure 19** [105].

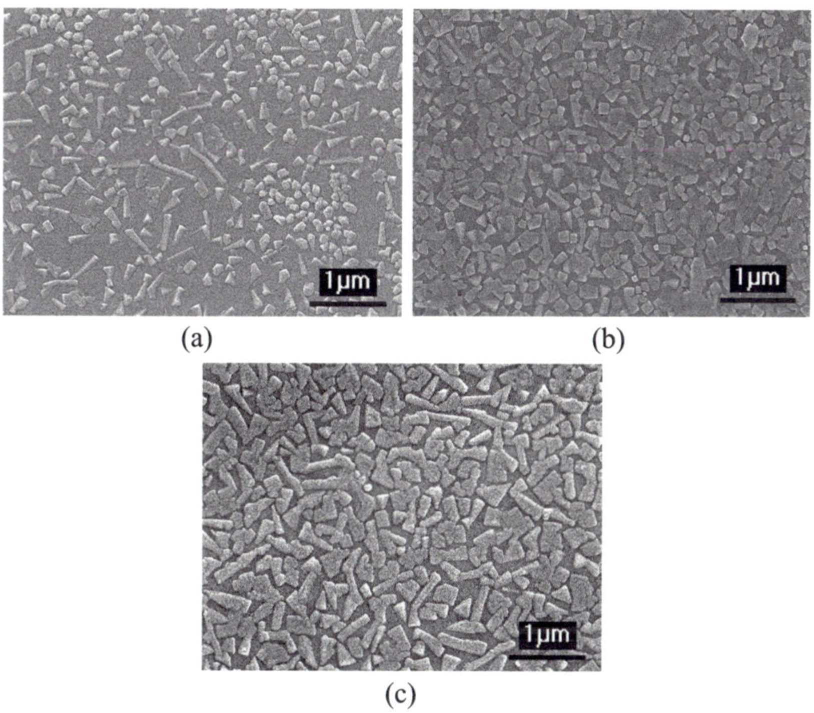

Figure 18.
SEM micrographs for glass-ceramic samples heated at different temperatures. (a) 525°C, (b) at 550°C, (c) 575°C [98].

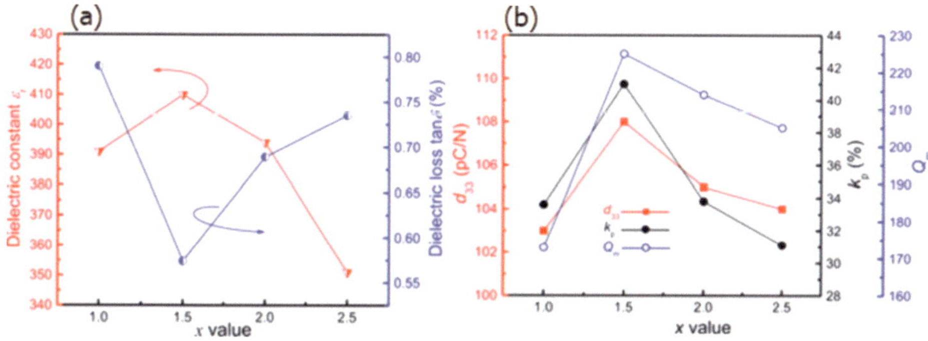

Figure 19.
(a) εr and tanδ (b) d33, kp and Qm of KNN + x wt% BBS ceramics as a function of the x value [105].

5. Conclusion

Several different ferroelectric glass-ceramics systems were discussed in this chapter. All glasses types were mixed with ceramics, resulting in much improved densification behaviour at reduced sintering temperatures. In addition, nanocrystalline glass-ceramics containing perovskite-structured ferroelectric phases have been researched. These studies indicate that the materials exhibit promising dielectric properties and good stability of relative permittivity values at low heat treatment temperatures comparison with conventional preparation methods, although relatively high dielectric losses were evident. This review suggests that ferroelectric glass-ceramics exhibit promising dielectric properties with good potential for use as energy storage dielectrics at high electric field levels as a result of their nanocrystalline microstructures.

Author details

Abdulkarim Ziedan Khalf
University of Kirkuk, College of Education for Pure Sciences Physics Department, Kirkuk, Iraq

*Address all correspondence to: karim_uot@yahoo.com

References

[1] Al-Assiri MS, El-Desoky MM. Grain-size effects on the structural, electrical properties and ferroelectric behaviour of barium titanate-based glass–ceramic nano-composite. Journal of Materials Science: Materials in Electronics. 2013;**24**:784-792

[2] Abhilash P et al. Facile Synthesis of "Quench-Free Glass" and Ceramic-Glass Composite for LTCC Applications. Journal of the American Ceramic Society. 2013;**96**(5):1533-1537

[3] Hsianga HI et al. Low temperature sintering and dielectric properties of BaTiO3 with glass addition. Materials Chemistry and Physics. 2009;**113**(2-3):658-663

[4] Su X, Tomozawa M, Nelson J. Effect of crystallizable glass addition on sintering and dielectric behaviors of barium titanate ceramics. Journal of Materials Science: Materials in Electronics. 2013;**24**:2135-2140

[5] Holand, W. and G.H. Beall, Glass-ceramic Technology 2012, John Wiley & Sons, Inc

[6] Zhang L, Zhai J, Yao X. Low-Sintering-Temperature Barium Titanate Thick Film Prepared by Electrophoretic Deposition Technique. Ferroelectrics. 2009;**384**:153-159

[7] Zhang H, Mak C. Preparation and characteristics of fine-grained ferroelectric glass-ceramic composites via a modified hybrid route at low temperature sintering. Journal of Electroceramics. 2011;**27**:126-133

[8] Kumar Yadav A, Gautam CR. A review on crystallisation behaviour of perovskite glass ceramics. Advances in Applied Ceramics. 2014;**113**(4):193-207

[9] Yongsiri P et al. Morphology of potassium sodium niobate based silicate glass system. Electronic Materials Letters. 2013;**9**(6):825-827

[10] Jain H. Transparent Ferroelectric Glass-Ceramics. Ferroelectrics. 2004;**306**(1):111-127

[11] Du J, Jones B, Lanagan M. Preparation and characterization of dielectric glass-ceramics in Na2O–PbO–Nb2O5–SiO2 system. Materials Letters. 2005;**59**(22):2821-2826

[12] Graça MPF et al. Electric and dielectric properties of a SiO2–Na2O–Nb2O5 glass subject to a controlled heat-treatment process. Physica B: Condensed Matter. 2007;**396**(1-2):62-69

[13] Graça MPF, Ferreira da Silva MG, Valente MA. NaNbO3 crystals dispersed in a B2O3 glass matrix – Structural characteristics versus electrical and dielectrical properties. Solid State Sciences. 2009;**11**(2):570-577

[14] Kim D et al. Dielectric properties and temperature stability of BaTiO3 co-doped La2O3 and Tm2O3. Current Applied Physics. 2012;**12**:952-956

[15] Zadeh HN et al. Low temperature sintering of barium titanate based ceramics with high dielectric constant for LTCC applications. Journal of the European Ceramic Society. 2011;**31**:589-596

[16] Carter, C.B. and M.G. Norton, Ceramic Materials Science and Engineering, M.G.N. C. Barry Carter, Editor 2013, Springer.

[17] Jaffe B et al. Piezoelectric Ceramics. London and New York: Academic Press; 1971

[18] Hao, X., A review on the dielectric materials for high energy-storage application. J. Adv. Dielect, 2013. 3(1): p. 1330001(1-14).

[19] Sarjeant WJ, Zirnheld J, MacDougall FW. Capacitors. IEEE Transactions on Plasma Science. 1998;**26**(5):1368-1392

[20] Moulson, A.J. and J.M. Herbert, Electroceramics, 2003, John Wiley & Sons Ltd.

[21] Kotz R, Carlen M. Principles and applications of electrochemical capacitors. Electrochimica Acta. 2000;**45**:2483-2498

[22] Fletcher NH, Hilton AD, Ricketts BW. Optimization of energy storage density in ceramic capacitors. Journal of Physics D: Applied Physics. 1996;**29**:253-258

[23] Patel S, Chauhan A, Vaish R. Enhancing electrical energy storage density in anti-ferroelectric ceramics using ferroelastic domain switching. Materials Research Express. 2014;**1**(4):045502

[24] Zhang Q et al. Improved Energy Storage Density in Barium Strontium Titanate by Addition of BaO-SiO2-B2O3 Glass. Journal of the American Ceramic Society. 2009;**92**(8):1871-1873

[25] Zheng P et al. Grain-size effects on dielectric and piezoelectric properties of poled BaTiO3 ceramics. Acta Materialia. 2012;**60**:5022-5030

[26] Wang X et al. Glass additive in barium titanate ceramics and its influence on electrical breakdown strength in relation with energy storage properties. Journal of the European Ceramic Society. 2012;**32**:559-567

[27] Kinoshita K, Yamaji A. Grain-size effects on dielectric properties in barium titanate ceramics. Journal of Applied Physics. 1976;**47**(1):371-373

[28] Arlt G, Hennings D, de With G. Dielectric properties of fine-grained barium titanate ceramics. Journal of Applied Physics. 1985;**58**(4):1619-1625

[29] Yamashita K et al. Statistical Analysis of Dielectric Strength of BaTiO3 Ceramic Films. Jap. J. Appl. Phys. 1980;**19**(5):867-871

[30] Tunkasiri T, Rujijangul G. Dielectric strength of fine grained barium titanate ceramics. Journal of Materials Science Letters. 1996;**15**:1767-1769

[31] Sarkar SK, Sharma ML. Liquid phase sintering of BaTiO3 by boric oxide (B2O3) and lead borate (PbB2O4) glasses and it's effect on dielectric strength and dielectric constant. Mat. Res. Bul. 1989;**24**:773-779

[32] Mahajan S et al. Effect of Nd Doping on Structural, Dielectric and Ferroelectric Properties of Ba(Zr0.05Ti0.95)O3 Ceramic. Integrated Ferroelectrics. 2010;**122**(1):83-89

[33] Puli VS et al. Structure, dielectric, ferroelectric, and energy density properties of (1 – x)BZT–xBCT ceramic capacitors for energy storage applications. Journal of Materials Science. 2013;**48**(5):2151-2157

[34] Wang Z et al. Synthesis, structure, dielectric, piezoelectric, and energy storage performance of (Ba0.85Ca0.15)(Ti0.9Zr0.1)O3 ceramics prepared by different methods. Journal of Materials Science: Materials in Electronics. 2016;**27**(5):5047-5058

[35] Puli VS et al. Structure and dielectric properties of BaO–B2O3–ZnO–[(BaZr0.2Ti0.8)O3]0.85–[(Ba0.7Ca0.3)TiO3]0.15 glass–ceramics for energy storage. Journal of Materials Science: Materials in Electronics. 2012;**23**(11):2005-2009

[36] Shirane G, Newnham R, Pepinsky R. Dielectric Properties and Phase Transitions of NaNbO3 and (Na,K)NbO3. Physical Review. 1954;**96**(3):581-588

[37] Safari A. and a.E.K. Akdogan, Piezoelectric and Acoustic Materials

for Transducer Applications, 2008. Springer.

[38] Egerton L, Dillon DM. Piezoelectric and dielectric properties of Ceramics in the System Potassium-Sodium Niobate. Journal of the American Ceramic Society. 1959;**42**(9):438-442

[39] Vousden P. The Structure of Ferroelectric Sodium Niobate at Room Temperature. Acta Cryst. 1951;**4**:545-551

[40] Qu B, Du H, Yangb Z. Lead-free relaxor ferroelectric ceramics with high optical transparency and energy storage ability. Journal of Materials Chemistry C. 2016;**4**:1795-1803

[41] Qu B et al. Enhanced dielectric breakdown strength and energy storage density in lead-free relaxor ferroelectric ceramics prepared using transition liquid phase sintering. RSC Advances. 2016;**6**:34381-34389

[42] Díaz, R.A. Thermal Analysis. Fundamentals and Applications to Material Characterization. in Proceedings of the International Seminar: Thermal Analysis and Rheology. 2003. Conference, Universidade da Coruña, Ferrol, Spain.

[43] Suzuki, Y., et al., Advanced Ceramic Technologies & Products, T.S.S. Y. Suzuki, K. Hirao, T. Tsuchiya, H. Nagata, J. S. Cross, Editor 2012, Springer.

[44] M. Nikl, et al., Silicate Glass-Based Nanocomposite Scintillators. in Advances in Nanocomposite Technology, A. Hashim, Ed. InTech, 2011(in Advances in Nanocomposite Technology, A. Hashim, Ed. InTech,).

[45] Shelby, J.E., Introduction to Glass Science and Technology, 2005, The Royal Society of Chemistry: The Royal Society of Chemistry. p. P001-P004.

[46] Gorzkowski EP et al. Glass-ceramics of barium strontium titanate for high energy density capacitors. Journal of Electroceramics. 2007;**18**(3-4):269-276

[47] McPherson J et al. Thermochemical description of dielectric breakdown in high dielectric constant materials. Applied Physics Letters. 2003;**82**(13):2121-2123

[48] Xin Zhou, et al., Electrical Energy Density and Discharge Characteristics of a Poly (vinylidene fluoride chlorotrifluoroethylene) Copolymer. IEEE Transactions on Dielectrics and Electrical Insulation, 2007. 14(5): p. 1133-1138.

[49] Smith NJ et al. Alkali-free glass as a high energy density dielectric material. Materials Letters. 2009;**63**(15):1245-1248

[50] Xue S et al. Dielectric properties and charge–discharge behaviors in niobate glass ceramics for energy-storage applications. Journal of Alloys and Compounds. 2014;**617**:418-422

[51] Abdel-Khaleka EK et al. Study of glass-nanocomposite and glass–ceramic containing ferroelectric phase. Materials Chemistry and Physics. 2012;**133**:69-77

[52] Shankar J, Deshpande VK. Electrical and thermal properties of lead titanate glass ceramics. Physica B. 2011;**406**:588-592

[53] Chaliha RS et al. Structure and dielectric properties of potassium niobate nano glass–ceramics. Journal of Materials Science: Materials in Electronics. 2010;**22**(7):728-734

[54] Yongsiri P et al. Fabrication of Ferroelectric Glass Ceramics from (K0.5Na0.5)NbO3-SiO2-Al2O3 Glass System. Ferroelectrics. 2011;**416**(1):144-150

[55] Chen S et al. Thermal and dielectric properties of the LTCC composites based on the eutectic

system BaO–Al2O3–SiO2–B2O3. Journal of Materials Science: Materials in Electronics. 2011;**22**:238-243

[56] Zhilin AA et al. Phase transformations in Na2O–K2O–Nb2O5–SiO2 glasses. Journal of Non-Crystalline Solids. 2004;**345-346**:182-186

[57] Hulsenberg, D., A. Harnisch, and A. Bismarck, Microstructuring of Glasses, A.H. D. Hulsenberg, A. Bismarck, Editor 2008, Springer-Verlag Berlin Heidelberg 2008.

[58] Shackelford, J.F. and R.H. Doremus, Ceramic and Glass Materials Structure, Properties and Processing, J.F.S.R.H. Doremus, Editor 2008, Springer Science+Business Media, LLC.

[59] Hwang G-H, Jeon H-J, Kim Y-S. Physical Properties of Barrier Ribs of Plasma Display Panels: I, Formation of Pores during Sintering of Lead Borosilicate Glass Frits. Journal of the American Ceramic Society. 2002;**85**:2956-2960

[60] Chang MS et al. Fabrication and characterization of dielectric materials of front and back panel for PDP. Journal of Information Display. 2001;**2**(3):39-43

[61] Lim E et al. Effect of BaO content on the sintering and physical properties of BaO–B2O3–SiO2 glasses. Journal of Non-Crystalline Solids. 2006;**352**:821-826

[62] Bobkova NM, Khot'ko SA. Low-melting glasses based on borate systems. Glass and Ceramics. 2004;**61**:175-177

[63] M. A. Vartanyan, E. S. Lukin, and N.A. Popova1, Low firing temperature ceramic for microcircut substrates. Glass and Ceramics, 2008. 65: p. 27-30.

[64] Chen S, Zhu D. Phase formation and properties of the BaO–B2O3–SiO2 and –Al2O3 ceramics prepared via an aqueous suspension route. Journal of Alloys and Compounds. 2012;**536**:73-79

[65] Thomas S, Sebastian MT. Effect of B2O3-Bi2O3-SiO2-ZnO glass on the sintering and microwave dielectric properties of 0.83ZnAl2O4-0.17TiO2. Materials Research Bulletin. 2008;**43**(4):843-851

[66] Hsiang H-I, Chen T-H. Electrical properties of low-temperature-fired ferrite–dielectric composites. Ceramics International. 2009;**35**(5):2035-2039

[67] Ling, W., et al., Effect of B2O3–Bi2O3–SiO2–ZnO glass on the dielectric and magnetic properties of ferroelectric/ferromagnetic composite for low temperature cofired ceramic technology. Journal of Applied Physics, 2010. 107(9): p. 09D911(1-3).

[68] Tulyaganov DU et al. Synthesis of glass-ceramics in the CaO–MgO–SiO2 system with B2O3, P2O5, Na2O and CaF2 additives. Journal of the European Ceramic Society. 2006;**26**:1463-1471

[69] Lim E et al. Characterization of the low temperature firing BaO–B2O3–SiO2 glass: The effect of BaO content. Journal of the European Ceramic Society. 2007;**27**:825-829

[70] Fedorov P, Kokh A, Kononova N. Barium borate β-BaB2O4 as a material for nonlinear optics. Russian Chemical Reviews. 2002;**71**(8):651-671

[71] Deshpande AM, Deshpande VK. Effect of SiO2 and Al2O3 addition on the density, Tg and CTE of mixed alkali-alkaline earth borate glass. IOP Conference Series: Materials Science and Engineering. 2009;**2**:012034

[72] Lim E, Kim B, Lee J. Dielectric, thermal and sintering behavior of BaO-B2O3-SiO2 glasses with the addition of Al2O3. Journal of Electroceramics. 2006;**17**:359-363

[73] Petrscu S et al. Vitreous and glass-ceramics materials in the SiO2-Al2O3-MeO-M2O type system. Journal of Optoelectronics and Advanced Materials. 2012;**14**:603-612

[74] Dutta A et al. Ac conductivity and dielectric relaxation in ionically conducting soda-lime-silicate glasses. Journal of Non-Crystalline Solids. 2008;**354**(33):3952-3957

[75] Wang T et al. Energy storage properties in Ba0.4Sr0.6TiO3 ceramics with addition of semi-conductive BaO–B2O3–SiO2–Na2CO3–K2CO3 glass. Journal of Alloys and Compounds. 2014;**617**:399-403

[76] Wang H, Wu J. Phase transition, microstructure, and electrical properties of Ca, Zr, and Sn-modified BaTiO3 lead-free ceramics. Journal of Alloys and Compounds. 2014;**615**:969-974

[77] Stookey SD. Photosensitive glass. Journal of Industrial and Engineering Chemistry. 1949;**41**(4):856-861

[78] Stookey SD. Catalyzed Crystallization of Glass in Theory and Practice. Journal of Industrial and Engineering Chemistry. 1959;**51**(7):805-808

[79] Herczog A. Microcrystalline BaTiO3 by Crystallization from Glass. American Ceramic Society. 1964;**47**(3):107-115

[80] McCauley D, Newnham RE, Randall CA. Intrinsic Size Effects in a Barium Titanate Glass Ceramic. Journal of the American Ceramic Society. 1998;**81**(4):979-987

[81] Takahashi J, Nakanoa H, Kageyama K. Fabrication and dielectric properties of barium titanate-based glass ceramics for tunable microwave LTCC application. Journal of the European Ceramic Society. 2006;**26**:2123-2127

[82] Lee T et al. Phase evolution of solid-state BaTiO3 powder prepared with the ultrafine BaCO3 and TiO2. Journal of Materials Research. 2012;**27**(19):2495-2502

[83] Liu G, Jiang Y, Button T. Low temperature sintering and dielectric properties of BaTiO3 ceramics incorporating nano-sized powders. Ferroelectrics. 2011;**421**:72-81

[84] Zadeh H et al. Low temperature sintering of barium titanate ceramics assisted by addition of lithium fluoride-containing sintering additives. Journal of the European Ceramic Society. 2010;**30**:81-86

[85] Jeon HP et al. Effects of BaO–B2O3–SiO2 glass additive on densification and dielectric properties of BaTiO3 ceramics. Materials Chemistry and Physics. 2005;**94**:185-189

[86] Lin JCC, Wei W-CJ. Low-temperature sintering of BaTiO3 with Mn-Si-O glass. Journal of Electroceramics. 2010;**25**(2-4):179-187

[87] Chen M-Y et al. Dielectric BaTiO3–BBSZ glass ceramic composition with ultra-low sintering temperature. Journal of the European Ceramic Society. 2015;**35**(1):139-144

[88] Choi S et al. Characteristics of BaO–B2O3–SiO2 nano glass powders prepared by flame spray pyrolysis as the sintering agent of BaTiO3 ceramics. Journal of Alloys and Compounds. 2011;**509**:7979-7984

[89] Lai Y et al. Effects of CaO–B2O3–SiO2 glass additive on the microstructure and electrical properties of BCZT lead-free ceramic. Ceramics International. 2016;**42**(11):12694-12700

[90] Zhang W et al. Structure and dielectric properties of BaxSr1−xTiO3-based glass ceramics for energy storage. Journal of Alloys and Compounds. 2014;**617**:740-745

[91] Chen J et al. Effect of the Ba/Ti Ratio on the Microstructures and Dielectric Properties of Barium Titanate-Based Glass-Ceramics. Journal of the American Ceramic Society. 2009;**92**(6):1350-1353

[92] Xue S et al. Improved discharged property for (BaO, Na2O)–Nb2O5–SiO2 glass-ceramics by La2O3 addition. Ceramics International. 2015;**41**:S441-S446

[93] Hanyu Z et al. Correlation between dielectric properties and crystallization treatment in potassium sodium niobate glass-ceramics for energy storage application. Journal of Alloys and Compounds. 2016;**674**:272-276

[94] Vernacotola DE, Chatlani S, Shelby JE. Ferroelectric Sodium Potassium Niobium Silicate Glass Ceramics. IEEE. 2000;**2**:829-832

[95] Aronne A et al. Crystallization and second harmonic generation in potassium–sodium niobiosilicate glasses. Journal of Solid State Chemistry. 2009;**182**(10):2796-2800

[96] Kioka K, Honma T, Komatsu T. Fabrication of (K,Na)NbO3 glass–ceramics and crystal line patterning on glass surface. Optical Materials. 2011;**33**(8):1203-1209

[97] Kongputhon P, Niyompan A, Tipakontitikul R. Dielectric Properties of Ferroelectric Glass-Ceramics of the Na2O-K2O-Nb2O5-Al2O3-SiO2 System with Partial Substitution of K2O for Na2O. Ferroelectrics. 2010;**403**(1):60-67

[98] Yongsiri P et al. Fabrication of transparent lead-free KNN glass ceramics by incorporation method. Nanoscale Research Letters. 2012;**136**(7):1-9

[99] Jarupoom P et al. Structures and Properties of Lead-Free NKN Piezoelectric Ceramics. Ferroelectric Letters. 2008;**35**:119-127

[100] Bomlai P et al. Effect of Calcination Conditions and Excess Alkali Carbonate on the Phase Formation and Particle Morphology of Na0.5K0.5NbO3 Powders. Journal of the American Ceramic Society. 2007;**90**(5):1650-1655

[101] Zhao Y et al. Low-temperature sintering of KNN with excess alkaline elements and the study of its ferroelectric domain structure. Current Applied Physics. 2013;**13**:2082-2086

[102] Bernard J et al. Low-Temperature Sintering of K0.5Na0.5NbO3 Ceramics. Journal of the American Ceramic Society. 2008;**91**(7):2409-2411

[103] Priya, S. and S. Nahm, Lead-Free Piezoelectrics, S.P.a.S. Nahm, Editor 2012, Springer New York Dordrecht Heidelberg London.

[104] Zhoua L et al. Low-temperature sintering of (K,Na)NbO3-based lead-free piezoceramicswith addition of LiF. Journal of the European Ceramic Society. 2014;**34**:1161-1167

[105] Li Y et al. Effect of BBS-Based Frit on the Low Temperature Sintering and Electrical Properties of KNN Lead-Free Piezoceramics. International Journal of Applied Ceramic Technology. 2013;**10**(5):866-872

Chapter 3

From the Laser Plume to the Laser Ceramics

Vladimir Osipov, Vyacheslav Platonov, Vladislav Shitov and Vladimir Solomonov

Abstract

The main stages of preparation of ceramic active elements of solid-state lasers are considered. The physical principles of laser synthesis of nanopowders are described. The features and processes taking place during compaction and compacts sintering are specified. Also we report on the investigation of characteristics of highly transparent ceramics on the basis of nanopowders synthesized in laser plume. It is shown that this approach enables to increase the "orange peel" formation threshold in the ceramics with strongly disordered crystalline structure. It opens the road to relatively simple synthesis technology from oxide materials and application of this ceramics as the gain media with oscillation efficiency higher than 50% and also leads to simplification of the synthesis technology of magnetoactive ceramics and to production of highly transparent YAG samples without the use of sintering heterovalent additives.

Keywords: laser plume, nanopowder, compact, sintering, highly transparent ceramics, laser ceramic elements

1. Introduction

In recent years, much attention has been paid to the developments aimed at creating solid-state lasers with a high average and peak power. This is primarily due to the wide range of applications of such laser systems: in the industry for remote cutting, welding, quenching, heat treatment and labeling of various materials [1–3], as well as in basic scientific research [4–6]. One of the key components of high power continuous and pulsed-periodic lasers is the active medium, where an inverse population of levels is created. In recent years, increasingly greater attention has been paid to the researches aimed at developing a technology to produce ceramic active elements for high-power laser systems. This is due to many advantages of optical ceramics over traditional media from single crystals and glasses: larger sizes, improved thermomechanical characteristics, the ability to synthesize composite samples, quick production, lower energy costs and price.

After pioneering work on synthesis of the laser ceramics and obtaining effective generation [7], a large amount of research was carried out in this direction. The requirements [8] are specified to achieve high-efficiency laser generation in ceramics: the thickness of the grain boundaries is of the order of 1 nm, the scattering loss per pass is less than 0.05–0.1% cm^{-1} (residual porosity at the level of 10^{-4} vol.%), optical uniformity with wavefront distortion of the $\lambda/19.5$. Using yttrium-aluminum garnet-based ceramics ($Y_3Al_5O_{12}$, YAG) with similar characteristics in the geometry of a thin disk (active medium Ø11 × 0.15 mm), an output power of 1.8 kW with a slope efficiency of 74.1% was implemented [9]. Moreover, a record output power of 6.5 kW with a slope efficiency of 57% was achieved in [10]. In a ceramic disc 8.5% Yb:LuAG with a thickness of 0.15 mm, an output power of 1.74 kW with a slope efficiency of 71.2% was demonstrated [11]. The most impressive output power values were achieved when using active elements of a sufficiently large volume. For example, in a ceramic plate of 1% Nd:YAG with a size of 89 × 30 × 3 mm^3, the power of continuous laser generation was 2.44 kW [12], and with increasing dimensions up to 120 × 50 × 3 mm^3—4.35 kW [13]. The cascade of several Nd:YAG ceramic elements sized 100 × 100 × 20 mm^3 allowed this value to increase to 67 kW [14], and further to 105.5 kW [15].

From the point of view of energy characteristics, the impulses with an energy of 105 J for a duration of 10 ns and an average power of 1 kW at a repetition rate of 10 Hz and cryogenic cooling of a Yb:YAG/Cr:YAG element of ceramics have been implemented to date [16].

One should also note the progress in the field of implementation of ultrashort laser pulses in ceramic active media. In this direction, laser pulses of 188 fs duration [17] and 152 fs [18] were demonstrated using $Yb:Y_2O_3$ ceramics. The shortest duration was achieved using composite ceramic $Yb:Y_2O_3/Yb:Sc_2O_3$ media with a total width of the amplification band of 27.3 nm, where a record low pulse duration of 53 fs was demonstrated [19].

When developing the technology to produce ceramic active elements, the main attention is paid to the formation of a nonporous microstructure of the material while maintaining the characteristic grain size in the range from several hundred nanometers to micrometers, which is important for reducing the local depolarization of laser radiation [20]. To meet these requirements, synthesis techniques were developed based on spark plasma sintering [21–23], hot isostatic pressing [24–26], and vacuum sintering with doping of heterovalent ions [27]. The latter option is more attractive due to the less expensive and uncomplicated technology. However, this approach, with a significant content of additives (more than 1 mol.%), is fraught with a significant disadvantage due to the release of heterovalent ions during sintering into the regions adjacent to the grain boundaries. In the synthesis of oxide ceramics, the possibilities of this approach can be expanded by using nanopowders obtained by laser evaporation, where the synthesis of nanoparticles proceeds at high temperatures and rapid cooling. This will ensure high uniformity of nanoparticles and ceramics based on them.

2. Preparing nanopowders

There are many methods for preparing nanopowders: mechanical crushing, precipitation from solutions, sol-gel, self-propagating high-temperature synthesis, physical vapor deposition. However, nanopowders prepared by the method of laser evaporation of a solid target in a gas atmosphere meet the above requirements to the fullest extent possible. Indeed, the radius of such particles (5–10 nm), the range of particle size distribution is rather narrow (5–40 nm), their purity is similar to the purity of the starting material, they usually have a spherical shape. The large capillary pressure and the significant surface energy due to the large surface of such nanopowders allow, under otherwise equal conditions, to reduce the duration or the sintering temperature. However, the most important advantage of the nanopowders thus prepared is that the doping takes place directly in the laser plume at high temperature and rapid cooling. This prevents segregation of the dopants and ensures high homogeneity of the ingredients in the nanoparticle, in the compact and, as will be shown, in samples of synthesized ceramics. In this connection, let us consider the process of laser nanopowder synthesis in more detail.

For the synthesis of oxide nanopowders by this method, a CO_2 laser (λ = 10.6 μm) and a 600 W fiber ytterbium laser (λ = 1.06 μm) were used. The average output power of the CO_2 laser was 550 W at a repetition rate of 650 Hz pulses with an energy of W = 1.4 J, a peak power of about 9 kW, and a duration of 500 μs at a power level of 0.1.

Figure 1a shows a block diagram of the laser complex for preparing nanopowders [28, 29]. Laser radiation was focused on the target with a lens, which also served as the entrance window of the evaporation chamber. The target was made from oxide micro-powder (or a mixture of them) by pressing and sintering it. As a result of the action of laser radiation, a laser plume consisting of target vapors appeared on the target near its surface. Mixing with ambient air or other buffer gas, the steam was cooled. The cooled vapor was condensed in the form of nanoparticles, which were in the evaporating chamber in a suspended state. A special drive rotated the target and moved it linearly in a horizontal plane so that the laser beam scanned the surface of the target at a constant linear velocity, thereby achieving uniform evaporation of the material from the surface. After evaporation of the surface, the target moved in a vertical direction. The fan pumped air through the chamber and transferred the powder to the cyclone and further into the electric filter where it was assembled. The air was cleaned additionally in a mechanical filter and returned to the chamber. The gas flow rate above the target surface was 15 m/s. **Figure 1b** (upper) shows photographs of the laser target before and after exposure of the CO_2 laser radiation for which the target material is opaque and the ytterbium laser radiation for which the target is semitransparent (lower). It can be seen that if the target is translucent for laser radiation, then it evaporates non-uniformly. Its surface consists of a number of needle formations 8 mm high and up to 1 mm thick.

The nanoparticles are formed in a laser plume. A laser plume is a flow of incandescent vapors of a solid target in the form of a weakly ionized plasma from the region of incidence of the laser beam on the target [30, 31]. In visible light, the plume is typically in the shape of a needle directed normal to the target surface, regardless of the angle of incidence of the laser beam (**Figure 2**). This tip is surrounded by a vortex structure, which is clearly manifested in shadow photography [32].

When exposed to single pulse or pulse-periodic laser radiation, the plume appears after the delay time td. During this time, the target substance is heated to

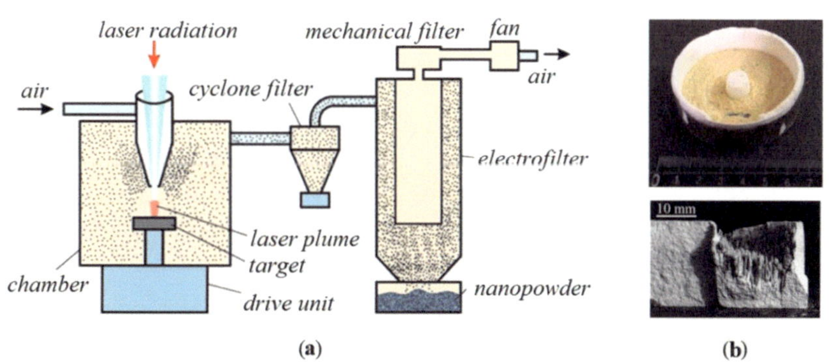

Figure 1.
Preparing a nanopowder: (a) block diagram of the laser complex for preparing a nanopowder, (b) image of laser target after exposure to radiation CO_2 laser (top), ytterbium fiber laser (bottom) [29].

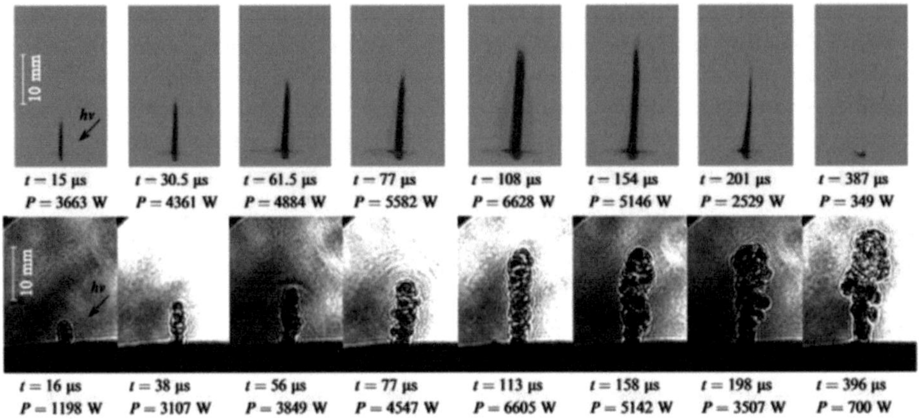

Figure 2.
Scanning of photographs (exposure 1 μs) of a laser plume (CO_2 laser pulse duration 200 μs, incident angle of 45°). Top row: visible light photography, bottom row: shadow photography. The captions below them indicate the shooting delay time relative to the start of the plume initiation (t) and the peak laser power (P) at the time of shooting [30].

the evaporation temperature in the area of the laser beam incidence. For a linear leading edge of a laser pulse, the delay time is defined as td = $2W_d/P_d$. Here, W_d is the energy required for preheating the target substance to the evaporation temperature, and P_d is the instantaneous power of laser radiation at the moment of the flare appearance. For different substances, due to the difference in W_d, the delay times of t_d differ. After the appearance of the glow (l = 0), the height of the plume (l) increases at a rate proportional to the square root of the peak power of the laser pulse. The maximum height of the plume (l_m) is reached at the moment of the maximum laser pulse. The diameter of the luminous zone of the plume is typically 0.5–1.0 mm, which approximately corresponds to the size of the laser spot on the target surface.

Over the entire length of the plume, its emission spectrum is represented by dominated structured molecular bands of radicals of cations of the target substance [31] against the background of a continuous band of recombination radiation (**Figure 3**). In this case, the short-wave part of the spectrum is well approximated by the Wien's curve, which makes it possible to determine the temperature of

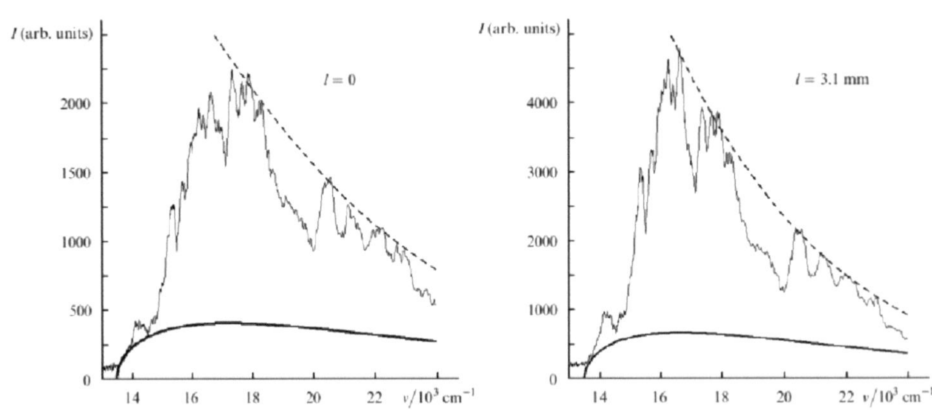

Figure 3.
The spectrum of the plume glow from a CO_2 laser at different distances (l) from the target of yttrium-stabilized zirconium oxide (solid curve), recombination radiation (bold curve), and approximation of the Wien's curve (dashed line) [31].

the luminous gas in the flare. Thus, when irradiated with a pulsed CO_2 laser, the maximum temperature close to the boiling point of the target material is reached at the target surface, and the flame temperature decreases nonmonotonically as it moves away from the target. When irradiated with pulses of a fiber ytterbium laser (1.07 μm), the temperature of the plume near the target slightly exceeds the melting point of the substance.

The transverse dimensions of the crater that appears on the target after exposure to a laser pulse almost coincide with the size of the laser spot on the target, and its depth depends on the wavelength of the laser radiation. For example, at the same pulse energy (1.0–1.4 J) after CO_2 laser irradiation, the crater depth is 5–10 μm, almost independently of the target substance. After a fiber laser pulse, the crater depth is 6–8 times greater, and with repeated exposure, the target surface becomes needle-like. These features are due to different mechanisms of absorption of radiation by the target of these. Thus, the frequency of a photon at the wavelength of a CO_2 laser is comparable to the frequency of optical phonons of oxide crystals. Therefore, in this case, such materials are almost opaque and the depth of penetration of laser radiation into them is only a few micrometers. For fiber ytterbium laser radiation, oxides we used are transparent – absorption is possible only on crystal and mechanical defects of the target. If these materials are single crystals with a minimum content of defects, the characteristic depth of penetration of laser radiation into them is tens of centimeters. This corresponds to the absorption index $\alpha \sim 10^{-2}$–$10^{-3} cm^{-1}$. If such defects are located inside the target in the area of the laser beam incidence, the initial heating also occurs inside the target (in the area of these defects). Then, due to the strong temperature dependence of the absorption coefficient, a heat wave is formed [33], which moves along the laser beam from the defect to the target surface, upon reaching which a laser plume is formed. This process is compounded by the fact that after repeated exposure, the surface of the initial target is covered with a layer of transparent melt 100–400 μm thick, in which the defect concentration is much lower than in the initial target made of sintered micro-powder.

This model is confirmed by the fact that the delay time for the appearance of a laser flare from the beginning of the laser pulse exposure has a large spread and on average increases with increasing transparency of the target. In particular, the delay in the appearance of a laser plume on the surface of a semitransparent $Nd:Y_2O_3$ ceramic with α – 23 cm^{-1} (an analog of the fused layer) averaged over several radiation pulses was 5–10 times greater than for the original sintered micro-powder target ($\alpha = 1.7 \times 10^3\ cm^{-1}$) at the same radiation intensity I = 0.4 MW/cm^2 [31]. The spread of the delay in the formation of a laser plume during evaporation of the same target is due to the stochastic nature of the depth of defects from the target surface at different points.

When nanopowders are obtained using both lasers, in addition to nanoparticles, spherical particles with sizes from 0.5 to 150 μm are also formed [29, 33], as well as shapeless target fragments of the same size. Fragments are formed after the end of radiation exposure to a specific part of the target due to thermal splitting of the cooling fused layer [30]. Spherical particles are liquid droplets of the melt, is sprayed by the vapor pressure of the laser crater.

Especially many drops are formed when the target is vaporized by continuous ytterbium laser radiation. At the same average radiation power (600 W) and the same intensity on the target as for the CO_2 laser ($\approx$1.3 MW/cm^2), the production capacity of the $Nd:Y_2O_3$ nanopowder decreased to 15 g/h, and its output during evaporation of one target to 9 wt.% [34]. High-speed shooting of the laser plume showed that this effect is due to the transition from steam to vapor-drop ablation. The latter becomes dominant ~500 μm after the start of the laser pulse. A similar pattern is observed in the evaporation of targets from YSZ and $FeMgAl_2O_4$. Theoretical analysis [33] allowed us to establish that one of the reasons for the appearance of drops in the laser flare is related to the presence of melt in the crater and is due to the development of the Kelvin-Helmholtz instability that is formed between the liquid wall of the crater and the flow of expiring vapor. This analysis made it possible to establish the characteristic size of the instability:

$$\lambda_e = \frac{2\pi\sigma}{\rho_2 V^2} = (20 \div 90) \cdot 10^{-6}\,\text{m} \tag{1}$$

and its development the increment

$$\tau = \frac{3\sqrt{3}\pi\sigma}{\rho_2 V^3}\sqrt{\frac{\rho_2}{\rho_1}} \approx 100\ \mu s, \tag{2}$$

where ρ_1 and ρ_2 are the melt and vapor densities, σ is the surface tension coefficient, V is the vapor flow rate.

Optimizing the duration (<00 μs) and radiation density, separation and trapping, it was possible to prepare high-quality nanopowders. **Figure 4** shows an example of a photo of YSZ nanopowder, and the distribution of particles of different composition in size is given as an example. Depending on the thermophysical properties of refractory oxides, the pressure and speed of the carrier gas, the productivity of producing a nanopowder using a CO_2 laser with an average radiation power of 600 W varies from 10 to 80 g/hour.

The distinguishing feature of nanoparticles synthesized in a laser plume, i.e. at a high temperature and rapid cooling, is a high homogeneity of the distribution of components in the volume. This is confirmed by the results of a study of the distribution of the concentration of dopant (Yb) in the Lu_2O_3 matrix, carried out in the scanning electron microscope (SEM) mode using the X-ray spectral microanalysis (X-ray SMA) method. The results of mapping the elemental composition of individual nanoparticles are shown in **Figure 5**. It follows from these images that the dopant is distributed uniformly over the Lu_2O_3 matrix, and there is no increased Yb concentration on the particle surface.

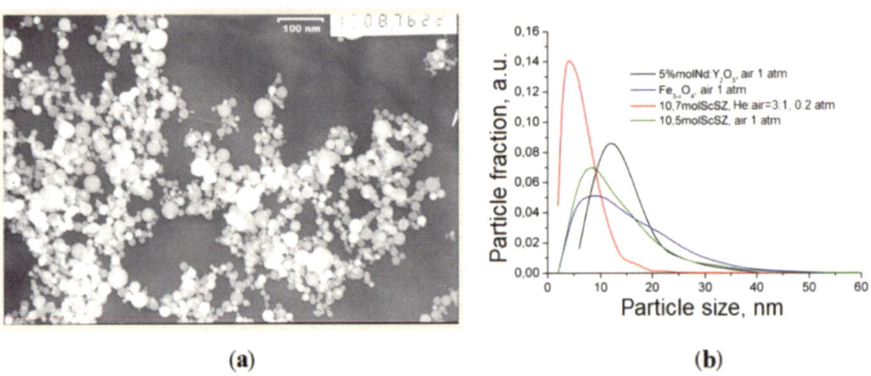

Figure 4.
A typical photo of YSZ nanoparticles (a) and the size distribution of nanoparticles of different compositions (b) [29].

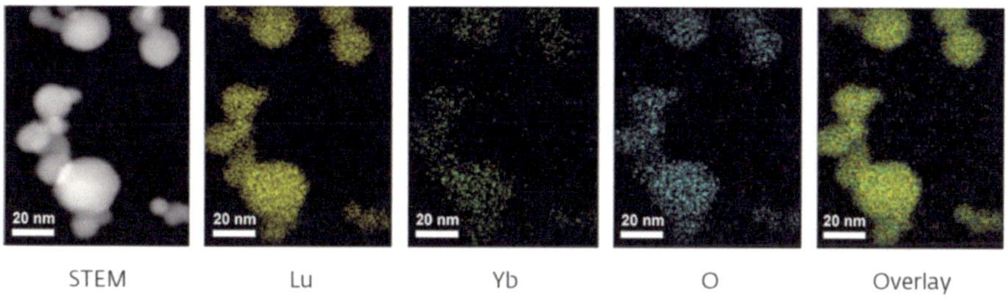

Figure 5.
Results of mapping the elemental composition of $Yb:Lu_2O_3$ nanoparticles with the use of SEM and X-ray SMA.

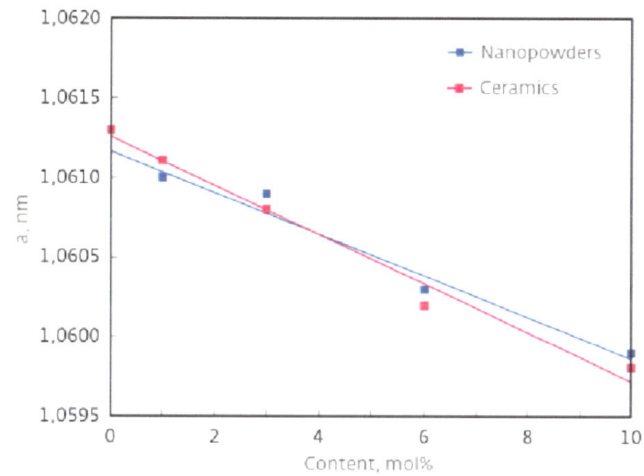

Figure 6.
Results of X-ray diffraction analysis of ceramics and nanopowders with different concentrations of HfO_2 [35].

This finding is supported by the results of X-ray diffraction analysis of $Nd:Y_2O_3$ nanopowders and ceramics doped with HfO_2 (**Figure 6**). It can be seen that the dependence of the parameters of the crystal lattice on the HfO_2 content is linear. This indirectly indicates a homogeneous occurrence of Hf in a Y_2O_3 matrix and the absence of second phases, both in a nanopowder and in ceramics.

A feature of the above method for producing nanopowders is that they crystallize in a laser plume, as a rule, in metastable phases. For example, yttrium oxide nanopowders crystallize in the monoclinic phase, while alumina nanopowders in the γ-phase. This effect is associated with very rapid cooling and quenching (within $\approx$1 ms) of the resulting nanoparticles during vortex mixing of the laser plume with air and, possibly, with the resulting oxygen deficiency in nanoparticles formed from trivalent cation radicals.

3. Compacting, annealing and sintering of nanopowders

The requirements that are imposed on the methods of "cold" pressing are, first of all, maximally possible compact density and uniformity of stacking of powders. To produce highly transparent ceramics, the following compacting methods are most often used: slip casting, slip casting under pressure, cold isostatic pressing, static pressing, static pressing with ultrasonic treatment on nanoparticles, magnetic pulse compacting.

In [36], the transparency of laser ceramics was investigated using compacts prepared by slip casting and dry pressing. It was shown that ceramics samples, whose compacts were prepared by cold isostatic pressing, have greater transparency than with slip casting. This difference is attributed to the high viscosity of the slip using nanoparticles, which prevented tight packaging. At the same time, when using hot pressing at 1750 °C and a pressure of 200 MPa, the samples prepared by slip casting have better characteristics than those based on the compacting of dry powders. However, the use of hot pressing is a complex and expensive step, therefore, there is a strong desire to create a technological chain of preparation of samples with theoretical transparency, without the use of hot pressing.

Given the above, the most studies are conducted using dry cold pressing of nanopowders. For these purposes, we have tested the method of static pressing of nanoparticles with and without ultrasonic treatment (UST), magnetic-pulse pressing and cold isostatic pressing. All of them showed rather close relative densities of compacts at the same pressures, which is confirmed by the results presented in [36, 37]. Nominally pure and neodymium-activated yttrium oxide nanopowders, designated by us as Y_2O_3, 8 NDY, 3NDY, and 1NDY (the number before the letter symbol NDY denotes the content of neodymium oxide in mole percent in nanopowder) were used in the experiments. For comparison, the dry nanopowders (without plasticizers) of all these types were pressed as uniaxial static pressing (without UST), and under the influence of ultrasonic vibrations. The pressures were 240, 480, and 720 MPa. The diameter of the pressed samples was 14 mm, the height of the samples was 2–4 mm. The experimental results in the form of the dependence of the relative density on the compacting pressure at a constant power of UST 3 kW and 0 kW (i.e. without UST) are shown in **Figure 7**.

According to the technique described in [37], the parameters of the pressing equation *b* and *Pcr* for each type of nanopowder were determined from the experimental compaction curve. The compression curves of the samples were described by the logarithmic compression equation in dimensionless form:

$$\rho / \rho_{theor}(P) = b \cdot Ln\left(\frac{P}{P_{cr}}\right) + 1 \quad (3)$$

where ρ is the density of the compact, ρ_{theor} is the theoretical density, b is the compaction rate, P_{cr} is the design pressure at which the theoretical density is reached. The results obtained show that the relative density of the compacts of the studied nanopowders is slightly dependent on the UST and is determined mainly by the compacting pressure, thereby confirming the findings obtained using other methods.

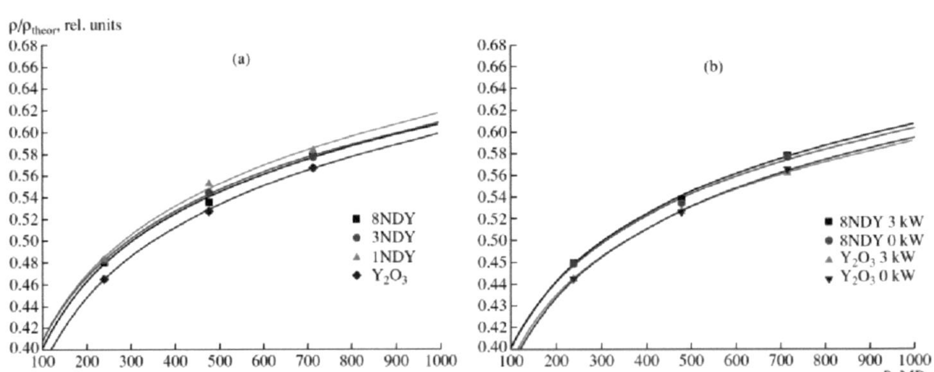

Figure 7.
Curves of nanopowder compaction: (a) 8NDY, 3NDY, 1NDY, Y_2O_3 with UST, W = 3 kW; (b) 8NDY, Y_2O_3 with UST W = 3 kW and without UST (W = 0 kW) [37].

The effect of nanoparticle size on compacts density is discussed in [38] using the above method, the granular dynamics. The calculations were carried out for nanopowders with particle sizes from 10 to 100 nm. Typically, deterioration of compressibility with decreasing particle sizes is associated with adhesion of the individual particles, which results in the formation of strong aggregates. As possible causes of the size effect are called Van der Waals forces of attraction, the absence of plastic deformation of nanoparticles, the formation of chemical bonds, electrostatic interaction, etc. The authors [38] sought to take into account the most important of these reasons. Their calculations of the dependence of the density of compacts on the axial pressure are shown in **Figure 8**.

Under the initial anisotropic configuration, the distribution of particles with the presence of vertical chains and a coordination number exactly equal to two accurately was adopted. It can be seen that as the particle size increases at the same pressing pressures, the density of the compacts increases substantially. We should also pay attention to the important role that the Van der Waals forces create (curve 4). Of course, there is no exact agreement with the experimental data, but the trend can be traced unequivocally. This fact raises the question of which nanopowders are most preferable for the synthesis of laser ceramics. On the one hand, small particles due to high surface energy provide high sinterability, and in the case of nanopowders, produced by laser evaporation, - greater solubility of ingredients in each other and particle uniformity, but poorer compressibility. This question remains open in relation to the synthesis of laser ceramics up to this point. Further, the results obtained using a nanopowder, obtained by laser evaporation of a solid target, with an average particle size of 10–20 nm and uniaxial static pressing will be presented for the preparation of compacts with dimensions less than 30 mm. Cold isostatic pressing was used for compacts of larger diameter. The prepared compacts with a relative density of 0.46–0.58 are usually air calcined to remove organic matter and to provide additional oxidation and phase transformations.

Figure 9 shows the dependence of the grain size on the calcination temperature. Each point on the graph corresponds to its own pattern. It can be seen that the grain sizes grow reasonably from 24 to 77 nm with an increase in temperature from 715 °C to 1300 °C, and the last point, apparently, is caused by a measurement error.

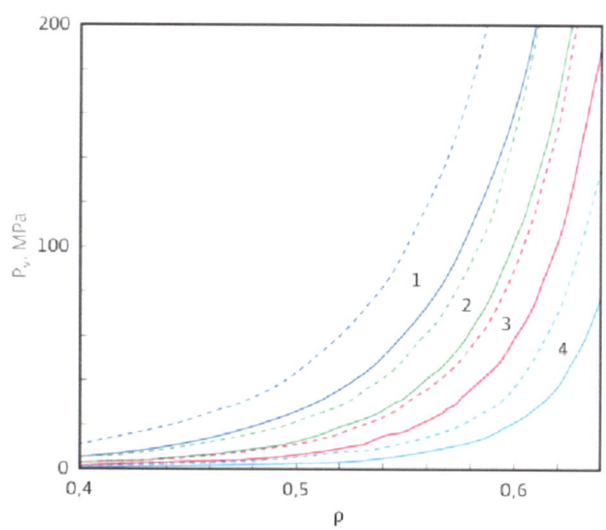

Figure 8.
Axial pressure as a function of the compact density for systems with a particle size d = 10 nm (1), 30 nm (2), 100 nm (3) and a system without Van der Waals forces (4). Solid lines are isotropic initial configurations; dashed lines are anisotropic configurations [38].

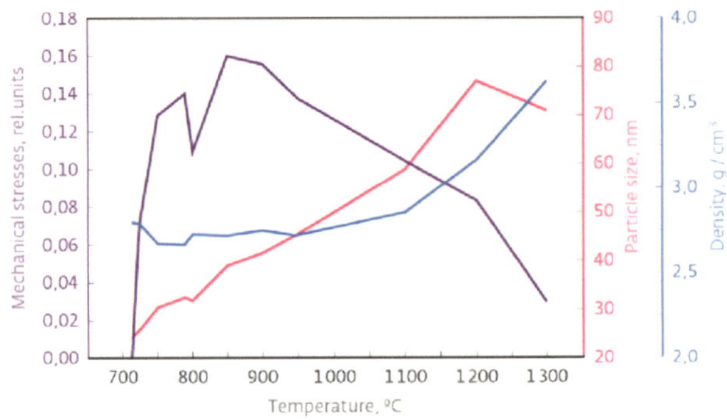

Figure 9.
Dependence of the grain size, mechanical stresses and densities of compacts on the calcination temperature of compacts from the nanopowder of the monoclinic phase [35].

The dependence of the mechanical stresses and density of compacts on temperature is also given there: after transformation at 715 °C into a cubic phase which parameters are greater than in the monoclinic one, mechanical stresses increase with the temperature raise, followed by a certain decrease, accompanied simultaneously by a shock of condensation of compacts, that we also interpreted as a mechanical ordering of grains. Further, the behavior of the curves is logical: the density of compacts increases, mechanical stresses decrease.

Sintering can be conditionally divided into three stages. The dependencies shown in **Figure 9**, characterize the processes in two of the three stages of sintering. In stage I (700–1200 °C), there is no shrinkage of the compact, but mass transfer from convex to concave surfaces occurs, mainly by near-surface diffusion. This leads to a decrease in the free surface of nanoparticles, which means that they smooth out, spheroidize and increase the size of contact spots between nanoparticles. In the case of nanopowders, the latter process leads to an increase in the dimensions of the nanoparticles, which is not observed for particles with dimensions of ~1 μm.

After 1200°C, a second stage is observed, characterized by rapid shrinkage of the sample. This is due to the diffusion sliding of the grains and the diffusion adjustment of their shape, as well as the "evaporation" of vacancies from the pore surface in the bulk of the particles, with their subsequent exit to the crystallite boundaries and displacement in the boundary layer. Since the particle sizes in our case are small, there are many grain boundaries, then the shrinkage process occurs quite intensively.

When the compacts are compacted, the diffusion processes are decisive. Therefore, an increase in these rates by introducing hetero- and isovalent additives that form solid solutions can significantly accelerate the compaction. In this case, heterovalent additives lead to the formation of vacancies that are much higher than their thermodynamic content in the unalloyed matrix. The introduction of isovalent additives leads to lattice distortion. Both these additives lead to an acceleration of mass transfer, release and filling of pores. When sintering with such additives, a situation may occur where the removal of pores outstrips the growth of crystallites. In this case, these processes are separated, and the crystallites grow non-porous, which facilitates the synthesis of high-transparency ceramics. Moreover, the introduction of additives changes the conditions for the transition of an atom across the boundary, which can significantly affect the final dimensions of the crystallites. We have investigated the replacement of the Y^{3+} cation in Nd^{3+}:Y_2O_3 with isovalent ions Lu^{3+} or Sc^{3+} ions or the Zr^{4+} and Hf^{4+} heterovalent ions, and also the Al^{4+} cations by Ce^{3+} in garnet ceramics. The compacts with a diameter of 15–32 mm, a thickness of 0.5–3.5 mm with a relative density of ~0.5 were sintered. The parameters of sintering varied over a wide range: the sintering temperature T = 1550–2050°C; sintering time t_s = 1–30 h; the rate of temperature rise v_T = 0.75 and 5.0 K/min. The influence of these factors on the characteristics of high-transparency ceramics will be discussed in the next section.

4. Highly transparent ceramics for lasers

Highly transparent ceramics are more commonly used as active elements of solid-state lasers intended for various purposes, optical armor, scintillation sensors, heat and mechanically resistant windows, bulbs for high-power high-pressure lamps, wide-angle lenses, etc. It was previously noted that ceramic samples of the highest optical quality are usually obtained using hot isostatic pressing. This is a rather expensive and complex technology. Therefore, numerous studies are being conducted to find technological solutions to avoid this operation. Here we present the results of only one of the ways to solve this problem, which is related to the use of nanopowders with a small average particle size and high uniformity of their composition within each of the nanoparticles. Let's consider the characteristics of a number of ceramics for various purposes prepared using nanopowders synthesized in a laser plume.

4.1 Ceramics with disordered crystalline structure

Such ceramics are formed by replacing matrix cations with impurity cations. This leads to a change in the local crystalline fields in the positions of the activator ions and, therefore, to broadening of the spectral lines and the gain band.

First the focused broadening of the laser transition band was implemented in ceramic yttrium-aluminum garnet [39], when a part of aluminum ions was replaced by scandium ions, i.e. ion of the same valence. In this ceramic, activated by Nd^{3+}, a laser pulse with a duration of 10 ps was obtained on its optical transitions in the 1 μm region, and when the neodymium was changed to ytterbium, it was reduced to 96 fs [40].

At the same time, it was found that the greatest broadening of the gain band is achieved in the ceramics based on yttrium oxide with the introduction of heterovalent ions. However, in [41] the doping with such ions did not allow achieving the transparency necessary for high-performance generation [35]. According to the authors, this was prevented by the formation of an "orange peel" due to the increased concentration of dopants near the intercrystalline boundaries. Since this class ceramic is important for the development of laser technology, we investigated its creation using two approaches. In the first case, the traditional approach [7] was implemented, i.e. the ceramics were synthesized from nanopowders of simple oxides Yb_2O_3, Nd_2O_3, Y_2O_3, HfO_2 and ZrO_2, mixed in the required ratio. We refer to them as to "mixed" powders. The second approach is original [42] and consists in the fact that the necessary components were mixed in the preparation of a laser target, and the synthesis of nanoparticles occurred in a laser plume, i.e. at high temperature and rapid (<1 ms) cooling. Let's refer to these powders as to "laser" ones.

Using these approaches, the samples of ceramics based on yttrium oxide with HfO_2 or ZrO_2 additives were prepared. The samples were 2–3 mm thick and 11 mm in diameter. Analysis of the appearance of the ceramics' samples based on yttrium oxide, obtained by different approaches, shows that they differ insignificantly. The differences are manifested in the study of their light scattering. **Figure 10** shows photographs of the initial radiation of the laser (λ = 633 nm) incident on the screen and of the radiation passing through samples of "mixed" and "laser" powders

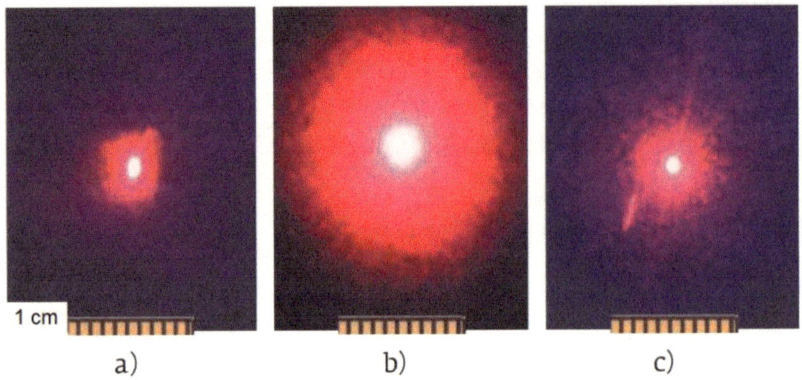

Figure 10.
The initial radiation of a semiconductor laser (λ = 633 nm) incident on the screen (a) and the radiation transmitted through the ceramic samples [(YbxLuyY1-x-y)$_2$O$_3$]1-z(ZrO$_2$)z from the "mixed" (b) and laser (c) powders [42].

having the same chemical composition $[(Yb_xLu_yY_{1-x-y})_2O_3]_{1-z}(ZrO_2)_z$. It can be seen that the ceramic made of "mixed" powders possesses a large light scattering and transparency by 15–20% lower than that made of laser powders [42], therefore it is not yet suitable for obtaining high-performance generation.

In this connection, the ceramics prepared from "laser" powders were investigated further. Their disordered crystalline structure manifests itself in the broadening of the emission bands at laser transitions between the Stark levels of the neodymium ion $^4F_{3/2} \leftrightarrow {}^4I_{11/2}$ and of the Yb^{3+} ion $^2F_{5/2} \leftrightarrow 2F_{7/2}$ (**Figure 11**). Moreover, it was found that the additives lead to a complete overlap (at a level less than 0.4 of the maximum intensity) of the contours of the two neodymium emission bands at λ = 1060 nm and 1075 nm (**Figure 11**, left). This leads to the formation of a continuous emission band with a width of up to 50 nm (on the base) in the range of 1040–1090 nm [42].

In the optical ceramics activated with ytterbium, the above additives also lead to broadening of the luminescence bands at λ = 1030 and 1075 nm on a laser transition between the Stark levels of the Yb^{3+} ion $^2F_{5/2} \leftrightarrow {}^2F_{7/2}$ (**Figure 11**, right). A complete overlap of the bands is observed at a minimum level of 0.25 of the maximum intensity with the width of the continuous band at this level reaching 100 nm on the base [42, 43].

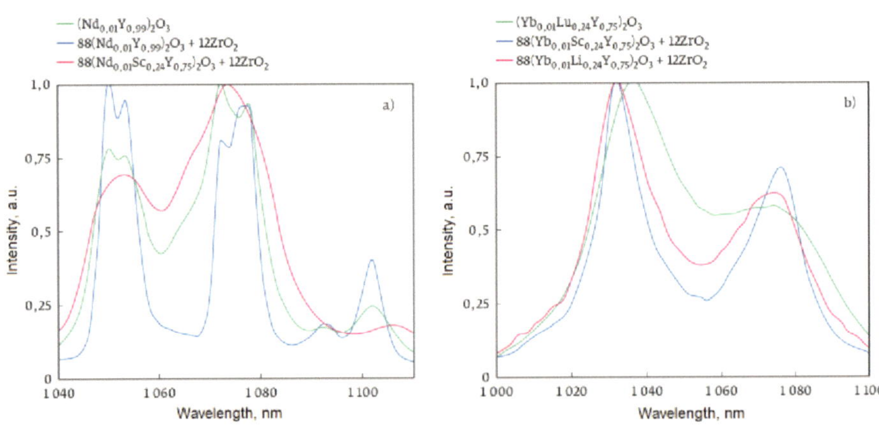

Figure 11.
IR spectra of luminescence of ceramic samples activated with Nd^{3+} ions (left) and Yb^{3+} ions (right) [42].

In the ceramics with additions of zirconium and hafnium, the trivalent Hf^{3+} and Zr^{3+} ions were found [42–44], which is confirmed by electron paramagnetic resonance spectra [44]. In the crystal, $3d^{10}4d^1$ Zr^{3+} and $4f^{14}5d^1$ Hf^{3+} ions form two Stark levels: the orbital doublet (E) and the triplet (T_2), with the energy gap equal to the strength of the crystal field in the positions of these ions. In yttrium oxide, these ions replace yttrium ions in two positions C_2 and C_{3i}, differing in symmetry and the strength of the crystalline field. Therefore, in the pulsed cathodoluminescence spectra of the ceramics containing zirconium or hafnium, both ions (Hf^{3+} and Zr^{3+}) emit two bands each, at λ = 818 nm and 900 nm about 30 nm wide [42, 43]. Furthermore, the energy of the radiative level of the short-wave band (12225 cm^{-1}) of the Hf^{3+} and Zr^{3+} ions coincides with the energy of the pumping level of $^4F_{5/2}$ (12138–12436 cm^{-1}) of the neodymium ion, and the energy of the radiative level of the second longer wavelength band (11100 cm^{-1}) – with that of the upper laser level $^4F_{3/2}$ (11208–11404 cm^{-1}) of the Nd^{3+} ion. It is because of the negative influence of the Hf^{3+} and Zr^{3+} ions on the inverse population of the laser levels caused by this coincidence, that we have not obtained laser generation on the neodymium ion ransitions in the ceramics with disordered crystalline structure with additions of hafnium or zirconium.

Another situation is observed for the activator Yb^{3+} ion. The energy of its upper laser level $^2F_{5/2}$ (10240–10673 cm^{-1}) is less than the energy of the radiative levels of Hf^{3+} and Zr^{3+} ions. Therefore, the Hf^{3+} and Zr^{3+} ions do not affect the population of the $^2F_{5/2}$ level of the Yb^{3+} ion, which allowed generation of laser radiation in disordered ceramic consisting of $0.88[(Yb_{0.01}Lu_{0.24}Y_{0.75})_2O_3]+0.12ZrO_2$ [42] obtained from "laser" nanopowders of a solid solution. The generation properties were investigated in a three-mirror V-shaped resonator formed by two spherical mirrors with radii of curvature of 100 mm and an output plane mirror with a transmittance of 1.2, 2.4 and 5.0%. The active element in the form of a polished ceramic disk 1.27 mm thick was installed in the resonator between spherical mirrors at the Brewster angle. Pumping was carried out through a dichroic spherical mirror with a reflection coefficient of 99.9% in the range of 1020–1100 nm and a transmittance factor of 98% in the range of 950–980 nm by a laser diode radiation with a fiber output of 9 W at a wavelength of 975 nm and a bandwidth of 3 nm. With an output mirror with a transmittance factor of 1.2, 2.4 and 5.0%, the slope efficiency was 16.5, 26.0 and 29.0% with an optical efficiency of 6.8, 7.0 and 9.5%, respectively.

Relatively low values of the laser generation parameters obtained are due to the presence of an "orange peel" in the ceramics with a high content (12 mol%) of zirconium. In the ceramic consisting of $0.95[(Yb_{0.05}Lu_{0.15}Y_{0.80})_2O_3] + 0.05ZrO_2$ with a content of the sintering additive ZrO_2 reduced to 5 mol%, the "orange peel" is not clearly manifested. While investigating the generation properties [45], it was found that the laser generation band on this ceramic (**Figure 12**) practically coincides with the IR-luminescence band (**Figure 11**, right), its width reaches 97 nm at the base, which is currently a record value in the visible and near-IR wavelengths. On this entire band, quasi-continuous generation with a slope efficiency equal to 49.3% and 51.2% in the band maxima at the wavelengths of 1077 and 1032 nm, respectively, was obtained. These factors provide good prospects for the development of lasers with ultrashort pulses and lasers with a wide range of smooth frequency tuning.

4.2 Ceramics of yttrium-aluminum garnet

Taking into account the importance for the creation of technological lasers and high-scale laser systems, the great attention has been paid to YAG ceramics, doped with Nd or Yb. Extensive studies have been carried out, the results of which have been presented in a number of reviews, for example [46], and monographs [8],

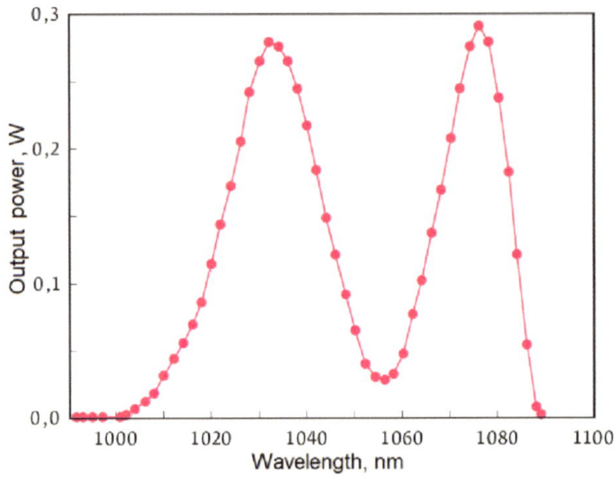

Figure 12.
Laser generation band of 0.95[$(Yb_{0.05}Lu_{0.15}Y_{0.80})_2O_3$] + 0.05$ZrO_2$ ceramics [45].

the methods for obtaining nanopowders, compaction and sintering have been developed that make it possible to synthesize samples with a transparency close to the theoretical one [8] and to generate a radiation with an efficiency of more than 74%. High-level results were obtained using both hot isostatic pressing (HIP) and vacuum sintering, but the presence of sintering additives in a mixture of nanopowders as TEOS [7] and MgO [47] was always mandatory. Using the nanopowders prepared by the laser synthesis method, we have studied the feasibility of synthesizing YAG ceramics without the use of these additives. Various approaches to the preparation of nanopowders were involved.

The successful attempt to produce highly transparent YAG ceramics without the use of sintering additives was associated with the mixing of separately obtained Nd:Y_2O_3 and Al_2O_3 nanopowders in the ratio of 3/5. Measured by the BET method the specific surface area of the Nd:Y_2O_3 powder was 50.7 m^2/g. It was a solid solution based on monoclinic yttrium oxide with crystalline lattice parameters a = 13.92 Å, b = 3.494 Å, c = 8.611 Å, β = 101.2°. After calcination at a temperature of 1000 °C for 30 minutes, the surface area of the powder was 25 m^2/g for conversion to the cubic phase, i. e. the particle size increased from 12 to 49 nm. Al_2O_3 nanopowder was also obtained by laser evaporation of a target followed by condensation of vapors in the air stream. Its specific surface, was 109.67 m^2/g. X-ray fluorescence analysis showed that the powder consists mainly of the γ-Al_2O_3 phase and the δ-phase content was less than 10%.

These powders were mixed in the indicated proportion in a drum mixer with an inclined rotation axis for 24 hours. Further, briquettes with a density of 20% compared to the theoretical were compacted from this mixture, which were then calcined at 1200 °C for 3 hours. As shown by X-ray fluorescence analysis, the YAG phase content in the briquettes was 96–98%. These briquettes were then milled by YSZ balls in a planetary mill for 48 hours.

The analysis of powder images after grinding showed that the agglomerates of the particles formed after calcination had an average size slightly less than 1 μm, but sometimes their size was close to 10 μm. The compacting of nanopowders into disks with a diameter of 15 mm and a thickness of 1.5–4.5 mm was carried out by the method of dry uniaxial static pressing without the use of any additives. The compacting pressure in these experiments was unchanged and was 200 MPa, which made it possible to obtain a density of 61.8%. Sintering was performed at a temperature of 1760 °C for 20 hours. The pore content in the samples was ~60 ppm, and the transparency was 83.28%. For the first time in the Nd: YAG ceramics that did not contain sintering additives, the generation was obtained with an average power of up to 4 W and a slope efficiency of 19% [48]. However, much better results were achieved when 0.5 wt% TEOS sintering additive was added to the nanopowder. In this case, the slightly agglomerated Nd:Y_2O_3 and Al_2O_3 nanoparticles of spherical shape with dimensions of 8–14 nm were calcined at a temperature of 900–1200 °C for transformation from the monoclinic to the cubic phase. These calcined nanopowders were weighed to ensure the $Nd_{0.03}Y_{2.97}Al_5O_{12}$ stoichiometry and mixed in a ball mill with an inclined axis of rotation in alcohol with the addition of 0.5 wt% TEOS for 48 hours.

Using the previously described approach, Nd(Yb):YAG ceramic samples were synthesized. **Figure 13** shows a photograph of a Nd:YAG ceramic sample, its transmission spectrum, and also the transmission spectrum of a single-crystal laser of the same composition, which has theoretical transparency. It can be seen that in the wavelength range of more than 450 nm, these spectra practically coincide. Compared with the above results, the optical quality of the resulting ceramic due to the presence of SiO_2 was improved due to a partial reduction in agglomeration of the powder during the calcination step, inhibition of crystallite growth and pore removal due to the formation of the liquid phase, which led to reducing their content to 17 ppm. Similar results were obtained by compacting the calcined Nd:Y_2O_3 and Al_2O_3 nanopowders into compacts with a relative density of 48% and reactive sintering at 1780 °C for 20 hours.

The comparative studies of our samples and samples by Konoshima Chemical [50] were carried out jointly with the National Institute of Optics (Florence, Italy). They had the same composition (1 at.% Nd:YAG) and a thickness of 1.5 mm. To obtain the generation, a V-shaped resonator was used (**Figure 14a**). Pumping was carried out through an end dichroic mirror having high transparency for pumping radiation and high reflection for the generated radiation and spaced from the sample by 4 mm. The distance from the end EM and the output mirror OC to the rotary mirror FM was 280 mm. The OC transmission varied between 2–20%. Pumping was carried out by rectangular pulses of a duration of 10 ms and a frequency of 12.5 Hz. Their peak power was 32 W, the radiation focusing spot was 0.8 mm.

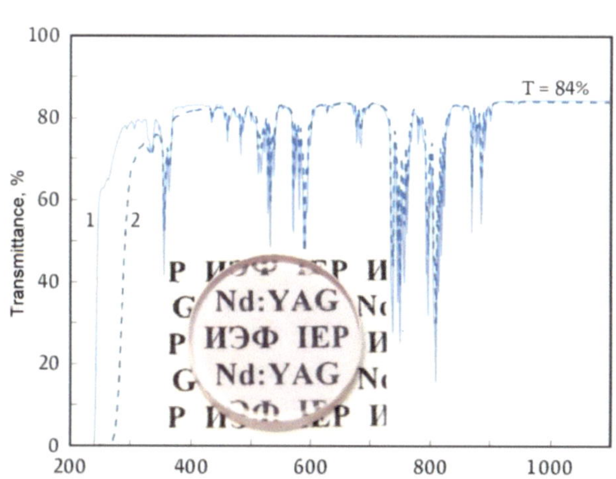

Figure 13.
Transmission spectra of Nd:YAG single crystal (1) and ceramics (2). The inset shows a photograph of ceramics [49].

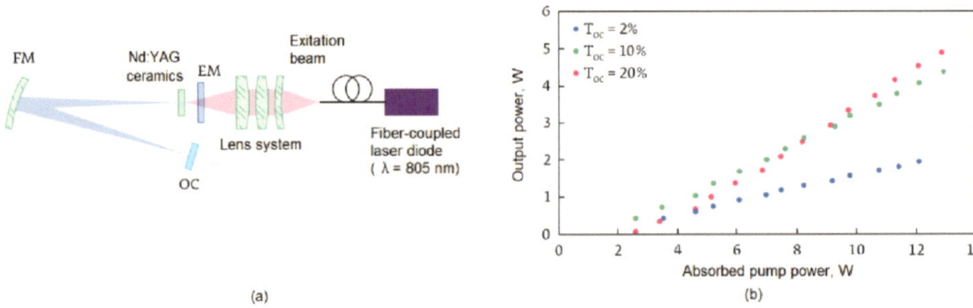

Figure 14.
Flow-chart of the experimental setting (a) and the dependence of the generation power on the pump power (b) [50].

	T_{oc} = 20%		T_{oc} = 10%		T_{oc} = 2%	
	P_{out}, W	η_S, %	P_{out}, W	η_S, %	P_{out}, W	η_S, %
IEP UrB RAS	4.91	52.7	4.38	35.5	2.07	16.7
Konoshima Chemical	5.29	53	4.69	40.4	2.49	16.6

Table 1.
Laser characteristics of Nd:YAG ceramics [50].

The dependence of the output power on the pump power is shown in **Figure 14b**. Similar results were obtained for the samples of Konoshima Chemical. Comparative data are given in **Table 1**. The best results were obtained with a transparency of the output mirror T_{oc} = 20%, when the radiation power was P_{out} = 4.91 W, and the slope efficiency η_{sl} = 52.7%. Thus, the introduction of a sintering additive in the form of TEOS had a significant effect on improving the characteristics of samples prepared from nanoparticles synthesized in a laser plume.

5. Conclusion

The main stages and processes taking place in the preparation of high-transparent ceramics, including laser ones. The optimal conditions at which the productivity of nanopowder production is realized, depending on the thermophysical properties of the material, were found to be 10–80 g/h. It is shown that the nanoparticles obtained are weakly agglomerated, have a spherical shape and an average size of ~10 nm. A feature of such nanoparticles is the high homogeneity of the composition at a high level of doping. It is shown that the density of compacts does not depend on the method of dry pressing and is determined by pressure, although the level of residual mechanical stresses differs. Pressing was carried out at diapason of pressures of 250–300 MPa, at which compact densities were ~50%.

The use of nanopowders synthesized in a laser plume for the preparation of highly transparent ceramics makes it possible to increase the threshold for the formation of an "orange peel". This opens the road to the use of sesquioxides with highly disordered crystalline structure as active elements of solid-state lasers using a relatively simple technology. In particular, this approach allowed to obtain the following.

1. In samples based on Y_2O_3 doped with Yb_2O_3 and ZrO_2, the slope efficiency of radiation generation can exceed 50%, and the band for smooth tuning of the radiation frequency can reach 100 nm.

2. Highly transparent YAG samples are prepared without the use of sintering additives, where the transparency and generation efficiency, however, is inferior to those realized when doping TEOS.

Acknowledgements

The research was carried out within the framework of the theme of state task No. 0389-2016-002 (2018–2020) and with the support of the project by UB of R AS No. 18-10-2-38.

Author details

Vladimir Osipov*, Vyacheslav Platonov, Vladislav Shitov and Vladimir Solomonov
Institute of Electrophysics of Ural Division of RAS, Yekaterinburg, Russia

*Address all correspondence to: osipov@iep.uran.ru

References

[1] Steen WM. 1 - 'Light' industry: an introduction to laser processing and its industrial applications. In: Lawrence J, Pou J, Low DKY, Toyserkani E, editors. Handbook of Woodhead Publishing Series in Welding and Other Joining Technologies, Advances in Laser Materials Processing. Woodhead Publishing; 2010. pp. 3-19. DOI: 10.1533/9781845699819.1.3

[2] Richardson M. Laser materials processing technologies and the future. In: Proceedings of the Conference on Lasers and Electro-Optics (CLEO Pacific Rim), 24-28 August 2015; Busan, South Korea. Paper 25D3-1

[3] Malinauskas M, Žukauskas A, Hasegawa S, Hayasaki Y, Mizeikis V, Buividas R, et al. Ultrafast laser processing of materials: from science to industry. Light Sci Appl. 2016 Aug 12;**5**(8):e16133. DOI: 10.1038/lsa.2016.133

[4] Evangelista B, NIF experiments draw interest at APS conference [Internet]. 2019. Available from: http://www.llnl.gov/news/nif-experiments-draw-interest-aps-conference [Accessed: 2020-09-29]

[5] Ribeyre X, Schurtz G, Lafon M, Galera S, Weber S. Shock ignition: an alternative scheme for HiPER. Plasma Physics and Controlled Fusion. 2009;**51**(1):015013. DOI: 10.1088/0741-3335/51/1/015013

[6] Arzeno H. Premier tir le 2 décembre au Laser Megajoule [Internet]. 2014. Available from http://www.sudouest.fr/2014/01/11/premier-tir-le-2-decembre-au-laser-megajoule-1425695-2941.php [Accessed: 2020-09-29]

[7] Ikesue A, Kinoshita T, Kamata K, Yoshida K. Fabrication and optical properties of high-performance polycrystalline Nd: YAG ceramics for solid-state lasers. Journal of the American Ceramic Society. 1995;**78**(4):1033-1040. DOI: 10.1111/j.1151-2916.1995.tb08433.x

[8] Ikesue A, Aung YL, Lupei V. Ceramic lasers. Cambridge, UK: Cambridge University Press; 2013 445 p

[9] Ikesue A, Aung YL. Synthesis of Yb:YAG ceramics without sintering additives and their performance. Journal of the American Ceramic Society. 2017;**100**:26-30. DOI: 10.1111/jace.14588

[10] Latham WP, Lobad A, Newell TC, Stalna KD, Phipps C. 6.5 kW, Yb:YAG ceramic thin disk laser. AIP Conf. Proc. 2010;**1278**:758-764. DOI: 10.1063/1.3507169

[11] Peng YH, Cheng J, Cheah YY, Lai K, Lau E, Ang S. High brightness continuous wave ceramic Yb:LuAG thin-disk laser. Optics Express. 2015;**23**:19618-19623. DOI: 10.1364/OE.23.019618

[12] Liu W, Li J, Jiang B, Zhang D, Pan Y. 2.44 kW laser output of Nd:YAG ceramic slab fabricated by a solid-state reactive sintering. Journal of Alloys and Compounds. 2012;**538**:258-261. DOI: 10.1016/j.jallcom.2012.05.050

[13] Chen J, Li J, Xu J. Liu W, Bo Y, Feng X, et. al. 4350 W quasi-continuous-wave operation of a diode face-pumped ceramic Nd: YAG slab laser. Optics & Laser Technology. 2014;63:50-53. DOI: 10.1016/j.optlastec.2014.03.005

[14] Yamamoto RM, Bhachu BS, Cutter KP, Fochs SN, Letts SA, Parks CW, Rotter MD, Soules TF. The use of large transparent ceramics in a high powered, diode pumped solid state laser. In Proceedings of Conference on Advanced Solid-State. Photonics. 27-30 January 2008; Nara, Japan. Paper WC5.

[15] Northrop Grumman Scales New Heights in Electric Laser Power, Achieves 100 Kilowatts From a Solid-State Laser [Internet]. 2018. Available from: http://news.northropgrumman.com/news/releases/photo-release-northrop-grumman-scales-new-heights-in-electric-laser-power-achieves-100-kilowatts-from-a-solid-state-laser. [Accessed: 2020-09-29]

[16] Mason P, Divoký M, Ertel K, Pilař J, Butcher T, Hanuš M, et al. Kilowatt average power 100 J-level diode pumped solid state laser. Optica. 2017;**4**(4):438-439. DOI: 10.1364/OPTICA.4.000438

[17] Tokurakawa M, Takaichi K, Shirakawa A, Ueda K, Yagi H, Yanagitani T, et al. Diode-pumped 188 fs mode-locked Yb^{3+}:Y_2O_3 ceramic laser. Applied Physics Letters. 2007;**90**. DOI: 071101(3), 10.1063/1.2476385

[18] Endo M, Ito I, Kobayashi Y. Direct 15-GHz mode-spacing optical frequency comb with a Kerr-lens mode-locked Yb:Y_2O_3 ceramic laser. Optics Express. 2015;**23**(2):1276-1282. DOI: 10.1364/OE.23.001276

[19] Tokurakawa M, Shirakawa A, Ueda K, Yagi H, Noriyuki M, Yanagitani T, et al. Diode-pumped ultrashort-pulse generation based on Yb^{3+}:Sc_2O_3 and Yb^{3+}:Y_2O_3 ceramic multi-gain-media oscillator. Optics Express. 2009;**17**(5):3353-3361. DOI: 10.1364/OE.17.003353

[20] Kagan MA, Khazanov EA. Compensation for thermally induced birefringence in polycrystalline ceramic active elements. Quantum Electronics. 2003;**33**(10):876-882. DOI: 10.1070/QE2003v033n10ABEH002516

[21] An L, Ito A, Goto T. Effect of sintering temperature on the transparency and mechanical properties of lutetium aluminum garnet fabricated by spark plasma sintering. Journal of the European Ceramic Society. 2012;**32**(12):3097-3102. DOI: 10.1016/j.jeurceramsoc.2012.04.020

[22] Kijko VS, Maksimov RN, Shitov VA, Demakov SL, Yurovskikh AS. Sintering of transparent Yb-doped Lu_2O_3 ceramics using nanopowder produced by laser ablation method. Journal of Alloys and Compounds. 2015;**643**:207-211. DOI: 10.1016/j.jallcom.2015.04.127

[23] Sokol M, Kalabukhov S, Kasiyan V, Dariel MP, Frage N. Functional Properties of Nd:YAG polycrystalline ceramics processed by high-pressure spark plasma sintering (HPSPS). Journal of the American Ceramic Society. 2016;**99**:802-807. DOI: 10.1111/jace.14051

[24] Zhang W, Lu T, Ma B, Wei N, Lu Z, Li F, et al. Improvement of optical properties of Nd:YAG transparent ceramics by post-annealing and post hot isostatic pressing. Optical Materials. 2013;**35**(12):2405-2410. DOI: 10.1016/j.optmat.2013.06.042

[25] Yang Zhang Y, Cai M, Jiang B, Fan J, Zhou C, Mao X, et al. Micro-structure of grain boundary in post-annealed sinter plus HIPed Nd:$Lu_3Al_5O_{12}$ ceramics. Optical Materials Express. 2014;**4**:2182-2189. DOI: 10.1364/OME.4.002182

[26] Chrétien L, Boulesteix R, Maître A, Sallé C, Reignoux Y. Post-sintering treatment of neodymium-doped yttrium aluminum garnet (Nd:YAG) transparent ceramics. Optical Materials Express. 2014;**4**:2166-2173. DOI: 10.1364/OME.4.002166

[27] Osipov VV, Solomonov VI, Orlov AN, Shitov VA, Maksimov RN, Spirina AV. Characteristics of yttrium oxide laser ceramics with additives. Quantum Electronics. 2013;**43**(3):276-281. DOI: 10.1070/QE2013v043n03ABEH015029

[28] Osipov VV, Kotov YuA, Ivanov MG, Samatov OM, Lisenkov VV,

Platonov VV, Murzakaev AM, Medvedev AI, Azarkevich EI. Laser synthesis of nanopowders. Laser Physics. 2006;16(1):116-120. DOI: 10.1117/12.633667

[29] Osipov VV, Platonov VV, Lisenkov VV. Laser ablation synthesis and properties of nanocrystalline oxide powders. in Handbook of Nanoparticles, ed.: Aliofkhazraei M, Springer International Publishing Switzerland. 2015. 1376 p. DOI 10.1007/978-3-319-13188-7_8-1.

[30] Osipov VV, Platonov VV, Lisenkov VV. Laser ablation plume dynamics in nanoparticle synthesis. Quantum Electronics. 2009;**39**(6):541-546. DOI: 10.1070/QE2009v039n06ABEH013981

[31] Osipov VV, Solomonov VI, Platonov VV, Snigireva OA, Ivanov MG, Lisenkov VV. Laser plume spectroscopy. 2. Graphite and yttrium-stabilized zirconium oxide targets. Quantum Electronics. 2005;**36**(7):633-637. DOI: 10.1070/QE2005v035n07ABEH004087

[32] Osipov VV, Lisenkov VV, Platonov VV, Orlov AN, Podkin AV, Savvin IA. Effect of pulses from a high-power ytterbium fiber laser on a material with a nonuniform refractive index. I. Irradiation of yttrium oxide targets. Technical Physics. 2014;**59**(5):716-723. DOI: 10.1134/S1063784214050223

[33] Platonov VV, Kochurin EA, Osipov VV, Lisenkov VV, Zubarev NM. Characteristic properties of laser ablation of translucent targets. Laser Physics. 2018;**28**(7). DOI: 076002 (8pp), 10.1088/1555-6611/aabdb4

[34] Osipov VV, Platonov VV, Lisenkov VV, Tikhonov EV, Podkin AV. Study of nanoparticle production from yttrium oxide by pulse-periodic radiation of ytterbium fiber laser. Applied Physics A. 2018;**124**:3. DOI: 10.1007/s00339-017-1348-9

[35] Osipov VV, Solomonov VI, Shitov VA, Maksimov RN, Orlov AN, Murzakaev AM. Optical ceramics based on yttrium oxide doped with tetravalent ions. Russian Physics Journal. 2015;**58**:107-116. DOI: 10.1007/s11182-015-0469-x

[36] Boltachev GSh, Lukyashin KE, Shitov VA, Volkov NB. Threedimensional simulations of nanopowder compaction processes by granular dynamics method. Physical Review E. 2013;88(1):012209(1-12). DOI: 10.1103/PhysRevE.88.012209

[37] Osipov VV, Khasanov OL, Shitov VA, Dvilis ES, Ivanov MG, et al. Optical $Nd^{3+}Y_2O_3$ ceramics of nanopowders compacted by static pressure using the ultrasonic method. Nanotechnologies in Russia. 2008;**3**(7-8):474-480. DOI: 10.1134/S1995078008070100045.

[38] Boltachev GS, Volkov NB. Size effect in nanopowder compaction. Technical Physics Letters. 2010;**36**:823-826. DOI: 10.1134/S1063785010090142

[39] Sato Y, Saikawa J, Taira T, Ikesue A. Characteristics of Nd^{3+}-doped $Y_3ScAl_4O_{12}$ ceramic laser. Optical Materials. 2007;**29**(10):1277-1282. DOI: 10.1016/j.optmat.2006.01.032

[40] Ma J, Wang J, Shen D, Ikesue A, Tang D. Generation of sub-100-fs pulses from a diode-pumped Yb:$Y_3ScAl_4O_{12}$ ceramic laser. Chinese Optics Letters. 2017;**15**(12):121403. DOI: 10.3788/COL201715.121403

[41] Greskovich C, Chernoch JP. Polycrystalline ceramic lasers. Journal of Applied Physics. 1973;**44**(10):4599-4606. DOI: 10.1063/1.1662008

[42] Bagayev SN, Osipov VV, Pestryakov EV, Solomonov VI, Shitov VA, Maksimov RN, et al. Laser ceramics with disordered crystalline structure. Journal of Appl. Mechanics

and Technical Physics. 2015;**56**(1):150-157. DOI: 10.1134/S0021894415010228

[43] Osipov VV, Solomonov VI, Orlov AN, Shitov VA, Maksimov RN, Spirina AV. Characteristics of yttrium oxide laser ceramics with additives. Quantum Electronics. 2013;**43**(3):276-281. DOI: 10.1070/QE2013v043n03ABEH015029

[44] Solomonov VI, Spirina AV, Konev SF, Cholach SO. Trivalent zirconium and hafnium ions in yttrium oxide ceramics. Optics and Spectroscopy. 2014;**116**(5):793-797. DOI: 10.1134/S0030400X14050257

[45] Toci G, Pirri A, Patrizi B, Maksimov RN, Osipov VV, Shitov VA, et al. High efficiency emission of a laser based on Yb-doped $(Lu,Y)_2O_3$ ceramic. Optical Materials. 2018;**83**:182-186. DOI: 10.1016/j.optmat.2018.06.006

[46] Wang SF, Zhang J, Luo DW, Gu F, Tang DY, Dong ZL, et al. Transparent ceramics: Processing, materials and applications. Progress in Solid State Chemistry. 2013;**41**(1-2):20-54. DOI: 10.1016/j.progsolidstchem.2012.12.002

[47] Li Y, Zhou S, Lin H, Hou X, Li W, Teng H, et al. Fabrication of Nd:YAG transparent ceramics with TEOS, MgO and compound additives as sintering aids. Journal of Alloys and Compounds. 2010;**502**(1):225-230. DOI: 10.1016/j.jallcom.2010.04.151

[48] Bagayev SN, Osipov VV, Solomonov VI, Shitov VA, Maksimov RN, Lukyashin KE, et al. Fabrication of Nd^{3+}:YAG laser ceramics with various approaches. Optical Materials. 2012;**34**(8):1482-1487. DOI: 10.1016/j.optmat.2012.03.004

[49] Osipov VV, Shitov VA, Solomonov VI, Lukyashin KE, Spirina AV, Maksimov RN. Composite Nd:YAG/Cr^{4+}:YAG transparent ceramics for thin disk lasers. Ceramics International. 2015;**41**(10):13277-13280. DOI: 10.1016/j.ceramint.2015.07.109

[50] Osipov VV, Maksimov RN, Shitov VA, Lukyashin KE, Toci G, Vannini M, et al. Fabrication, optical properties and laser outputs of Nd:YAG ceramics based on laser ablated and pre-calcined powders. Optical Materials. 2017;**71**:45-49. DOI: 10.1016/j.optmat.2016.06.021

Section 3

Fabrication Processes

Chapter 4

The Investigation on the Fabrication and Characterization of the Multicomponent Ceramics Based on PZT and the Relaxor PZN-PMnN Ferroelectric Materials

Le Dai Vuong, Vo Thanh Tung and Phan Dinh Gio

Abstract

This chapter presents the investigation of fabrication and the physical properties of the $Pb(Zr_{1-x}Ti_x)O_3$-$Pb(Zn_{1/3}Nb_{2/3})O_3$-$Pb(Mn_{1/3}Nb_{2/3})O_3$ multicomponent ceramics. The multicomponent $yPb(Zr_{1-x}Ti_x)O_3$-$(0.925-y)Pb(Zn_{1/3}Nb_{2/3})O_3$-$0.075Pb(Mn_{1/3}Nb_{2/3})O_3$ (PZT-PZN-PMnN) ceramics were synthesized by conventional solid-state reaction method (MO) combined with the B-site oxide mixing technique (BO). Research results show that the electrical properties of PZT-PZN-PMnN ceramics are optimal at a PZT content of 0.8 mol and Zr/Ti ratio of 48/52. At these contents, the ceramics have the following optimal properties: electromechanical coupling factor, $k_p = 0.62$ and $k_t = 0.51$; piezoelectric constant (d_{31}) of 130 pC/N; mechanical quality factor (Q_m) of 1112; dielectric loss ($\tan\delta$) of 0.005; high remanent polarization (P_r) of 30.4 μC.cm^{-2}; and low coercive field (E_C) of 6.2 kV.cm^{-1}. Investigation of the domain structure of the two ferroelectric phases (tetragonal and rhombohedral) in the ZnO-doped PZT-PZN-PMnN with compositions at near the morphotropic phase boundary is described as follows: the 90 and 180° domains exist in the tetragonal phase, while the 71, 109, and 90° domains are located in the rhombohedral phase, and the widths of these domains were about 100 nm. Besides, the ceramics exhibited excellent temperature stability, which makes them a promising material for high-intensity ultrasound applications.

Keywords: ceramics, the multicomponent ceramics, PZT-PZN-PMnN, ZnO nanoparticles, the ultrasonic transducers

1. Introduction

Over the years, piezoelectric materials have been heavily investigated for ultrasonic device applications. Of the many piezoelectric materials, $Pb(Zr_{1-x}Ti_x)O_3$ (PZT)-based materials are more attractive for these applications, such as piezoelectric actuators, ultrasonic motors, and piezoelectric transformers [1–9].

As $Pb(Mn_{1/3}Nb_{2/3})O_3$ (PMnN), $Pb(Zn_{1/3}Nb_{2/3})O_3$ (PZN) have been found to be promising ferroelectric ceramics with good piezoelectric characteristics, high Curie temperature, they meet well with the requirements of ultrasonic transducer applications [6–8]. They are ferroelectric materials that have characteristics such as: high dielectric constant, the temperature at the phase transition point between the ferroelectric and paraelectric phase is broad (the diffuse phase transition), and a strong frequency dependency of the dielectric properties [6, 10–12]. The PZT-PZMnN ceramics, as one of PZT-Pb(B′, B″)O_3 solid solutions, received more attention due to their high piezoelectric properties [6, 10–14]. So far, the sintering temperature of PZT-based ceramics is usually too high, approximately 1200°C [9]. To improve the sinterability and properties of lead piezoelectric ceramics, on the basis of the conventional solid phase sintering method, various advanced manufacturing techniques have been applied to the fabrication of lead ceramics such as the two-stage calcination method [15], high energy mill [16] and liquid phase sintering [9, 15, 17–20], hot isostatic pressing, hot pressing, microwave sintering, and spark plasma sintering [17] has been used effectively. Among them, the liquid phase sintering is a simple and effective method of improving the properties of PZT-based ceramics, which is currently attracting the interests of many scientists [15, 16]. By using various additives, such as NiO, B_2O_3, Bi_2O_3, Li_2CO_3, $BiFeO_3$, ZnO, CuO, and Bi_2O_3, many researchers have successfully decreased the sintering temperature of PZT-based ceramics [5, 6, 13, 14, 18–23]. We also attempted decreasing sintering temperatures from 1150 to 930°C, which significantly improved the electrical properties of the ceramics. In these ceramics, Li_2CO_3 is considered as a liquid-phase sintering aid [5, 21, 24]. The addition of Li_2CO_3 improved the sinterability of the $Bi_{0.5}(Na_{0.8}K_{0.2})_{0.5}TiO_3$ ceramic samples and caused an increase in the density and grain size at a sintering temperature of 1100°C [19]. With increasing Li_2CO_3 content, the phase structure of the ceramics changed from rhombohedral to tetragonal, indicating that it is close to the morphotropic phase boundary (MPB) of this system.

In this chapter, in order to develop the composition ceramics for high-intensity ultrasound applications, $xPb(Zr_yTi_{(1-y)})O_3$-$(0.925\text{-}x)Pb(Zn_{1/3}Nb_{2/3})O_3$-$0.075Pb(Mn_{1/3}Nb_{2/3})O_3$ (PZT-PZN-PMnN) ceramics were fabricated by the B-site oxide mixing technique. The aim of the chapter was, first, to carry out a phase formation, piezoelectric, ferroelectric, and dielectric characteristics in a solid solution of PZT-PZN-PMnN. The compositions synthesized in this study were x = 0.65, 0.70, 0.75, 0.80, 0.85, and 0.90 in the ternary system, $xPb(Zr_{0.47}Ti_{0.53})O_3$-$(0.925 - x)Pb(Zn_{1/3}Nb_{2/3})O_3$-$0.075Pb(Mn_{1/3}Nb_{2/3})O_3$. Then detailed systematic structural analysis and the study of physical properties were carried out for x = 0.8 compositions by varying the value of y in the Zr/Ti ratio. This will help to better determine how variations in the phase content affect local atomic arrangements and hence the electrical properties; and, second, to study the effect of ZnO nanoparticles on the sintering behavior and electrical properties of $0.8Pb(Zr_{0.48}Ti_{0.52})O_3$-$0.125Pb(Zn_{1/3}Nb_{2/3})O_3$-$0.075Pb(Mn_{1/3}Nb_{2/3})O_3$ piezoelectric materials; the application, fabrication of ultrasonic transducers are reported and discussed.

2. Synthesis of PZT-PZN-PMnN ceramics by the B-site oxide mixing technique

Lead-based mixed B-site cation perovskites of (B′, B″)O_3-type exhibit diffuse phase transition (DPT) behaviors of broad dielectric constant spectra in contrast to the sharp phase transitions of $Pb(Zr,Ti)O_3$ and $PbTiO_3$ [25]. The complex perovskite compounds are difficult to synthesize by conventional solid-state reaction method owing to the formation of pyrochlore phases and reduction of desirable properties, such as the electromechanical coupling factor and dielectric constant, which originate from the perovskite structure [25]. The B-site oxide mixing technique (BO) [26, 27] (formation of a B-site precursor of (B′, B″)O_2-type, followed by a reaction with PbO) has been applied to several complex perovskite compositions and the results are quite successful [22, 28, 29]. In the conventional method (MO), oxide powders of PbO, ZrO_2, TiO_2, ZnO, MnO_2, and Nb_2O_5 were weighed and milled for 8 h. To identify the temperature for calcining PZT-PZN-PMnN, we investigated the data of thermal gravimetric (TG) and thermal analysis (DTA) of PZT-PZN-PMnN powder (**Figure 1**). As per the above results, the TG curve exhibits a linear decrease in the total mass of the studied powder. However, the DTA curve shows an endothermic peak from 739 to 840°C, corresponding to the ion evaporation. To ensure the phase creation in the sample, the mixture powder was calcined at temperatures a little higher than 850°C after being milled for 8 h and pressed into pellets [30]. Afterward, the calcined PZT-PZN-PMnN pellets were continued to be milled for 16 h and pressed into disk 12 mm in diameter and 1.5 mm in thick under 100 Tan/cm^2.

In the B-site oxide mixing technique, in order to identify the temperature for calcining of (Zn,Mn)Nb_2(Zr,Ti)O_6 (BO), we investigated the data for thermal gravimetric (TG) and thermal analysis (DTA) of (Zn,Mn)Nb_2(Zr,Ti)O_6 powders, as shown in **Figure 2**. As per results, the TG curve of the mixture powder shows that the total mass of the studied powder decreases linearly. However, the DTA curve shows the endothermic peak at 978°C, corresponding to the temperature of powder evaporation. In order to ensure that the temperature is at least above 978°C for each powder grain, the mixture powder was calcined at little higher temperature of 1100°C [6, 11, 21, 22] after the powders of BO and PbO were weighed and milled for 8 h.

The powders were calcined at a temperature of 850°C for 2 h, producing the PZT-PZN-PMnN compound. The samples were sintered at 950°C for 2 h. **Figure 3** shows the X-ray diffraction (XRD) patterns of the PZT-PZN-PMnN ceramics prepared by different methods. From X-ray diagrams, we can see that the BO sample has only pure perovskite phase with rhombohedral structure, and this was determined by the (*200*) diffraction peak observed near 44° with no splitting. As seen in **Figure 4**, the BO sample was almost fully dense, and the average grain size of the

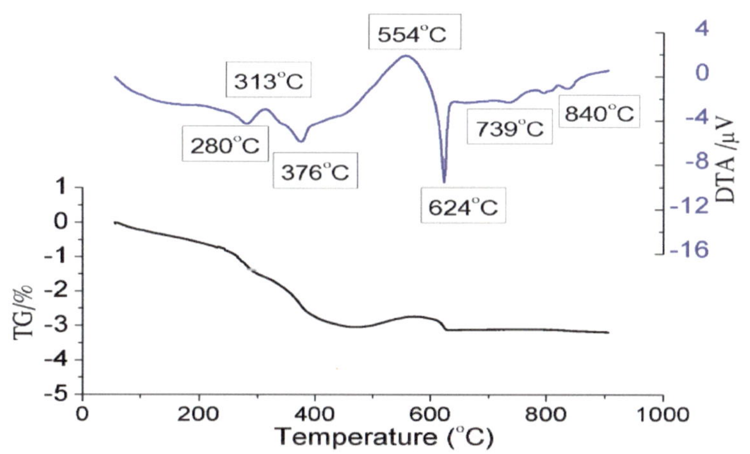

Figure 1.
TG/DTA curves for the powder mixture compositions.

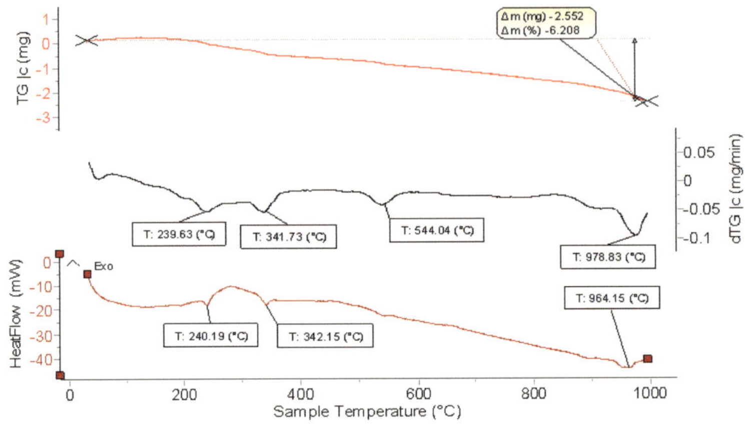

Figure 2.
TG and DTA curves of (Zn,Mn)Nb_2(Zr,Ti)O_6 power at 10°C/min heating rate.

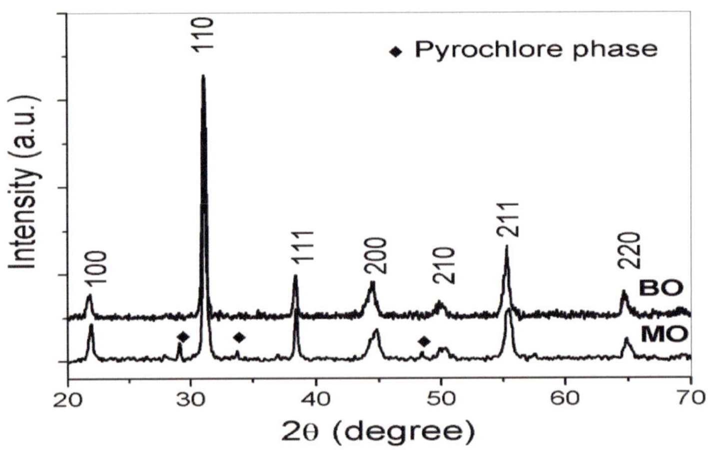

Figure 3.
X-ray diffraction diagram of the BO and MO samples.

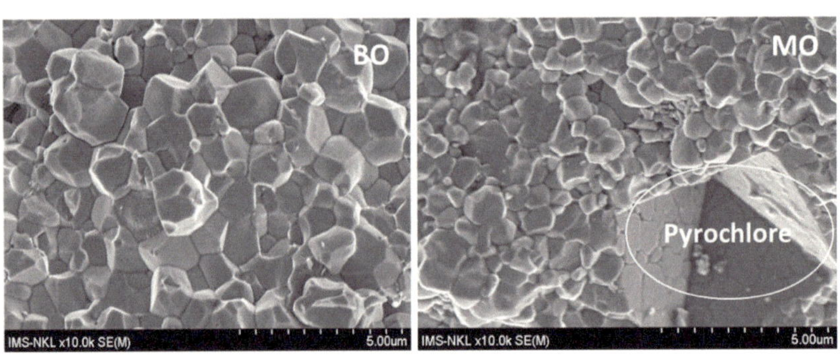

Figure 4.
Microstructures of BO and MO samples.

sample was large. However, in the MO sample, besides the perovskite phase with rhombohedral structure (**Figure 3**), there is a small pyrochlore phase. This is in good accordance with the microstructure (**Figure 4**). It is said that formation of the pyrochlore phase may be due to the steric and electrostatic interaction between the high polarization of the Pb^{2+} cation and the B-sites cation (Zn^{+2}, Nb^{+5}) [24, 27, 31].

3. Characterization of ceramics

The crystalline structure of the sintered ceramics was analyzed by X-ray diffraction (XRD) analysis at room temperature. The surface morphology was examined using field emission scanning electron microscopy (SEM), X-ray energy dispersive spectra (EDS) was measured using a Hitachi S-3400N scanning electron microscope with an EDS system Thermo Noran, and the densities of the ceramic samples were measured by the Archimedes method from the ceramic samples weighed in air, in water and the density of water. The grain size is determined from SEM micrographs by a linear intercept method. The dielectric properties of ceramics (relative dielectric constant and dielectric loss) were measured with a HIOKI 3532 impedance analyzer. The electromechanical coupling factors k_p, k_t and piezoelectric constant (d_{31}) were determined by the resonance method according to the IEEE Standard 61 using an impedance analyzer Agilent 4196B and RLC HIOKI 3532 [32, 33]. The ferroelectric properties were measured by applying the Sawyer-Tower method [34].

4. The effects of $Pb(Zr_{0.47}Ti_{0.53})O_3$ on the structure, microstructure, and the dielectric properties of $xPb(Zr_{0.47}Ti_{0.53})O_3$-(0.925 – x) $Pb(Zn_{1/3}Nb_{2/3})O_3$-0.075$Pb(Mn_{1/3}Nb_{2/3})O_3$ ceramics

Lead-zinc niobate $Pb(Zn_{1/3}Nb_{2/3})O_3$ (PZN) materials were first synthesized in the 1960s [35, 36]. It is one of the well-known relaxor perovskite ferroelectrics exhibiting a diffused phase transition with a phase transition temperature around 140°C (T_m) [6, 12]. However, pure perovskite lead-zinc niobate ceramics are relatively difficult to prepare by conventional solid-state reaction method [37]. The addition of other perovskite materials such as $PbTiO_3$, $BaTiO_3$, and $PbZr_{0.47}Ti_{0.53}O_3$ (PZT) is necessary to stabilize the perovskite structure for PZN ceramics [12, 25, 38, 39]. The B-site ions in the PZT perovskite structure (Zr^{4+}, Ti^{4+}) might have been partially substituted by the B-site ions of the relaxor-type PZN structure (Zn^{2+}, Nb^{5+}), which allowed the PZT-PZN solid-solution system to retain the perovskite structure and the high sinterability of lead-based relaxor ceramics [12, 25, 38, 39]. Based on the preparation of pyrochlore-free $Pb(Ni_1/_3Nb_2/_3)O_3$ (PNN), Vittayakorn et al. [40] studied the effects of PZT contents on the dielectric and ferroelectric properties of 0.5PNN-(0.5 – x) PZN-xPZT ceramics. The results showed that the dielectric constant (ε_r), the remanent polarization (P_r), and Curie temperature (T_c) increase with the increase in PZT content. With the combination of the preeminent properties between PZT, PZN, and PMnN, the PZT-PZN solid-solution systems hope to achieve the prominent properties of normal ferroelectric PZT and relaxor ferroelectric PZN and PMnN, which could exhibit better piezoelectric and dielectric properties simultaneously. In this section, in order to improve electrical properties, we have prepared xPZT-(0.925 – x)PZN-0.075PMnN ceramics with the content of PZT from 0.65 to 0.90. The $xPb(Zr_yTi_{(1-y)})O_3$-(0.925 – x)$Pb(Zn_{1/3}Nb_{2/3})O_3$–0.075$Pb(Mn_{1/3}Nb_{2/3})O_3$ ceramic samples have been fabricated by the B-site oxide mixing technique as described in Section 2.

Figure 5 shows XRD patterns of the PZT-PZN-PMnN ceramics at various contents of PZT. As observed, all ceramics have pure perovskite phase with dominantly tetragonal structure. The lattice parameters (*a*, *c*) of the samples have been evacuated from the (002) and (200) peaks of diffraction patterns, which are shown in the inset of **Figure 5**. When PZT content increases, the tetragonality *c*/*a* ratio increases. According to the $PbZrO_3$-$PbTiO_3$ phase diagram, $Pb(Zr_{0.47}Ti_{0.53})O_3$ is the tetragonal phase (space group P4mm) near the morphotropic phase boundary region at room temperature (RT) [41, 42]. While $Pb(Mn_1/_3Nb_2/_3)O_3$ is a cubic structure and the PZN composition was determined to be the rhombohedral (space group R3m) [36, 38]. Therefore, with increasing the molar fraction of PZT, the crystal symmetry of the PZT-PZN-PMnN should change due to the tetragonal distortions of PZT [6, 25, 40]. In order to determine what chemical composition of the PZT-PZN-PMnN ceramic changes during sintering, the EDS analysis is performed and is shown in **Figure 6**. The presence of lithium (Li) is not plotted here because its atomic number is low and the mass percentage is too small [43]. **Table 1** also showed the comparison in the mass of Pb, Zr, Ti, Nb, Zn, and Mn elements between before and after sintering of the PZT-PZN-PMnN ceramics. It is quite clear that the chemical composition of the synthesized ceramic obtained by EDS analysis can roughly accord with

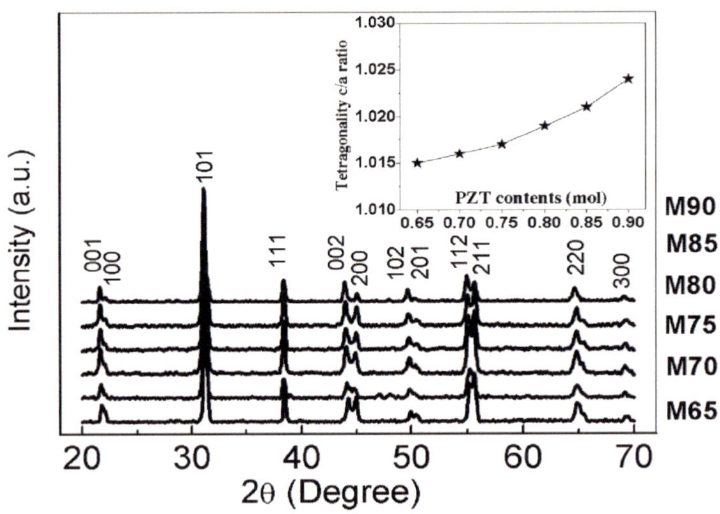

Figure 5.
XRD patterns of PZT-PZN-PMnN ceramics at various contents of PZT.

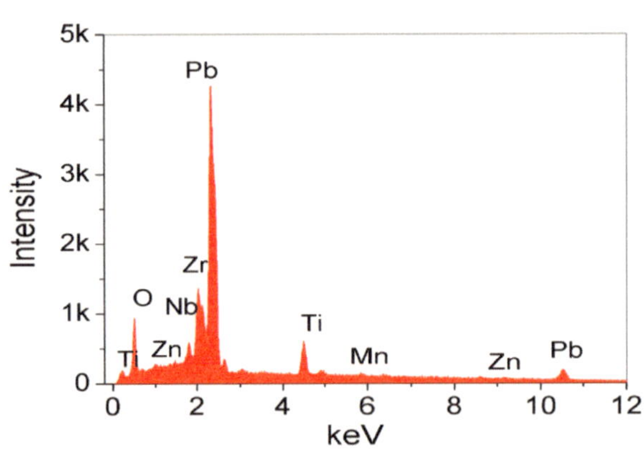

Figure 6.
EDS spectrum of 0.8PZT-0.125PZN-0.075PMnN ceramics.

Elements	The mass percentage of elements from the precursors	Mass percentage of elements from the synthesized ceramic
Pb	64.4	57
Zr	10.72	10.24
Ti	6.10	4.97
Nb	3.79	3.06
Zn	0.83	0.59
Mn	0.42	0.43

Table 1.
The chemical composition of the PZT-PZN-PMnN ceramics.

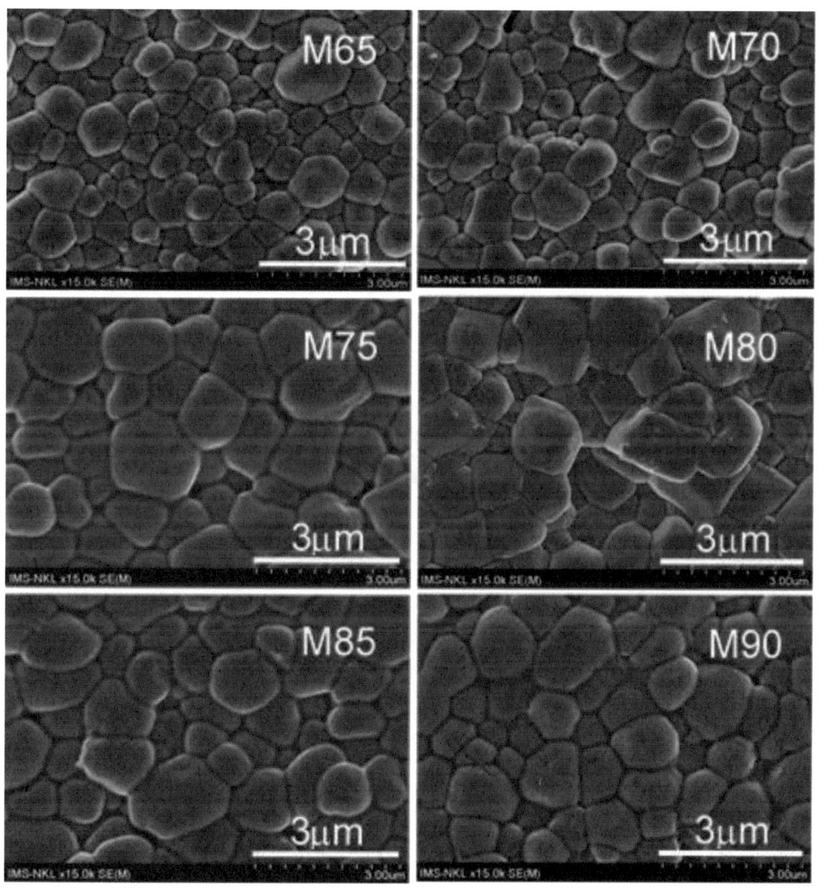

Figure 7.
Surface morphologies observed by the SEM of PZT-PZN-PMnN ceramics at various contents of PZT.

the general formula of the material without Pb. The reason could be explained by the evaporation of PbO during sintering [6, 25, 31]. Therefore, it is necessary to add excess 5 wt% PbO to compensate for lead loss during sintering.

Figure 7 shows microstructures of the PZT-PZN-PMnN ceramics at various contents of PZT. The average grain size of these samples is increased with the increase of PZT content in **Table 2**. On the other hand, the average grain size is reduced when x increases above 0.8. These results are obviously consistent with the change in the density of PZT content of PZT-PZN-PMnN ceramics, as shown in **Table 2**.

Samples	M65	M70	M75	M80	M85	M90
Average grain size (μm)	0.56 ± 0.02	0.66 ± 0.02	0.90 ± 0.02	1.04 ± 0.01	0.85 ± 0.02	0.83 ± 0.03
The average density of ceramics (g/cm^3)	7.77 ± 0.02	7.78 ± 0.01	7.80 ± 0.01	7.81 ± 0.01	7.72 ± 0.01	7.69 ± 0.02

Table 2.
The average grain size of PZT-PZN-PMnN ceramics.

Table 2 shows the density of the PZT-PZN-PMnN ceramics as a function of the PZT content. With the increase of PZT content up to 0.8, the mass density of PZT-PZN-PMnN ceramics increases. It achieves a maximum value (ρ = 7.81 g/cm^3, 96% of the theoretical density in which the theoretical density of ceramic is calculated using Eq. (1):

$$\rho = \frac{nA}{V_C N_A} \tag{1}$$

where n = number of atoms associated with each unit cell in ABO_3, A = atomic weight, V_C = volume of the unit cell in ABO_3, and N_A = Avogadro's number.

This is explained by the content of PZT was added to the ceramic system is less than 0.8 mol, a large number of pores were present, indicating insufficient densification of the sample (**Figure 7**: some SEM for M70, M90 are missing and M75 is not good). As the PZT content increases, the ceramics became denser, and the sample was almost fully dense at a PZT content of 0.8 mol.

The PZT content dependence of the dielectric constant (ε_r), dielectric loss (tan δ), and mass density (ρ) of the PZT-PZN-PMnN ceramics at 1 kHz and RT is illustrated in **Figure 8**. It can be seen that dielectric properties are strongly influenced by the composition of the ceramics. When the content of PZT increases from 0.65 to 0.8 mol, values of ε_r increase and reach to the maximum of 1230 at 0.8 mol of PZT. Then, these rapidly decrease with increasing x, while tan δ decreases with increasing PZT content. The minimum tan δ of 0.005 is obtained at x = 0.8 and then increased. It could be explained by the combination of a large and homogeneous grain size and the highest densification for the composition of 0.8PZT-0.125PZN-0.075PMnN ceramic [22].

In order to characterize the dielectric loss of all samples, the measurement of dielectric constant dependent on temperature is carried out at 1 kHz, as shown in **Figure 8**. With increasing PZT content, the dielectric constant peak increases and becomes sharpened. Hence, the material properties change from relaxor ferroelectricity to normal ferroelectricity. The permittivity and the maximum temperature (T_m) of the ceramics are shown in **Figure 9**. It shows that the T_m increases with increasing PZT content and is in the range of 206–275°C. There is a difference between the phase transformation temperatures of PZN (T_m ~ 140°C) [36, 38, 40] and PZT (T_C ~ 390°C) [25, 35], so it is significant to study the dependence of phase transition temperature of the PZT-PZN-PMnN ceramics on PZT content [40]. When the temperature is higher than T_m, the function $\varepsilon(T)$ is out of order the Curie-Weiss law in the normal ferroelectric materials. The fact the relationship between dielectric constant (ε) and temperature (T) above T_m can be complied by the modified Curie-Weiss law for analyzing of experimental data [44] is shown as follows:

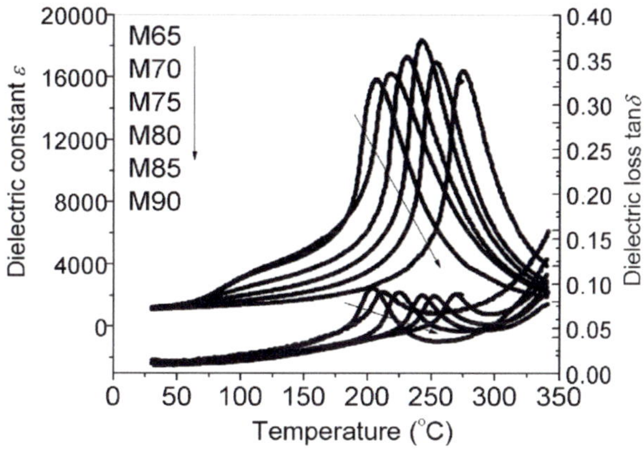

Figure 8.
Temperature dependence of the dielectric constant and dielectric loss tan δ at 1 kHz of samples.

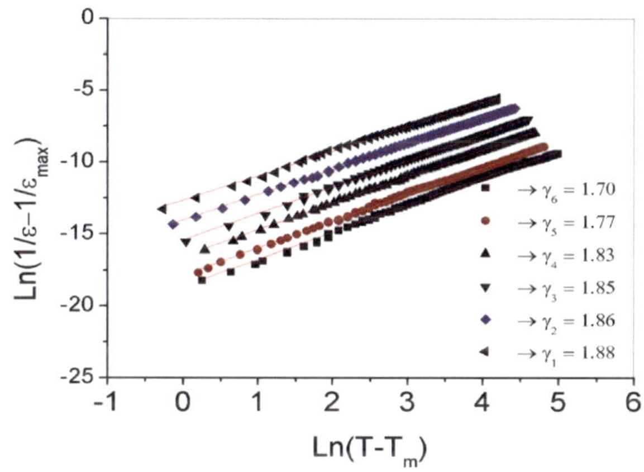

Figure 9.
The plot of $ln(1/\varepsilon - 1/\varepsilon_m)$ versus $ln(T\text{-}T_m)$ of PZT-PZN-PMnN ceramics at 1KHz.

$$\frac{1}{\varepsilon} - \frac{1}{\varepsilon_{\max}} = \frac{(T - T_m)^{\gamma}}{C'} \tag{2}$$

where C is the modified Curie–Weiss constant and γ is the diffuseness exponent, which changes from 1 to 2 for normal ferroelectrics to fully disorder relaxor ferroelectrics, respectively [44].

The slopes of the fitting curves (**Figure 9**) are used to determine the γ value of xPZT-(0.925 - x)PZN-0.075PMnN ceramics at 1 kHz. As can be seen in **Figure 9**, the γ changes from 1.70 to 1.88. Thus, it is indicated that the transitions are of a diffuse type and the ceramics are highly disordered.

To analyze the frequency dependence of T_m, it is necessary to use Vogel-Fulcher law [6, 45]:

$$F = f_o \exp\left(-T_o / \left(T - T_f\right)\right) \tag{3}$$

$$Ln(f) = f_o - T_o\left(T_m - T_f\right) \tag{4}$$

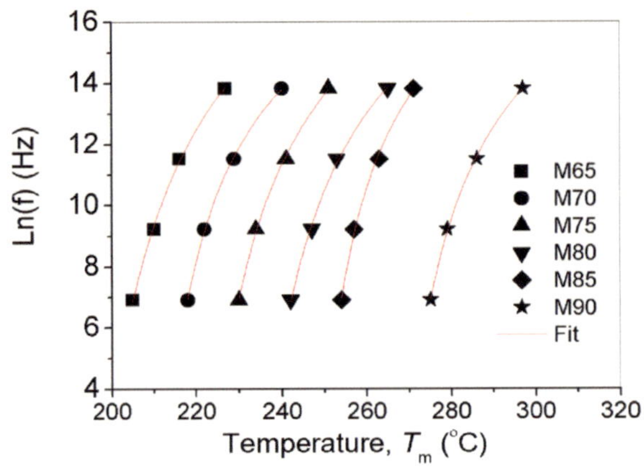

Figure 10.
The plot of ln(f) versus T_m as a function of the measured frequency of PZT-PZN-PMnN.

Samples	M65	M70	M75	M80	M85	M90
T_0 (°C)	398	254	302	331	217	253
T_f (°C)	179	198	208	218	237	255

Table 3.
The value of fitting parameters to Vogel-Fulcher relationship.

where T_f is the freezing temperature, E_a is the activation energy for polarization fluctuation of a polar nanoregion, f_o is a characteristic frequency or Debye frequency, and k_B is the Boltzmann constant = 1.38 × 10^{23} J/K, and $T_o = E_a/k_B$. **Figure 10** shows the plot of ln(f) versus T_m as a function of the measured frequency of PZT-PZN-PMnN. The symbols are the experimental points, and the line is the corresponding fitting to the Vogel-Fulcher relationship as listed in **Table 3**.

5. The effects of Zr/Ti ratio on the structure, microstructure, and the electrical properties of $0.8Pb(Zr_yTi_{1-y})O_3$–$0.125Pb(Zn_{1/3}Nb_{2/3})O_3$–$0.075Pb(Mn_{1/3}Nb_{2/3})O_3$ ceramics

The influence of Zr/Ti ratio on the structure of PZT-PZN-PMnN ceramics has been analyzed through the X-ray diffraction patterns (**Figure 11**). The patterns reveal a pure perovskite phase for all ceramic samples.

As can be seen, the tetragonality of PZT-PZN-PMnN ceramics decreased with increasing Zr/Ti ratio content through the c/a ratio decreases. According to Dixit et al. [46] and Kahoul et al. [47], the morphology of $Pb(Zr,Ti)O_3$ ceramics is strongly dependent on the Zr and Ti content. The content of the rhombohedral phase gradually increases within decreasing the Zr content simultaneously, and the tetragonal phase gradually decreases. The morphological evolution with Zr contents in this work may be attributed to the increase of a rhombohedral phase in these ceramics [46, 47]. This may be because the large Zr^{+4} (0.86 Å) ions diffuse into the PZT-PZN-PMnN lattice to replace Ti^{4+} (0.61 Å), resulting in the increase in the lattice constant and a shift in the XRD peak position toward lower 2θ values, similar to our recent research [48].

Effects of the contents of Zr/Ti ratio on the microstructure development of the ceramics are shown in **Figure 12**. In general, surface ceramics with large grains and

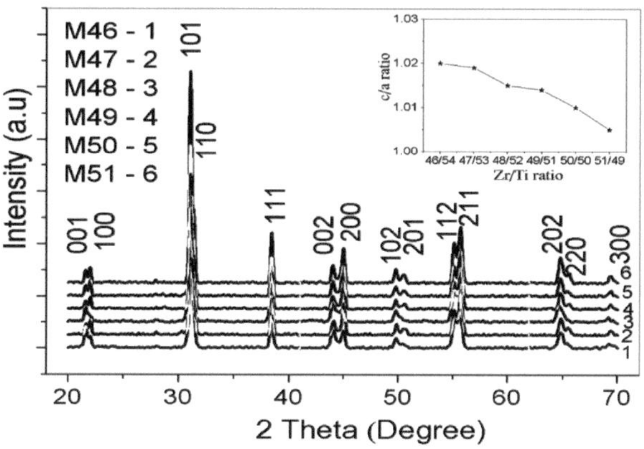

Figure 11.
X-ray diffraction patterns of ceramics with different Zr/Ti ratio contents: M46 (Zr/Ti = 46/54), M47 (Zr/Ti = 47/53), M48 (Zr/Ti = 48/52), M49 (Zr/Ti = 49/51), M50 (Zr/Ti = 50/50), and M51 (Zr/Ti = 51/49).

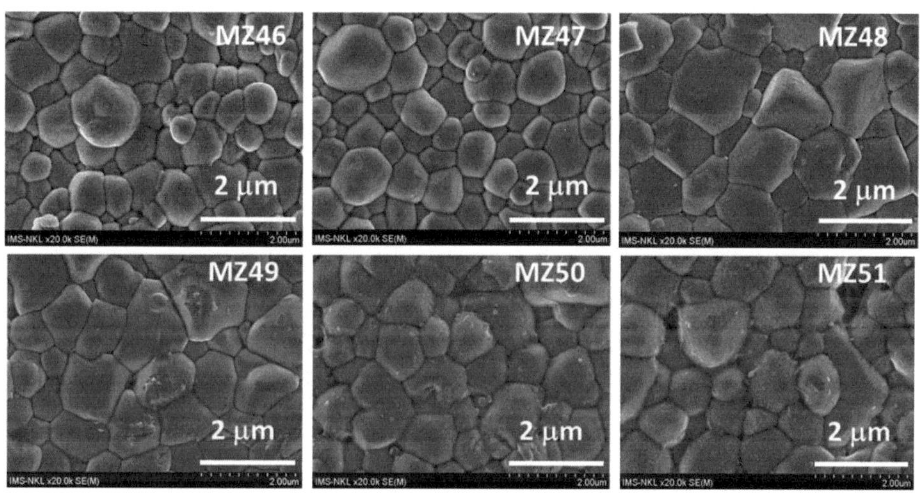

Figure 12.
Microstructures of samples with the different Zr/Ti ratio contents: MZ46 (Zr/Ti = 46/54), MZ47 (Zr/Ti = 47/53), MZ48 (Zr/Ti = 48/52), MZ49 (Zr/Ti = 49/51), MZ50 (Zr/Ti = 50/50), and MZ51 (Zr/Ti = 51/49).

uniform microstructure were obtained in all samples, and the average grain size of samples is increased with the increasing content of Zr/Ti ratio. In conformity with the previous densification results, highly dense samples exhibited high degrees of grain close packing. However, some pores and abnormal grain boundaries were observed in **Figure 12** (MZ50 and MZ51) and the average grain size is reduced.

Figure 13 shows the temperature dependence of ε and tan δ of the ceramic samples measured at different frequencies (1 kHz 1 MHz). It can see that with the increase in Zr amount, the T_m temperature of ceramics decreases as indicated in **Figure 14**. This may explain that the Curie temperature of $PbZrO_3$ ceramics is about 232°C [25], and it is lower than that of $PbTiO_3$ ceramics, 490°C [49, 50]. It is due to the decrease of lattice parameters and bond lengths [46, 47].

In order to determine the piezoelectric properties of ceramics, resonant vibration spectra of the PZT-PZN-PMnN samples were measured at room temperature (**Figure 15**), and from these resonant spectra, the piezoelectric parameters of the samples, such as electromechanical coefficients k_p, piezoelectric coefficients d_{31}, mechanical quality coefficient Q_m, and dielectric loss tan δ were determined (**Figure 16**).

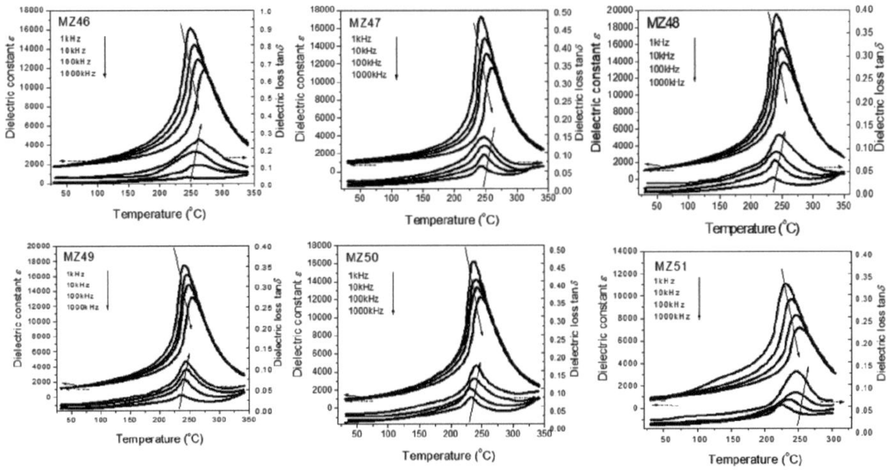

Figure 13.
Temperature dependence of relative dielectric constant ε and dielectric loss tan δ of samples at different frequencies: MZ46 (Zr/Ti = 46/54), MZ47 (Zr/Ti = 47/53), MZ48 (Zr/Ti = 48/52), MZ49 (Zr/Ti = 49/51), MZ50 (Zr/Ti = 50/50), and MZ51 (Zr/Ti = 51/49).

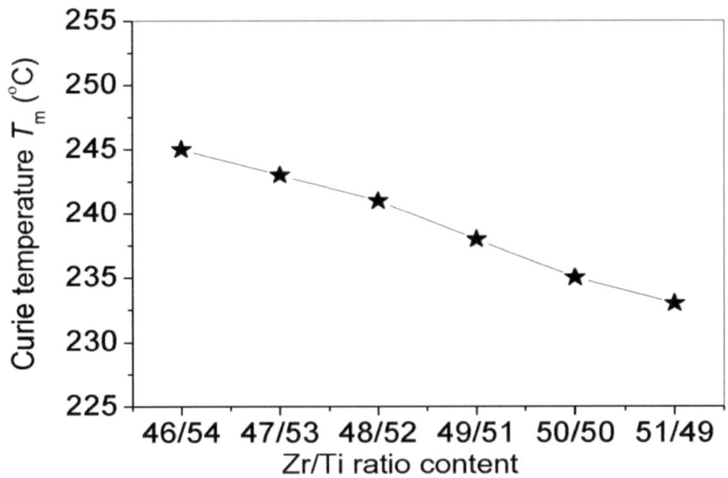

Figure 14.
The Curie temperature T_m of PZT-PZN-PMnN ceramics with different amounts of Zr/Ti ratio.

It can be observed that the k_p, d_{31}, Q_m, and tan δ depend on the amount of the Zr/Ti ratio content. The piezoelectric properties of ceramics are markedly improved. The following optimized values were obtained at Zr/Ti = 48/52, k_p = 0.62, d_{31} = 140 pC N^{-1}, Q_m = 1112, and tan δ = 0.005. This fact can be explained by the increased grain size effect and better modification of microstructure in ceramics [10, 25, 49, 50]. However, with the further increasing the Zr/Ti ratio content, the electrical properties of PZT-PZN-PMnN ceramics are reduced. The cause is due to an abnormal grain boundary, and the average grain size is also reduced, as shown in **Figure 12**.

The *P-E* hysteresis loops of PZT-PZN-PMnN at room temperature are displayed in **Figure 17(a)**, and P_r and E_c are presented in **Figure 17(b)**. With the increase in P_r and the decrease in E_c, the ferroelectric properties of PZT-PZN-PMnN ceramics improve. With increasing of Zr/Ti ratio content, the value of P_r increases and reaches the highest value of 34.5 μC/cm^2) at the Zr/Ti ratio of 48/52, and then decreases. The coercive field E_C decreases slightly with the increasing of Zr/Ti ratio content and reaches the smallest value of 9.0 kV/cm at Zr/Ti ratio of 48/52.

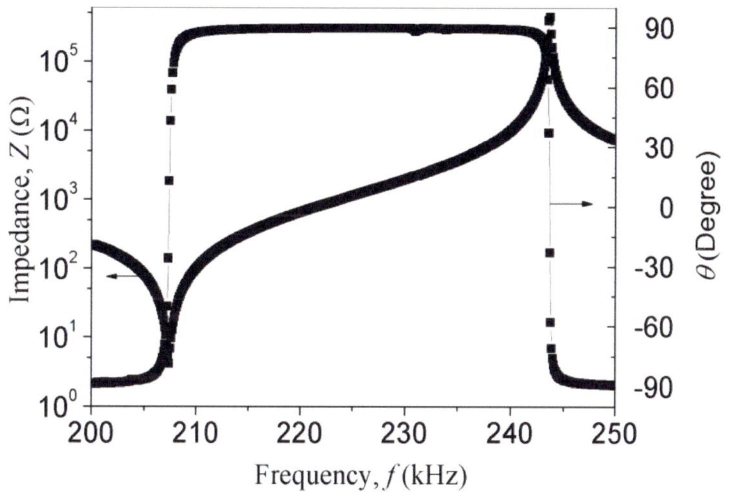

Figure 15.
The spectrum of radial resonance of MZ48 sample (at room temperature).

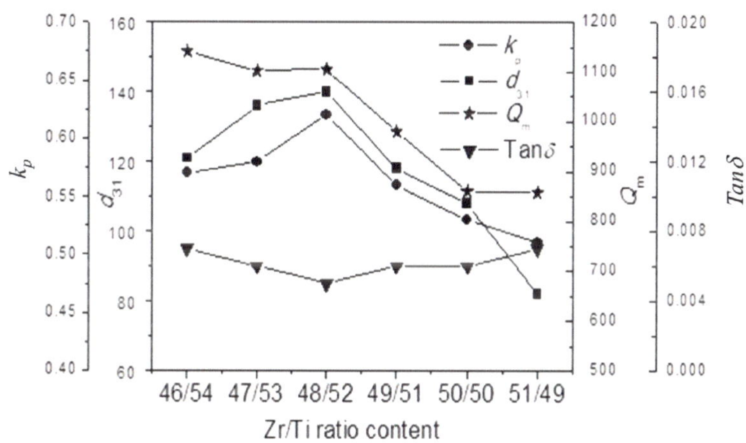

Figure 16.
The values of k_p, d_{31}, Q_m, *and tan δ of the PZT-PZN-PMnN ceramic samples.*

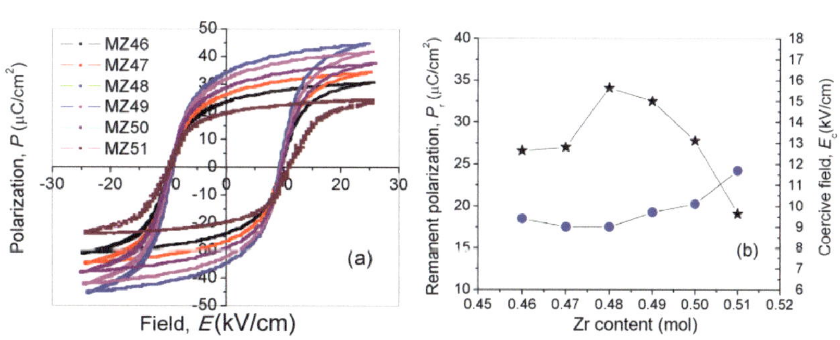

Figure 17.
(a) Hysteresis loops of samples and (b) P_r *and* E_c *as a function of Zr/Ti ratios.*

The effect of temperature on the ferroelectric properties of ceramics was studied by the hysteresis loops of the $0.8Pb(Zr_{0.48}Ti_{0.52})O_3$–$0.125Pb(Zn_{1/3}Nb_{2/3})O_3$–$0.075Pb(Mn_{1/3}Nb_{2/3})O_3$ sample in the temperature range of 30–280°C

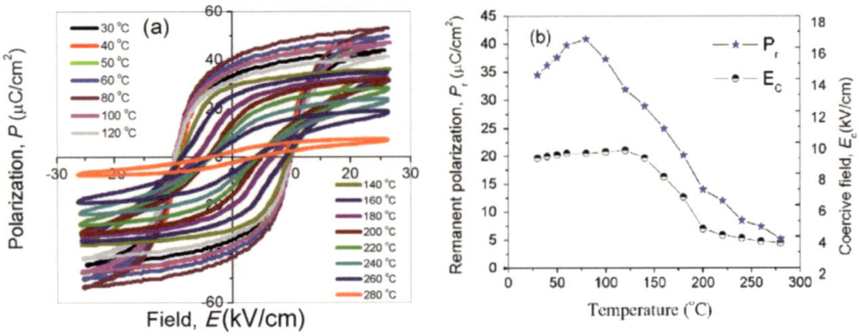

Figure 18.
(a) Hysteresis loops and (b) temperature dependence of P_r and E_C of MZ48 sample at a different temperature.

(**Figure 18(a)**). The hysteresis loops of the ceramics exhibited excellent temperature stability due to the broad diffusive phase transition between the nonergodic and ergodic relaxor states that coexisted over a wide temperature range [51]. When the temperature increased from room temperature to 120°C, the remanent polarization and the coercive field increased (**Figure 18(b)**). The reason is when the temperature increases, the oxygen vacancies in the perovskite structure will move and significantly increase the conductivity of the material, which should increase the dielectric loss. However, when the temperature rises above 120°C, the remanent polarization P_r and the coercive field E_c decreased (**Figure 18(b)**). Generally, the size of the hysteresis loops depends on the dielectric loss of the material and the metastable macro-domain structure and the immobilizations of the domain walls [52]. Therefore, when the temperature increased, large thermal motion energy caused an increase in bipolar disorder, narrowed the hysteresis loops, and decreased the remanent polarization and the coercive field. In addition, the hysteresis loops showed that the P_r is nonzero at T_m but decays to zero at temperatures above T_m. These results are consistent with the literature [40].

6. Ferroelectric domain structures around the morphotropic phase boundary of the $0.8Pb(Zr_{0.48}Ti_{0.52})O_3$-$0.125Pb(Zn_{1/3}Nb_{2/3})O_3$-$0.075Pb(Mn_{1/3}Nb_{2/3})O_3$ ceramics

In this section, in order to develop the composition ceramics for high-intensity ultrasound applications, $0.8Pb(Zr_{0.48}Ti_{0.52})O_3$–$0.125Pb(Zn_{1/3}Nb_{2/3})O_3$–$0.075Pb(Mn_{1/3}Nb_{2/3})O_3$ + z wt% ZnO nanoparticles ceramics were fabricated by the B-site oxide mixing technique with the variations of z and then the phase formation, piezoelectric and dielectric characteristics were investigated with the variations of z. The general formula of the studied materials is $0.8Pb(Zr_{0.48}Ti_{0.52})O_3$–$0.125Pb(Zn_{1/3}Nb_{2/3})O_3$–$0.075Pb(Mn_{1/3}Nb_{2/3})O_3$ + z wt% ZnO, where z = 0.0, 0.20, 0.25, 0.30, 0.35, 0.40, and 0.45. The obtained ZnO nanoparticles are spherical in shape, with their average diameter about 27 nm [9, 18]. On the other hand, reagent-grade oxide powders of PbO, ZnO, MnO_2, Nb_2O_5, ZrO_2, and TiO_2 (purity ≥99%) were used as starting raw materials for the fabrication of the PZT-PZN-PMnN ceramics.

Figure 19 shows X-ray diffraction patterns of the PZT-PZN-PMnN ceramics at various contents of ZnO nanoparticles. All the compositions have demonstrated pure perovskite phases and no trace of the second phase. Further XRD analysis is performed in the 2θ ranges from 43 to 46°, as shown in the inset of **Figure 19**. It can be seen that a phase transformation from the rhombohedral structure to the tetragonal structure occurs with increasing ZnO content. The samples with z = 0.0 and 0.2

have the rhombohedral structure characterized by a peak $(200)_R$ at $2\theta \approx 44.5°$. With $z = 0.40$ and 0.45, the ceramics exist as a tetragonal phase which is indicated by the splitting of $(002)_T$ and $(200)_T$ peaks in the 2θ range from 44 to 45° [23, 53, 54]. In the z range from 0.25 to 0.35, the ceramics coexist as rhombohedral and tetragonal phase, which is revealed by the coexistence of $(002)_T$, $(200)_T$, and $(200)_R$ peaks. Therefore, it could be said that the composition z from 0.25 to 0.35 is close to the morphotropic phase boundary (MPB) of this system. The phenomenon can be explained by the penetration of Zn^{2+} ions into the grains to substitute for B-site ions due to the similar radii of Zn^{2+} (0.074 nm), Ti^{4+} (0.0605 nm), Zr^{4+} (0.072 nm), and Nb^{5+} (0.064 nm) at the octahedral sites of the perovskite lattice, forming additional anionic vacancies. This causes a distortion in the lattice; therefore, the substitution of Zn^{2+} ions at Ti^{4+} or Zr^{4+} sites caused the c-axis to be lengthened and changed in lattice parameters and degree of tetragonality (c/a). These results are consistent with the literature [5, 19, 21, 48, 55].

Effects of the contents of ZnO nanoparticles on the microstructure development of the ceramics are shown in **Figure 20**. As can be described in the microstructure of

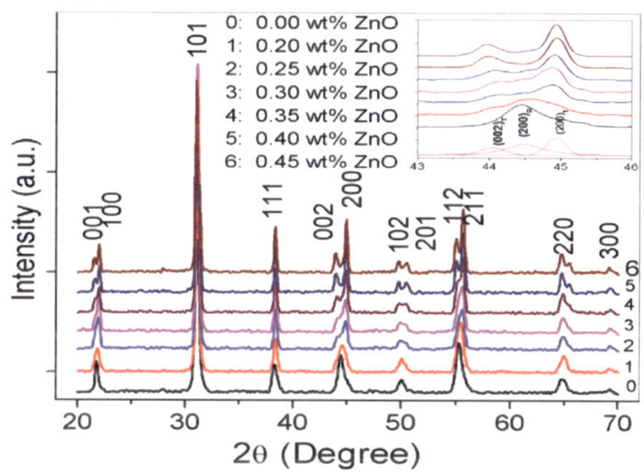

Figure 19.
X-ray diffraction patterns of PZT-PZN-PMnN ceramics at various contents of ZnO nanoparticles [9].

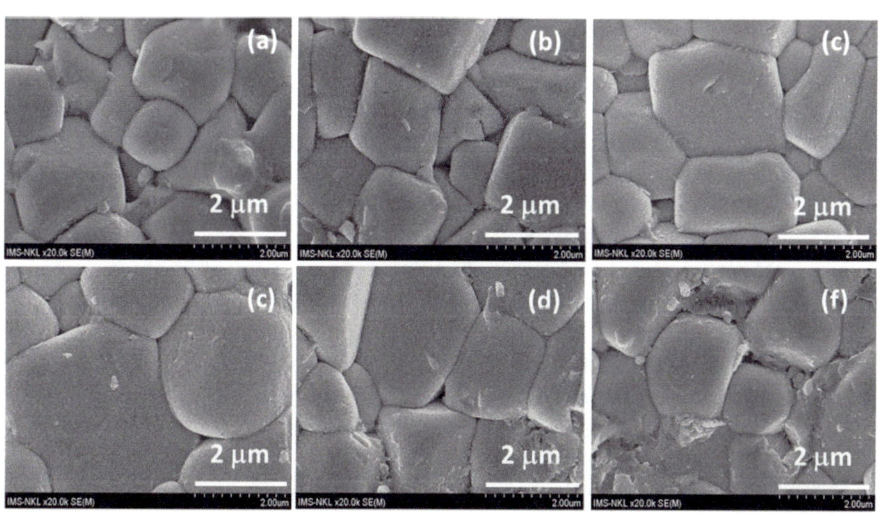

Figure 20.
Microstructures of PZT-PZN-PMnN ceramics at different contents of ZnO nanoparticles: (a) 0.20 wt%, (b) 0.25 wt%, (c) 0.3 wt%, (d) 0.35 wt%, (e) 0.40 wt%, and (f) 0.45 wt%.

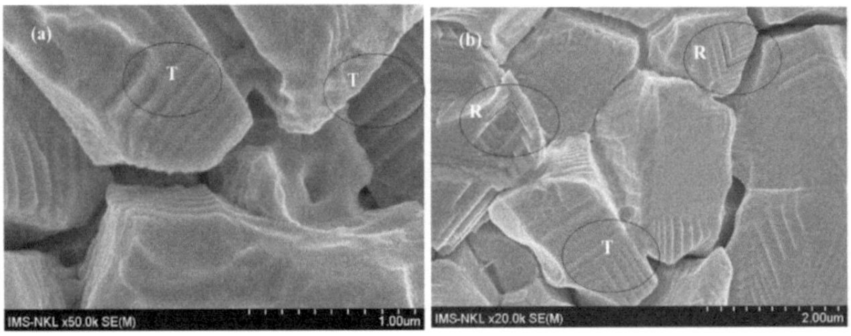

Figure 21.
The domain structures micrographs of the PZT-PZN-PMnN ceramics; (a) 0.4 wt% ZnO; (b) 0.35 wt% ZnO [9].

ceramics, the grain size of PZT-PZN-PMnN samples is increased with the increasing content of ZnO nanoparticles. This may explain that the low melting point of ZnO nanoparticles is beneficial to generate eutectic liquid phase at low temperature, and it can act as lubrication during the sintering process, wetting solid particles and providing capillary pressure between them, thus resulting in faster grain growth of PZT-PZN-PMnN ceramics [56, 57]. However, when the ZnO concentration is large, it exceeds the solubility limit of ZnO into the ceramics, and they will be located at grain boundaries preclude the grain growth process, as shown in **Figure 20(d)–(f)**.

Figure 21 shows evolution examples of the ferroelectric domain with the rhombohedral to tetragonal phase transformation and the grain size of the PZT-PZN-PMnN samples of about 2 μm. The SEM images of the domain structure suggest the presence of 90 and 180° domains in the tetragonal phase (**Figure 21(a)**), whereas the 71, 109, and 90° domains are located in the red-bordered region and primarily viewed in **Figure 21(b)**), and the widths of these domains were about 100 nm. Inspection of SEM images acquired at lower magnifications showed that the abundance and scale of these microtwin structures varied with location both within and between ceramic grains, with abrupt changes in the domain structure occurring at the grain boundaries [58]. One of the important contributions from our experimental works is the confirmation of the SEM images by corrosion method as a valid method for domain size assessment in bulk ceramics.

7. Conclusions

This chapter presents the investigation on the fabrication and characterization of sample groups of PZT-based ceramics and the relaxor PZN-PMnN ferroelectric materials with perovskite structure. The B-site oxide mixing technique reported in this study is simple, produces large quantities, and is easy to reproduce. Experimental results showed that the electrical properties of $x\text{Pb}(\text{Zr}_y\text{Ti}_{(1-y)})\text{O}_3$-$(0.925 - x)\text{Pb}(\text{Zn}_{1/3}\text{Nb}_{2/3})\text{O}_3$-$0.075\text{Pb}(\text{Mn}_{1/3}\text{Nb}_{2/3})\text{O}_3$ ceramics are optimal at a $\text{Pb}(\text{Zr}_y\text{Ti}_{(1-y)})\text{O}_3$ content of 0.8 mol and Zr/Ti ratio of 48/52. At these contents, the ceramics have good electrical properties: d_{31} = 140 pC/N, k_p = 0.62, k_t = 0.51, Q_m = 1112, tan δ = 0.005, and P_r = 34.5 μC/cm^2. Investigation of the domain structure of the two ferroelectric phases (tetragonal and rhombohedral) in the PZT-PZN-PMnN with compositions at near the MPB is described as follows: the 90 and 180° domains exist in the tetragonal phase, while 71, 109, and 90° domains are located in the rhombohedral phase. The widths of these domains were about 100 nm. The hysteresis loops of the ceramics exhibited excellent temperature stability due to the broad diffusive phase transition between the nonergodic and ergodic relaxor states that coexisted over a wide temperature range, which makes it as a promising material for high-intensity ultrasound applications.

Acknowledgements

This research is funded by the Vietnam National Foundation for Science and Technology Development (NAFOSTED) under Grant Number 103.02-2017.308.

Author details

Le Dai Vuong[1]*, Vo Thanh Tung[2] and Phan Dinh Gio[2]

1 Faculty of Chemical and Environmental Engineering, Hue Industrial College, Hue City, Vietnam

2 Department of Physics, College of Sciences, Hue University, Vietnam

*Address all correspondence to: ledaivuongqb@gmail.com

References

[1] Huang X, Zeng J, Ruan X, Zheng L, Li G. Structure, electrical, and thermal expansion properties of PZnTe–PZT ternary system piezoelectric ceramics. Journal of the American Ceramic Society. 2018;**101**(1):274-282

[2] Hong C-H, Kim H-P, Choi B-Y, Han H-S, Son JS, Ahn CW, et al. Lead-free piezoceramics – Where to move on? Journal of Materiomics. 2016;**2**(1):1-24

[3] Kour P, Pradhan S, Kumar P, Sinha S, Kar M. Study of ferroelectric and piezoelectric properties on Ca doped PZT ceramics. Materials Today: Proceedings. 2017;**4**(4):5727-5733

[4] Yue Y, Zhang Q, Nie R, Yu P, Chen Q, Liu H, et al. Influence of sintering temperature on phase structure and electrical properties of $0.55Pb(Ni_{1/3}Nb_{2/3})O_3$-$0.45Pb(Zr_{0.3}Ti_{0.7})O_3$ ceramics. Materials Research Bulletin. 2017;**92**:123-128

[5] Qaiser MA, Hussain A, Xu Y, Wang Y, Wang Y, Yang Y, et al. CuO added $Pb_{0.92}Sr_{0.06}Ba_{0.02}(Mg_{1/3}Nb_{2/3})_{0.25}(Ti_{0.53}Zr_{0.47})_{0.75}O_3$ ceramics sintered with Ag electrodes at 900°C for multilayer piezoelectric actuator. Chinese Physics B. 2017;**26**(3):037702

[6] Vuong LD, Gio PD, Tho NT, Chuong TV. Relaxor ferroelectric properties of PZT-PZN-PMnN ceramics. Indian Journal of Engineering & Materials Sciences. 2013;**20**(Dec-2013):555-560

[7] Zhang S, Li F, Jiang X, Kim J, Luo J, Geng X. Advantages and challenges of relaxor-$PbTiO_3$ ferroelectric crystals for electroacoustic transducers – A review. Progress in Materials Science. 2015;**68**:1-66

[8] Irzaman H, Darvina Y, Fuad A, Arifin P, Budiman M, Barmawi M. Physical and pyroelectric properties of tantalum-oxide-doped lead zirconium titanate [$Pb_{0.9950}(Zr_{0.525}Ti_{0.465}Ta_{0.010})O_3$] thin films and their application for IR sensors. Physica Status Solidi (a). 2003;**199**(3):416-424

[9] Vuong LD, Gio PD, Quang NDV, Dai Hieu T, Nam TP. Development of $0.8Pb(Zr_{0.48}Ti_{0.52})O_3$-$0.2Pb[(Zn_{1/3}Nb_{2/3})_{0.625}(Mn_{1/3}Nb_{2/3})_{0.375}]O_3$ ceramics for high-intensity ultrasound applications. Journal of Electronic Materials. 2018;**47**(10):5944-5951

[10] Vuong LD, Gio PD. Structure and electrical properties of Fe_2O_3-doped PZT-PZN-PMnN ceramics. Journal of Modern Physics. 2014;**5**(14):1258-1263

[11] Vuong LD, Gio PD, Kieu VTT. Raman scattering spectra and dielectric relaxation behavior of PZT-PZN-PMnN ceramics. International Journal of Chemistry and Materials Research. 2014;**2**(6):48-55

[12] Gao F, Cheng L-H, Hong R-Z, Liu J, Wang C-J, Tian C. Crystal structure and piezoelectric properties of $xPb(Mn_{1/3}Nb_{2/3})O_3$–$(0.2-x)Pb(Zn_{1/3}Nb_{2/3})O_3$-$0.8Pb(Zr_{0.52}Ti_{0.48})O_3$ ceramics. Ceramics International. 2009;**35**(5):1719-1723

[13] Tsai C-C, Chu S-Y, Hong C-S, Chen S-F. Effects of ZnO on the dielectric, conductive and piezoelectric properties of low-temperature-sintered PMnN-PZT based hard piezoelectric ceramics. Journal of the European Ceramic Society. 2011;**31**(11):2013-2022

[14] Gio PD, Vuong LD, Hoa HTT. Electrical properties of CuO-doped PZT-PZN-PMnN piezoelectric ceramics sintered at low temperature. Journal of Materials Science and Chemical Engineering. 2014;**2**(11):20-27

[15] Bakhsh N, Khalid F, Hakeem AS, editors. Effect of sintering temperature on densification and mechanical properties of pressureless sintered CNT-alumina nanocomposites. IOP Conference Series: Materials Science and Engineering; 2014: IOP Publishing.

[16] Zheng T, Wu J. Enhanced piezoelectricity over a wide sintering temperature (400-1050 C) range in potassium sodium niobate-based ceramics by two step sintering. Journal of Materials Chemistry A. 2015;**3**(13):6772-6780

[17] Lee G, Ji JH, Koh JH. Enhanced piezoelectric properties of (Bi, Na)TiO_3–(Bi, K)TiO_3 ceramics prepared by two-step sintering process. International Journal of Applied Ceramic Technology. 2018;**15**(2):531-537

[18] Vuong LD, Truong-Tho N. Effect of ZnO nanoparticles on the sintering behavior and physical properties of $Bi_{0.5}(Na_{0.8}K_{0.2})_{0.5}TiO_3$ lead-free ceramics. Journal of Electronic Materials. 2017;**46**(11):6395-6402

[19] Vuong LD, Tho NT. The sintering behavior and physical properties of Li_2CO_3-doped $Bi_{0.5}(Na_{0.8}K_{0.2})0.5TiO_3$ lead-free ceramics. International Journal of Materials Research. 2017;**108**(3):222-227

[20] Wang J, Wang G, Nie H, Chen X, Cao F, Dong X, et al. Low-temperature sintering and electric properties of $Pb_{0.99}(Zr_{0.95}Ti_{0.05})0.98Nb_{0.02}O_3$ ferroelectric ceramics with CuO additive. Journal of the American Ceramic Society. 2013;**96**(8):2370-2373

[21] Kang SH, Ahn CW, Lee HJ, Kim IW, Park EC, Lee JS. Dielectric and pyroelectric properties of Li_2CO_3 doped $0.2Pb(Mg_{1/3}Nb_{2/3})O_3$-$0.5Pb(Zr_{0.48}Ti_{0.52})O_3$-$0.3Pb(Fe_{1/3}Nb_{2/3})O_3$ ceramics. Journal of Electroceramics. 2008;**21**(1-4):855-858

[22] Yoo J, Lee I, Paik DS, Park Y-W. Piezoelectric and dielectric properties of low temperature sintering $Pb(Mn_{1/3}Nb_{2/3})O_3$–$Pb(Zn_{1/3}Nb_{2/3})O_3$–$Pb(Zr_{0.48}Ti_{0.52})O_3$ ceramics with variation of sintering time. Journal of Electroceramics. 2009;**23**(2-4):519-523

[23] Kim Y-J, Yoo J, Lee J-Y. Piezoelectric properties of PCW-PNN-PZT ceramics sintered at low temperature. Ferroelectric Letters Section. 2017;**44**(1-3):1-7

[24] Yoon M-S, Kim Y-M, Kweon S-Y, Hong T-W, Lee Y-G, Ryu S-L, et al. Effects of ZnO on the piezoelectric properties of Pb $(Mn_{1/3}Sb_{2/3})O_3$-$Pb(Zr, Ti)O_3$ ceramics. Journal of Electroceramics. 2007;**18**(1-2):73-75

[25] Yuhuan X. Ferroelectric Materials and Their Applications. North Holland; 1991

[26] Vuong LD, Gio PD, Lien NTK. Physical properties of PZT-PZN-PMnN ceramics were fabricated by B-site oxide mixing technique. Hue University Journal of Science (HU JOS). 2013;**84**(6):93-99

[27] Fang B, Sun R, Shan Y, Tezuka K, Imoto H. On the feasibility of synthesizing complex perovskite ferroelectric ceramics via a B-site oxide mixing route. Journal of Materials Science. 2007;**42**(22):9227-9233

[28] Tsai C-C, Chu S-Y, Liang C-K. Low-temperature sintered PMnN-PZT based ceramics using the B-site oxide precursor method for therapeutic transducers. Journal of Alloys and Compounds. 2009;**478**(1-2):516-522

[29] Lee S-H, Yoon C-B, Lee S-M, Kim H-E. Reaction sintering of lead zinc niobate–lead zirconate titanate ceramics. Journal of the European Ceramic Society. 2006;**26**(1-2):111-115

[30] Necira Z, Boutarfaia A, Abba M, Menasra H, Abdessalem N. Effects

of thermal conditions in the phase formation of undoped and doped Pb(Zr1-xTix)O_3 solid solutions. Materials Sciences and Applications. 2013;**4**(05):319-323

[31] Fan H, Kim HE. Effect of lead content on the structure and electrical properties of Pb(($Zn_{1/3}Nb_{2/3}$)0.5 ($Zr_{0.47}Ti_{0.53}$)0.5)O_3 ceramics. Journal of the American Ceramic Society. 2001;**84**(3):636-638

[32] Kwok KW, Chan HLW, Choy CL. Evaluation of the material parameters of piezoelectric materials by various methods. IEEE Transactions on Ultrasonics, Ferroelectrics, and Frequency Control. 1997;**44**(4):733-742

[33] Baghai-Wadji A, Gharb NB, Bouakaz A, Damjanovic D, Dayton P, Degertekin L, et al. IEEE ultrasonics, ferroelectrics, and frequency control society. Aurora. 1987;**51**:60504

[34] Sawyer CB, Tower C. Rochelle salt as a dielectric. Physical Review. 1930;**35**(3):269

[35] Lee S-H, Kim H-G, Choi H-I, Sa-Gong G, editors. Dielectric and piezoelectric properties of PNN-PZN-PZT ceramics for actuator application. In: Proceedings of the 5th International Conference on Properties and Applications of Dielectric Materials; 1997. IEEE; 1997

[36] Halliyal A, Safari A. Synthesis and properties of lead zinc niobate: Pb($Zn_{1/3}Nb_{2/3}$)O_3 based relaxor ferroelectrics. Ferroelectrics. 1994;**158**(1):295-300

[37] Pakawanit P, Ngamjarurojana A, Prasatkhetragarn A, Ananta S. Characterization of 0.93Pb($Zn_{1/3}Nb_{2/3}$)O_3-0.07BaTiO_3 ceramics derived from a novel $Zn_3Nb_2O_8$ B-site precursor. Ceramics International. 2013;**39**:S325-S329

[38] Fan H, Kim H-E. Perovskite stabilization and electromechanical properties of polycrystalline lead zinc niobate–lead zirconate titanate. Journal of Applied Physics. 2002;**91**(1):317-322

[39] Seo SB, Lee SH, Yoon CB, Park GT, Kim HE. Low-temperature sintering and piezoelectric properties of 0.6Pb($Zr_{0.47}Ti_{0.53}$)$O_3$0.4Pb($Zn_{1/3}Nb_{2/3}$)O_3 ceramics. Journal of the American Ceramic Society. 2004;**87**(7):1238-1243

[40] Vittayakorn N, Rujijanagul G, Tunkasiri T, Tan X, Cann DP. Perovskite phase formation and ferroelectric properties of the lead nickel niobate–lead zinc niobate–lead zirconate titanate ternary system. Journal of Materials Research. 2003;**18**(12):2882-2889

[41] Hou Y-D, Chang L-M, Zhu M-K, Song X-M, Yan H. Effect of Li_2CO_3 addition on the dielectric and piezoelectric responses in the low-temperature sintered 0.5 PZN–0.5 PZT systems. Journal of Applied Physics. 2007;**102**(8):084507

[42] Yogaraksa T, Hikam M, Irzaman H. Rietveld analysis of ferroelectric $PbZr_{0.525}Ti_{0.475}O_3$ thin films. Ceramics International. 2004;**30**(7):1483-1485

[43] Prakash BJ, Buddhudu S. Synthesis and analysis of $LiNbO_3$ ceramic powders by co-precipitation method. Indian Journal of Pure and Applied Physics. 2012;**320-324**(5):320-324

[44] Uchino K. Relaxor ferroelectrics. Journal of the Ceramic Society of Japan. 1991;**99**(1154):829-835

[45] Rubio-Marcos F, Romero J, Martín-Gonzalez M, Fernández J. Effect of stoichiometry and milling processes in the synthesis and the piezoelectric properties of modified KNN nanoparticles by solid state reaction. Journal of the European Ceramic Society. 2010;**30**(13):2763-2771

[46] Dixit A, Majumder S, Dobal P, Katiyar R, Bhalla A. Phase transition studies of sol–gel deposited barium zirconate titanate thin films. Thin Solid Films. 2004;**447**:284-288

[47] Kahoul F, Hamzioui L, Boutarfaia A. The influence of Zr/Ti content on the morphotropic phase boundary and on the properties of PZT–SFN piezoelectric ceramics. Energy Procedia. 2014;**50**:87-96

[48] Vuong LD, Gio PD. Enhancement in dielectric, ferroelectric, and piezoelectric properties of $BaTiO_3$-modified $Bi_{0.5}(Na_{0.4}K_{0.1})TiO_3$ lead-free ceramics. Journal of Alloys and Compounds. 2020;**817**:152790

[49] Yoo J, Lee Y, Yoon K, Hwang S, Suh S, Kim J, et al. Microstructural, electrical properties and temperature stability of resonant frequency in $Pb(Ni_{1/2}W_{1/2})O_3$–Pb $(Mn_{1/3}Nb_{2/3})O_3$–$Pb(Zr, Ti)O_3$ ceramics for high-power piezoelectric transformer. Japanese Journal of Applied Physics. 2001;**40**(5R):3256-3259

[50] Vuong LD, Gio PD, Van Chuong T, Trang DTH, Hung DV, Du NT. Effect of Zr/Ti ratio content on some physical properties of low temperature sintering PZT-PZN-PMnN ceramics. International Journal of Materials and Chemistry. 2013;**3**(2):39-43

[51] Jaita P, Sanjoom R, Lertcumfu N, Rujijanagul G. Improvement of electric field-induced strain and energy storage density properties in lead-free BNKT-based ceramics modified by BFT doping. RSC Advances. 2019;**9**(21):11922-11931

[52] Fan H, Zhang L, Zhang L, Yao X. The effect of defect field on dielectric ageing of lead magnesium niobate-lead titanate relaxor ferroelectrics. Journal of Physics: Condensed Matter. 2000;**12**(19):4381-4390

[53] Luan NDT, Vuong LD, Van Chuong T, Truong TN. Structure and physical properties of PZT-PMnN-PSN ceramics near the morphological phase boundary. Advances in Materials Science and Engineering. 2014;**2014**(821404):8

[54] Truong-Tho N, Vuong LD. Effect of sintering temperature on the dielectric, ferroelectric and energy storage properties of SnO_2-doped $Bi_{0.5}(Na_{0.8}K_{0.2})_{0.5}TiO_3$ lead-free ceramics. Journal of Advanced Dielectrics. 2020;**10**(4):2050011-2050019

[55] Li H-B, Li Y, Wang D-W, Lu R, Yuan J, Cao M-S. Effects of ZnO nanoneedles addition on the mechanical and piezoelectric properties of hard PZT-based composites. Journal of Materials Science: Materials in Electronics. 2013;**24**(5):1463-1468

[56] German RM. Liquid Phase Sintering. Springer Science & Business Media; 2013

[57] Luo CY, Hu MZ, Huang Q, Fu Y, Gu HS. Influence of ZnO and Nb_2O_5 additions on sintering behavior and microwave dielectric properties of $(Mg_{0.95}Ca_{0.05})TiO_3$ ceramics. Key Engineering Materials. 2012;**512**:1184-1188

[58] Rossetti GA Jr, Popov G, Zlotnikov E, Yao N. Domain structures and nonlinear mechanical deformation of soft Pb $(Zr_xTi_{1-x})O_3$ (PZT) piezoelectric ceramic fibers. Materials Science and Engineering A. 2006;**433**(1-2):124-132

Chapter 5

Self-Healing of Concrete through Ceramic Nanocontainers Loaded with Corrosion Inhibitors and Microorganisms

George Kordas

Abstract

Self-healing was considered for repairing of destruction of reinforced concrete on bridges, houses, etc., that comes from corrosion of reinforcement metals as well as cracking in cement. The work was accomplished at producing and assessing through incorporate ceramic nanocontainers loaded with microorganisms. We produced various types of organic and inorganic nanocontainers that were incorporated into the cement that can act as carriers for the transport of bacteria. The microorganisms used in the work are *Escherichia coli* and *Staphylococcus aureus*. Precipitation of $CaCO_3$ was observed by both bacteria. As microspheres do not affect the submersion of the mineral by the microorganism, additional studies were carried out to assess the interaction between transmission microsystems and bacterium. The mechanism of self-healing of building materials in this work was based on $CaCO_3$ precipitation, through the ureolytic action of bacteria. When a crack appears in the cement, then the bacterium trapped in a nanocontainers is released and comes into contact with the water. In this way, the microorganism begins to metabolize and precipitate the mineral, in a way that eventually observes healing of the crack. $CaCO_3$ microbial precipitation was based on the breakdown of urea ($CO\ (NH_2)_2$) into CO_3^{2-} and NH_3. Due to the high pK value of the NH_3/NH_4^+ system (pKa = 9.2), the breakdown reaction led to an increase in pH, favoring the release of carbonated ions (CO_3^{2-}), and in an environment rich in calcium ions (Ca^{2+}), $CaCO_3$ precipitation took place.

Keywords: nanocontainers, self-healing, Concrete, corrosion

1. Introduction

Reinforced concrete is a composite material resulting from the strengthening of concrete with steel rods with greater strength. The combination of these materials leads to a new one that will meet the needs of the construction. The reinforcement is done to increase the tensile strength of concrete. On the other hand, steel rods are sensitive to corrosion. Concrete, though, forms an alkaline environment that hinders the oxidation of steel. There are many factors that affect the strength of concrete and contribute to its early wear. A key cause that activates various wear

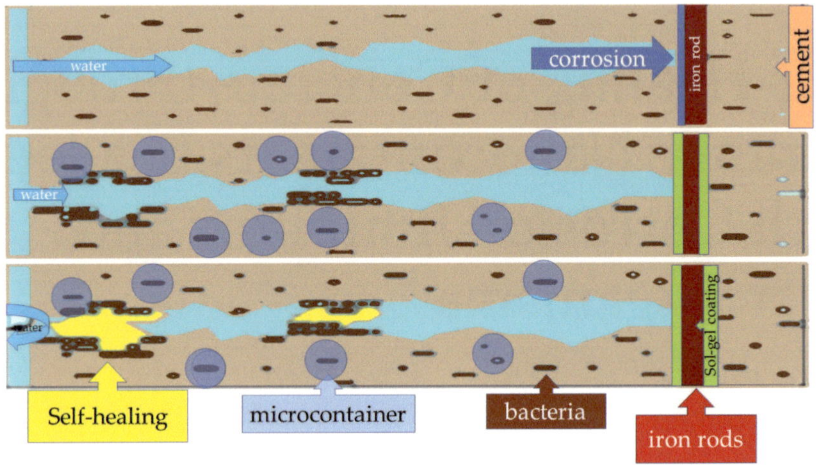

Figure 1.
Schematic representation of the healing of cracks in cement with integrated microorganisms in microcontainers and protective sol-gel coating against corrosion of metal rods.

mechanisms is cracks that dramatically increase the permeability of cement. The microstructure of hardened cement is porous, isolated, and interconnected. The interconnected pores determine the permeability of the material as they allow water and chemical compounds to penetrate the concrete matrix. Permeability essentially increases as the crack connects isolated resources to resource networks. In most concrete wear mechanisms, permeability plays an important role. Specifically, interconnected pores determine the permeability of the material as they allow water and chemical compounds to penetrate the concrete matrix. In the same way, CO_2 diffuses through the pores, which reacts with the alkaline components of cement (e.g. Ca $(OH)_2$) in the process called carbonation.

The above points make it clear that cracks in concrete must be reduced and that ideally a mechanism should lead to the sealing or clogging of newly formed cracks, in order to limit the permeability of the uterus. A healing mechanism is ideal when you do not need constant testing for repair, and it is financially lucrative [1]. In recent times, concrete self-healing is proposed to be done using a biological restoration technique through the introduction of bacteria into the concrete. **Figure 1** presents such technology. The idea is based on the incorporation of a bacterium that metabolizes urea and sinks $CaCO_3$ into the crack environment. $CaCO_3$ microbial submersion is certified by a number of factors, such as the concentration of dissolved inorganic carbonate ions and Ca^{2+} ion concentration. The protection of the bacterium in the cement is done by locking them in micro-containers, the incorporation of which can reduce the strength of the concrete.

The present work deals with the development of coatings to protect steel rods from corrosion and nanocontainers filled with bacteria to induce self-healing.

2. Materials and methods

2.1 Chemicals

The reagents we used are methyl methacrylate (MMA), poly(ethylene glycol) methacrylate (PEG), ammonia solution (30%), 2,2-azobis(2-methylpropionitrile) (AIBN), sodium dodecyl sulphate (SDS), ethylene dimethacrylate (EGDMA), toluene diisocyanate (TDI), tetraethyl orthosilicate(TEOS), ethylene diamine (EDA), diethylene triamine (DETA), polyvinylpyrrolidone (PVP), chloride ammonium, tryptone, yeast extract, and sodium chloride. The solvents we used were distilled water, acetonitrile, acetone, ethanol, 1-octadecene and paraffin. All chemicals were obtained by commercial sources.

2.2 Instruments

FEI's scanning electronic microscope (SEM) was used with tungsten filament at 25 kV. A Perkin Elmer's FT-IR Spectrum100 spectrometer was used where the scanning range was from 4000 cm^{-1} to 400 cm^{-1}. X-ray diffraction (XRD) was measured using a Siemens D-500 apparatus. The thermal analyses were carried out with Perkin Elmer analyst Pyris Diamant S II. Dynamic light scattering (DLS) was performed at Malvern's Zetasizer Nano and each measurement represents the average of three measurements. A Sorvall Evolution RC centrifuge was used for container separation. An Elma Sonic ultrasonic bath, S 30H, was used to disperse the samples. The sterilizations were carried out with steam sterilizer of Trade Raypa. Any utensils and solutions to be used for the development of microorganisms were sterilized at 120°C for 20 min.

2.3 Nutrient (solutions and LB agar)

The microorganisms used to carry out the experiments were grown in an aqueous culture medium Luria Broth (LB). Recommendation of the culture medium LB was NaCl 0.5% w/v, tryptone 1% w/v, and yeast extract 0.5% w/v. In a Petri dish, nutrient material is placed in which the cells of the microorganisms of the experiments grow. The composition of the nutrient is agar (gel agent) 1.5 g, NaCl (electrolyte) 0.5 g, tryptone (source of proteins and nitrogen) 0.5 g, fungal extract (source of metals and carbohydrates) 0.3 g and distilled water 100 ml. Each dish is partially filled with warm LB agar containing a mixture of substances described in the table above. The investigation of cell vitality is a characteristic indicator and a necessary condition in cases of cytotoxic action of polymer microspheres where they will be used in materials that will come into direct contact with humans. It is performed by identifying vital cell functions. Cell vitality refers to the percentage of healthy cells in a culture. This indicator, by definition unclear, is usually determined by controlling vital functions such as cell metabolic activity (MTT).

2.4 Sterilization of glassware and nutrients

All tools used in microorganism experiments were sterilized to avoid any contamination of microbial solutions. The same procedure was carried out for their nutrients. All glassware and nutrients were sterilized for 20 min at 120°C. For the development of the microorganism, the experimental process of cultivation in nutrient solutions is described as follows: the reculture of the microorganism takes place after recovery from ampoules stored at −80 °C. The nutrient medium is vaccinated with a small amount of the ampoule of the microorganism. The procedure takes place under aseptic conditions, conditions achieved by the use of a reducing flame in order to avoid contamination of the sample. Vaccination of the microorganism is done with a sterile Pasteur pipette, where its nose has been sterilized in flame. The nutrient solution with the microorganism is incubated in a special stirring chamber at 37°C, at approximately 100 rpm for 24 hours in order to optimally develop the microorganism. For the development of the microorganism used each time, the experimental process of growing it in a Petri dish is as follows: ampoules of the microorganism are inoculated in a Petri dish forming zeta as shown in the following image. The procedure takes place under aseptic conditions using a reducing flame to avoid any complications of the sample. The microorganism is vaccinated in the dish with sterile Pasteur pipette, where its nose has been sterilized as above. The vaccinated dish is then placed in an incubation oven at 37°C for about 24 hours.

2.5 Cell cultures

The type of cells that covered the needs of this work are THE-293. They're human cells of fetal tissue. For their maintenance they contain adenovirus and therefore their management must be very careful. They are used for efficacy tests, host contamination tests and iodide tests [2]. Their development conditions depend on the specific characteristics of the cells. Cell culture materials consist of a nutrient cell growth solution DMEM (Dulbecco's Modified Eagle Medium) enriched with 2 mM glutamine, 0.85 g/L NaHCO3, 25 mM HEPES, 10% FBS (Fetal Bovine Serum), $6.8 < pH < 7.2$. 0.2% w/v streptomycin, 2 × 103 U/mL penicillin, in PBS (1×) and cell separation solution (0.02% EDTA-thrypsin). Cell culture is suspended with fresh nutrient solution, at a final concentration of 2–4 × 10^5 cells/mL, and maintained at 37°C, removal of nutrients from cultivation, addition of cell separation solution (4 mL/bottle 75 mL) and incubation at 37°C for 4 min, removal of the solution, re-addition of 0.4 mL of cell separation solution and incubation at 37 °C for 15–20 min, in order to detach the liver and separate the cells and add a DMEM nutrient solution and cell growth at 37 °C. Cell lines can undergo uncontrolled changes related to their morphology, growth, vitality, and karyotype due to prolonged recultures or any unfortunate infections. This risk is avoided by creating a renewable cell bank, after cooling the cells and keeping them in liquid nitrogen for long periods of time (years). Cooling is led to cells that are in a logarithmic phase of growth or are close to filling a single carpet. The methodology followed is (a) cell implantation at a concentration of 4 × 10^5 cells/mL, (b) at 48 h, where the cells are at the completion of the logarithmic growth phase; detachment and centrifugation at 1000 rpm for 5 min, (c) re-dissolution of cells in DMEM in the presence of 10% DMSO in FBS (cryoprotective substance), at a final concentration of 4 × 106–10 × 10^6 cells/ mL (depending on the cell line), (d) transfer to ampoules which are then gradually placed in a freezer; so as to avoid the creation of crystals inside the cells and therefore cellular solution, and (e) finally is placed in a cell retention device (−196 °C, liquid nitrogen) for several years. The restoration of cells stored in liquid nitrogen into current culture (36.5–37°C) is carried out by quickly defrosting the ampoule sample at 37°C (avoid denaturing the protein content of the cell). Centrifuge to remove cold protective material (DMSO) from cultivation. Rapid removal of DMSO is particularly important because it acts as an inhibitor in the development of cell proliferation and in some cases triggers differentiation, apoptosis, or even necrosis; depending on the cell type, redialysis of cells in nutrient solution and growth of crops at 37°C, frequent cell recultures at the beginning are necessary, to fully restore the normal growth rate of the crop. Infectious environmental factors, the most common of which are fungi and bacteria, are often an obstacle to maintaining an in vitro cell culture. For this reason, all cell cultures are handled within a nematically flow chamber in order to achieve sterile conditions for the cells. Sterilization of the cell area is achieved on a daily basis by exposure to ultraviolet radiation for at least 15 minutes, while ethanol solution (70% v/v) is used for local sterilization of the site. The materials used are sterilized in a special liquid sterilization furnace (automatically) at 120°C and pressure 1 kp/cm^2 for 20 minutes.

3. Synthesis and characterization of organic hybrid microspheres

The purpose of this work on the synthesis of organic-inorganic microcontainers is to create a material where it can be used by industry for the ultimate purpose of self-healing of concrete. In other words, a study was carried out to find a synthetic path that is easy, fast, but also easily adaptable to the final product. For these reasons, polymer microspheres were combined, changing parameters in the experimental process to achieve a size of microspheres, approximately 1 μm. The purpose of using microspheres at this size was to interact with *St. aureus*, the size of which is 0.5–1.5 μm.

3.1 Water traps

The purpose of producing water traps with a desired size of 1 μm and then incorporating them into the material is when they come into contact with the bacterium and their nutrient, to adsorb and retain the bacterium on their surface. Spherical water traps were made up by a two-step process, which initially includes the cross-linked polymethacrylic acid (PMAA) spheres through submersion polymerization after distillation, and then the conversion of carboxylic groups into corresponding calcium salts through the treatment of microtraps with Ca $(OH)_2$ solution. Methacrylic acid (MAA) and dimethacrylate glycolic acid ester (EGDMA) were used as monomers in a crusader role, in acetonitrile solvent with azodisisobutyronitrile (AIBN) beginning. Acetonitrile, being a nonprimary, polar solvent, favors polymer-polymer interactions, that is, hydrogen bonds between the carboxylic edges of polymer chains. MAA together with EGDMA are dissolved in acetonitrile in a 250 ml triple-blooded spherical flask. A freezer, thermometer, and N_2 supply are installed in this bottle. After an hour of stirring, the AIBN is added, which has been dissolved in a solvent quantity before. Stirring continues at the same temperature for another hour. In order to distill the solvent, the temperature of the experiment increases to the boiling point of acetonitrile (80°C). After a certain amount of solvent has been collected, the reaction is terminated, and the resulting final solution is in the form of an emulsion. Finally, after the solution is allowed to reach ambient temperature, the sample is centrifuged twice, rinsed with acetonitrile, and then dried. PMAA spheres acquire the ability to absorb water when the carboxylic groups are converted into the corresponding calcium salts. The experimental procedure is as follows: 2 g PMAA spheres dissolve in acetonitrile and spread through ultrasound. After the sample has been dispersed homogenously, add 0.74 g of 0>1 M Ca $(OH)_2$ solution. The mixture becomes clear (depending on the ratio of monomers—the less EGDMA is added, the clearer the solution becomes). After stirring for 30 minutes, the sample is centrifuged, rinsed with acetonitrile, and left to dry. Four different sized water traps were made up, changing the ratio of monomers in each case (**Figure 2**).

One can conclude from **Table 1** that the submersion polymerization after distillation gives uniform polymeric microspheres with different functional groups. **Figure 3** shows the SEM image of sample 2 together with the elemental analysis 2 before **Figure 3A** and after **Figure 3B** treatment with Ca $(OH)_2$.

According to the SEM images and the data in **Table 1**, the experimental process of composing water traps shows repeatability by giving spherical particles, with a similar diameter and small size distribution. The diameter shown in **Table 1** was calculated from the average of the diameters of hundreds of water traps with the corresponding standard deviation of each. The dispersion indicator is close to one and proves that there is a good size distribution. Also, the elemental analysis

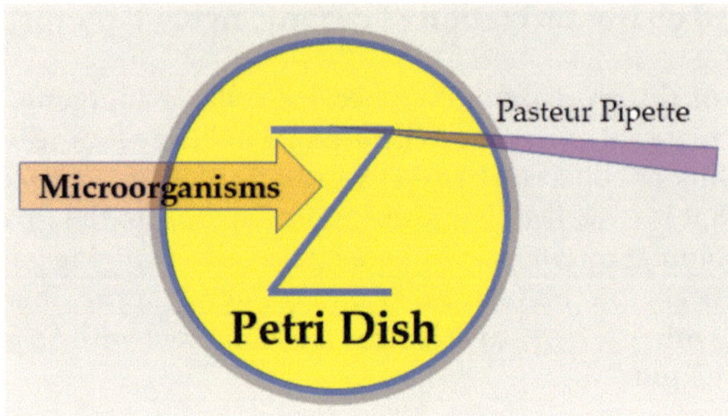

Figure 2.
Vaccination of microorganisms in petri dish.

Samples	n = [EGDMA]/[MAA] Molar ratio	Size (nm) (determined by TEM)
1	0.05	1004 ± 55
2	0.056	1150 ± 72
3	0.06	830 ± 40
4	0.085	626 ± 58

Table 1.
n = [EGDMA]/[MAA] molar ratio and size of water traps for 4 n compositions.

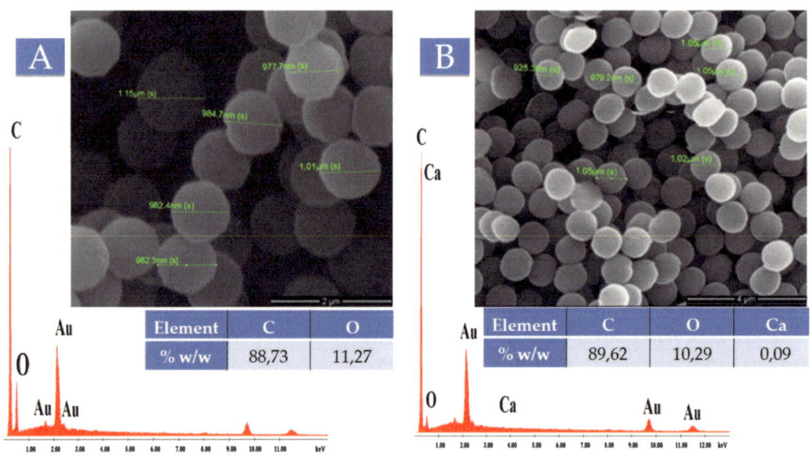

Figure 3.
(A) SEM image and EDS analysis of sample 2 as well as %w/w of data before modification with Ca (OH)2. (B) SEM and EDS analysis of sample 4 as well as the % w/w of data after modification with Ca $(OH)_2$.

(**Figure 3B**) in each sample case confirms the existence of Ca, after its modification, which means that the water microtraps are now capable of absorbing water. **Figure 3** confirms that statement via dynamic light scattering measurements.

For small amounts of water, up to 10% v/v of the total amount of solvent, the differences in hydrodynamic diameter are not significant, but as the percentage increases, there is a fairly large difference in size. This difference is evident when the percentage of water exceeds 15%, and this is probably due to the interaction of trap-water with water. These interactions may relate to the development of hydrogen bonds between particles and water but also between the particles themselves. The elucidation of the water absorbing mechanism needs further work (**Figure 4**).

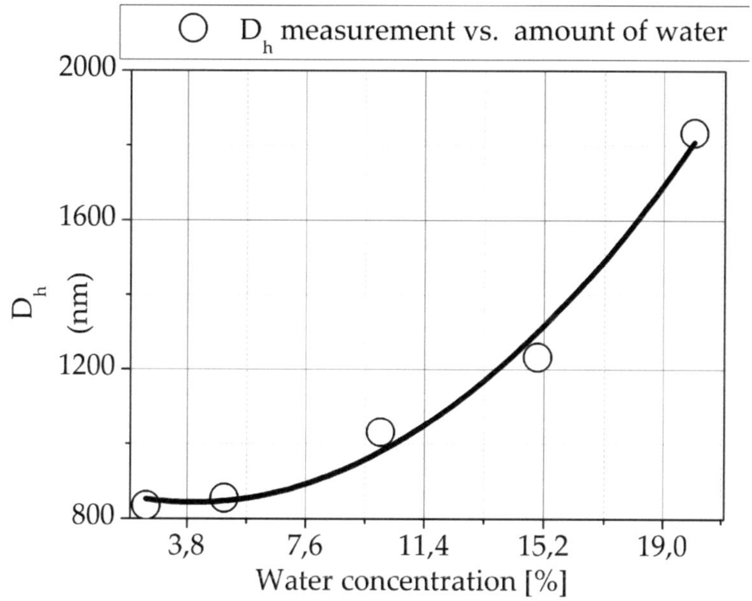

Figure 4.
Change in hydrodynamic diameter depending on different quantities of water.

3.2 $Ca(OH)_2$ @SiO_2 modified water traps

The purpose of the Ca $(OH)_2$@SiO_2 modification of the water traps is to give the adsorbed water traps a durable shell that will protect the bacterium and LB when it is introduced into cement. The SiO_2 shell synthesis process was done using the sol-gel method. For the more durable and complete coating of activated water traps, six experiments were carried out, each increasing the amount of the TEOS reagent. The experimental procedure is the same in all six experiments carried out and is as follows: in 150 ml boiling glass, the amount of traps is dissolved in acetonitrile, and the solution is left to ultrasound for 20 minutes and then for another while stirring, in order to spread the sample well in the solvent. Then the ammonia is added, and the stirring continues for another 15 minutes. TEOS is then added to the sample and the solution is left stirring for about 20 hours. Finally, each sample is centrifuged, rinsed, and left to dry. The whole experimental process takes place at ambient temperature. To make up the SiO_2 shell around the water traps, a water emulsion in oil (W/O emulsion) was created. Due to the organic solvent used (acetonitrile), when aqueous solution is added to it, water drops are created inside the solvent. In this case, due to the ability of water traps to retain water, the role of the aqueous phase in a W/O emulsion is made up of particles that have adsorbed water. The water comes from the solution of NH_3 30%. Thus, in the middle of the water traps and acetonitrile, when the precursor TEOS compound is added, hydrolysis and condensation take place there. This overlays silicon-shelled water traps (**Figure 5**).

Figure 6 shows the size of the nanospheres as a function of R_w (H_2O/ Si ratio). This graph was determined by keeping all parameters constant except the quantity of the added water in the experiments. Note that the amount of modified water microtraps covered was the same in all experiments, (0.5 g).

Figure 7 shows the SEM micrographs of the modified nanotraps. The SEM image and the corresponding EDS elemental analysis are reported for the sample of the experiment corresponding to the one with the largest amount of TEOS added.

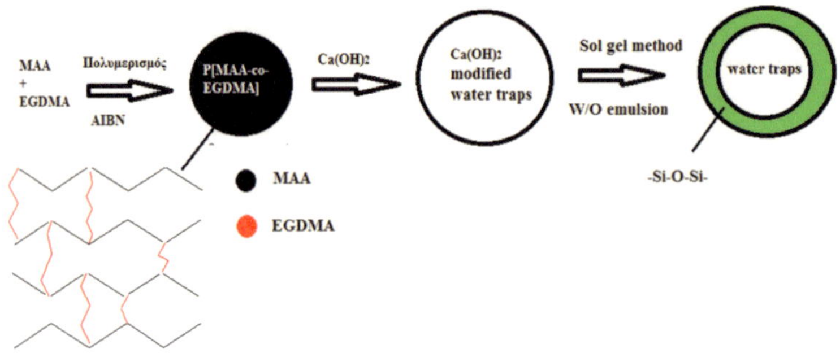

Figure 5.
Schematic representation of the coating of modified water traps with SiO_2 in a W/O emulsion.

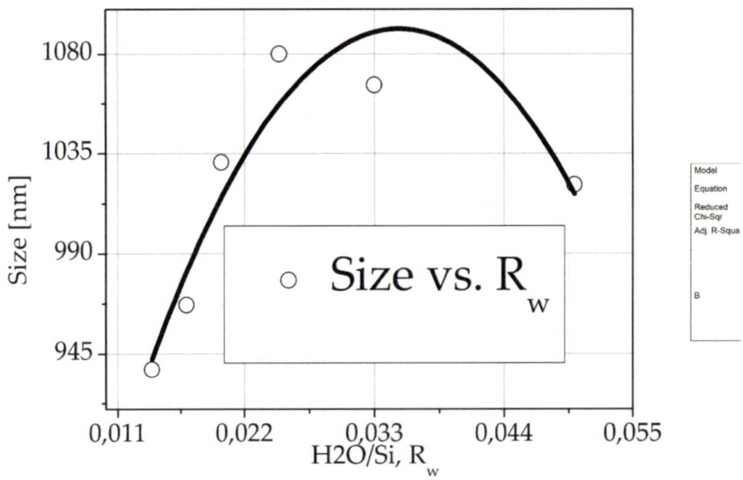

Figure 6.
The corresponding Rw ratio for each case of SiO_2@Ca $(OH)_2$ modified water traps.

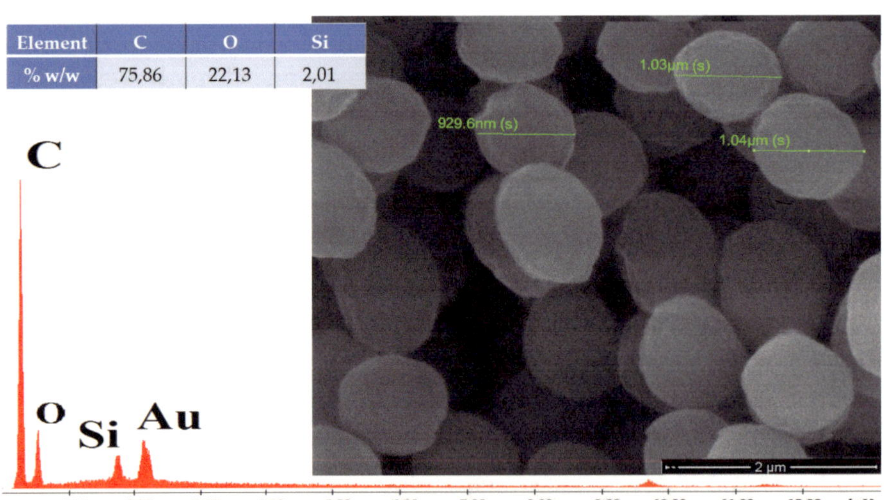

Figure 7.
SEM micrograph. EDS analysis of the experiment with 21 ml of TEOS as well as the %w/w of each element.

3.3 SiO_2 microspheres

The purpose of the study is to learn the behavior of bacteria with an inorganic microsphere of appropriate size. As mentioned above, silica is well compatible with cement and is therefore used as a small transport system. The synthesis process is relatively cheap and easily manageable. The desired ball size is about 1 μm. Through the sol-gel method, SiO_2 micro-containers were combined into two series of experiments by changing the parameters of the experiment while keeping the remaining constant. In the 1st series of experiments, we increased the amount of TEOS added. In the second series of experiments, the volume of the solvent increased. In both cases, the ratio of H_2O/Si, R_w and the change in size of the microspheres was investigated. We used ethanol, water, TEOS, and 30% NH_3 as the catalyst. The experimental SiO_2 ball synthesis process was done using 150 ml boiling glass in which EtOH and H_2O were mixed and left to stir for about 10 minutes. After that, we added NH_3 and stirred for 20 minutes. The reaction starts when the TEOS drip is added. Leave the final solution for about 20 hours under the same conditions, then centrifuge and dry. The whole experimental process takes place at ambient temperature. **Table 2** gives the ratios H_2O/Si, Rw for each case. This ratio results from the added water in each experiment. The water comes from the 30% NH_3 solution. That is, 0.3 ml H_2O is contained in 1 ml of NH_3. **Table 3** gives the exact amounts of chemicals used. The resulting SEM micrographs are shown in **Figure 8**. According to the SEM image, the SiO_2 microsphere synthesis process shows repeatability by giving spherical particles, with a similar diameter and small size distribution. The dispersion proves that there is a good size distribution.

The characterization of the size of SiO_2 microspheres was done with DLS. The study was carried out to find the optimal sample concentration for the correct measurement in DLS. In this case, it was found that the optimal concentration is 10 mg/L. According to the literature [3], a suspension is considered stable (i.e. not cracked) when it has a potential positive of +30 mV and more negative than -30 mV. According to the DLS measurement, the size of the SiO_2 microspheres is 782 ± 46 nm, which is also in line with the measurement of SEM. The ζ-potential is –56 ± 0.361, which attests to the stability of the sample composed. In summary, the synthesis of an inorganic microspheres of about 1 μm size was achieved by the sol-gel.

Experiment	TEOS (ml)	$[H_2O/Si]$ (R_w)
1	4	3.2
2	8	1.6
3	10	1.28
4	12	1.06

Table 2.
The ratio R_w resulting by changing the amount of TEOS in the first case of experiments, keeping the remaining parameters constant.

Experiment	EtOH (ml)	H_2O (ml)	$[H_2O/Si]$ (R_w)
1	100	11	1.28
2	200	22	2.38

Table 3.
The ratio Rw, which results by changing the amount of solvents in the second case of experiments, keeping the remaining parameters constant.

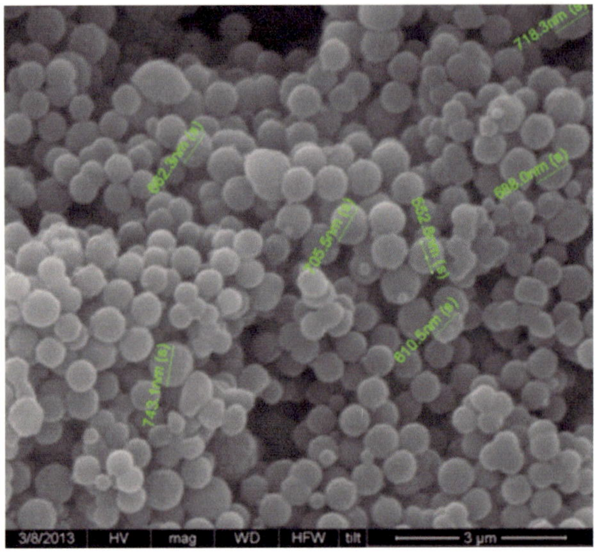

Figure 8.
SEM image of 725 ± 84 nm size SiO2 microspheres.

3.4 Polymethyl methacrylate microspheres

The purpose of the study is to verify the behavior of bacteria with an organic microcontainers of appropriate size. The method is economical and easily manageable. Desired sphere size is about 1 μm. Polymeric methyl polymethacrylate microspheres (PMMA) were made up with radical polymerization and in particular with emulsion polymerization. The reagents we used are methyl methacrylate monomer (MMA), KPS start-up and water solvent. As the desired sphere size is approximately 1 μm, experiments were carried out changing the following parameters of the experiment: (a) the amount of solvent, (b) the amount of the authority, and (c) the amount of water. The experimental procedure was carried out in a 100 ml spherical flask, the solvent was added where, with the help of a heating plate under stirring, the temperature was maintained at 70°C. Then the monomer was added, and after 20 min stirring, the starter was added. The colloidal solution was left stirring for approximately 20 h. Polymerization took place under a nitrogen atmosphere. The following describes the puma synthesis mechanism consisting of (1) the starting stage where the authority gives free radicals and reacts with the monomer, (2) the phase of propagation is observed polymer development, and (3) the stage of termination at which two roots react with each other, and the polymerization is terminated by taking PMMA as a product. **Figure 9** summarizes the procedure (**Table 4**).

Table 5 gives the quantities of reagents that led to PMMA microsphere synthesis with the largest size, in each of a series of experiments.

Figure 10 shows the size of the microcontainers as a function of water (A), MMA (B), and C starter.

According to the above diagrams, it is observed that the largest size of PMMA microbeads synthesized results from the second series of experiments, i.e., by changing the amount of monomer by keeping the remaining parameters of the experiment constant. **Figure 11** shows the micrograph of the samples by SEM.

The SEM image above shows the size of the PMMA micro-containers with the largest diameter achieved. As observed from the above image, the sample shows a large multi-dispersion. The characterization of the size of PMMA micro-containers was subsequently made with the dynamic scattering of light. As described in the

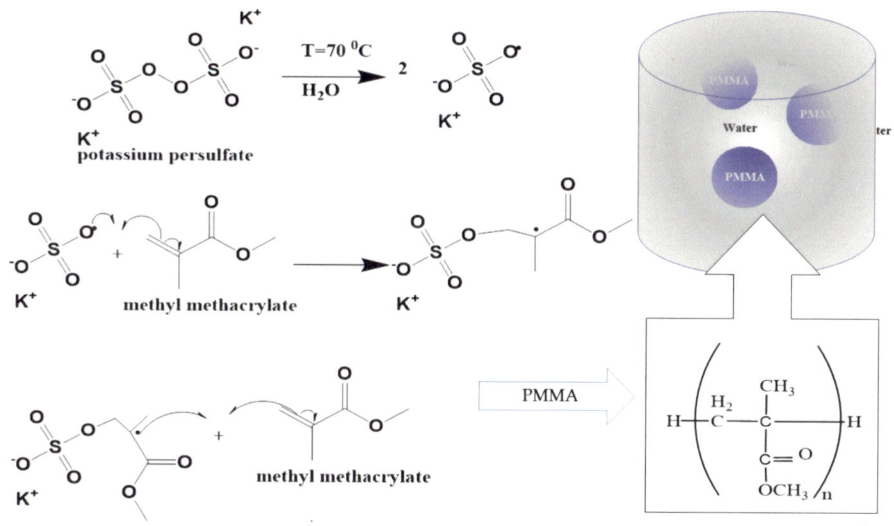

Figure 9.
PMMA microcontainer synthesis mechanism.

	Size (nm) Determined by SEM
$1^{η}$ series of experiments	725 ± 84
$2^{η}$ series of experiments	587 ± 38

Table 4.
The size of microspheres resulting from both series of experiments.

Change	MMA (ml)	KPS (mg)	Water (ml)	Size (nm, TEM)
Solvent	4	40	100	417 ± 23
Monomer	6	40	30	711 ± 90
Starter	4	40	30	299 ± 26

Table 5.
Quantities of reagents in each series of experiments with the corresponding size resulting each time.

case of SiO_2 micro-containers, a study was carried out to find the optimal sample concentration for the appropriate measurement of DLS. In this case, it was found that the optimal concentration is 10 mg/L. The following diagram, in addition to the size of the micro-containers resulting from the measurement, also lists the z-potential of the sample. This studies whether or not the suspension is stable. According to the DLS measurement, the resulting PMMA size is 780 ± 25 nm, which is in line with the SEM measurement. The ζ-potential is −38.6 ± 0.651, which attests to the stability of the samples. PMMA microsphere synthesis is economical, easily manageable and takes place in aqueous environment. For the production of a product, in industry, the above factors are very important.

3.5 Microcontainers with polyurea shell

The ultimate purpose of this study is to trap the bacteria spores in the microcontainers with polyurea shells. This can be done if the spores of the microorganism are scattered in the drops of the oily phase, and the shell forms around them.

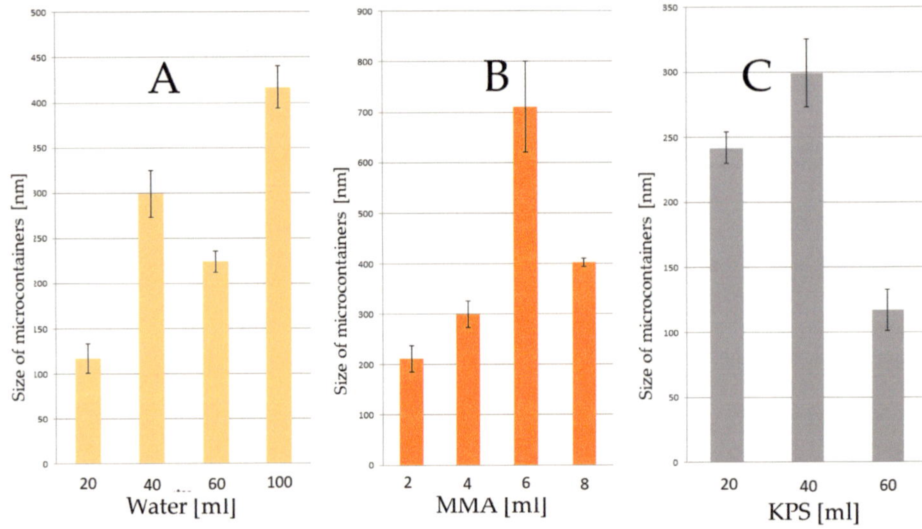

Figure 10.
(A) PMMA microsphere size changing the amount of solvent, (B) PMMA microsphere size changing the amount of monomer, and (C) PMMA microsphere size changing the amount of the beginning.

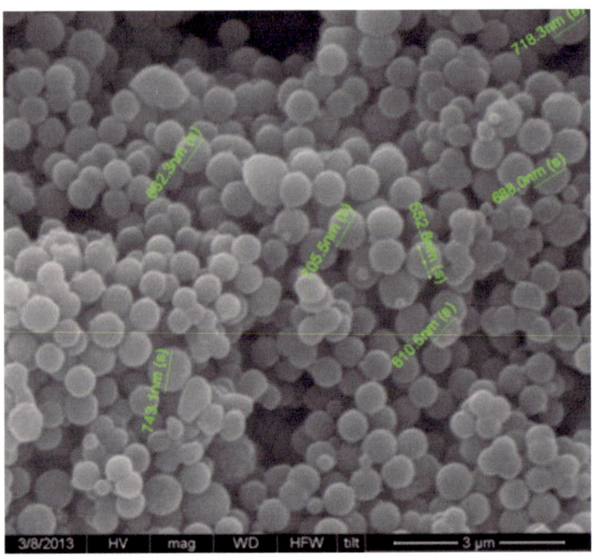

Figure 11.
SEM image of the PMMA micro-container with a size of 711 ± 90 nm.

The synthesis of polyurethane-coated microcontainers is achieved through interfacial polymerization (IP) in an oil emulsion in water (O/W emulsion). The synthesis of polyurea shell micro-containers is achieved by polymerization on the middle surface of two liquids that are not mixed together. This study investigated an oil emulsion in water (O/W emulsion). The reagents used are organic soluble monomer (toluene diisocyanate [TDI]), water-soluble monomer (ethylenediamine, EDA, diethylenediamine, DETA), ethylenediamine (EDA), oil phase (1-octadecene, paraffin), water phase (H_2O), emulsifier (sodium sulphonate dodecyl, SDS), methyl ester poly ethylene glycol (PEG), polyvinylpyrrolidone (PVP), Triton x-100, catalyst (NH_4Cl) and solvent (acetone). Four series of experiments were carried out by changing some parameters, keeping the remaining constants, in order to synthesize polyuria shell containers. The experimental procedure followed in all four series of experiments is as follows: For the composition of the oily solution in a boiling glass, mix the organic soluble monomer with the organic phase and acetone. The stirring lasts a few minutes. The oil emulsion in water is formed when the oil solution is added to 50 ml of aqueous solution containing 5% w/w emulsifier. Stirring of the system takes place at ambient temperature and at 300 rpm/min for 5 min. The water-soluble monomer solution in the emulsion is added drip and stirring at 600 rpm. Then add 5% w/w NH_4Cl and stir continuously for 3 h at 60°C. Finally, the sediment is centrifuged, washed with 30% ETOH, and left for drying [4]. In the first series of experiments, the possibility of forming polyurea shell micro-containers using two different water-soluble monomers (EDA and DETA) with an emulsifier SDS is studied. All other factors in the experiment remain the same. In the second series of experiments, the possibility of different emulsifiers (SDS, PEG, PVP, and Triton x-100) for the formation of polyuria shell micro pots is studied, using DETA as a water-soluble monomer. All other factors in the experiment remain the same. The third series of experiments studies the action of different phases of oil (1-Decaoctene and paraffin) using Triton x-100 as an emulsifier and as a water-soluble monomer DETA. All other factors in the experiment remain the same. The fourth series of experiments studies the dependence of the composition of micro pots on the quantity of peg emulsifier. Paraffin was used as an oily phase and DETA was used as a water-soluble monomer. All other factors in the experiment remain the same. **Figure 12** shows the synthesis of micro-containers of polyurea with different amines.

The mixture of the oil phase, containing the oil and TDI, is dispersed in the aqueous phase with the emulsifier, thus forming an oil emulsion in/water (O/W emulsion). Initially, in the aqueous phase, the hydrophobic groups of the emulsifier are covered by its hydrophilic groups, in such a way that micelles are formed (**Figure 10**). Adding the oil solution, the hydrophilic groups are diffused into

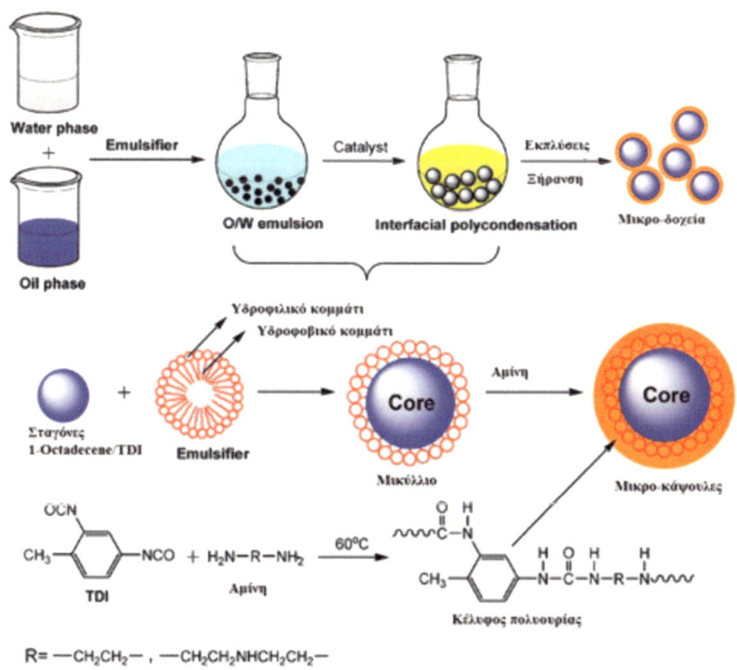

Figure 12.
Representation of the composition of polyurea micro-containers with different amines.

the oil, while the hydrophilic along the drop covering it. The formation of the shell begins when regional isocyanate groups of TDI are hydrolyzed in the oil-water middle surface forming amines, according to the reaction:

$$RNCO + H_2O \rightarrow RNH_2 + CO_2 \tag{1}$$

These amines react with non-hydrolyzed isocyanate groups, thereby forming a polyuria network. When the original shell is formed, the added amine (the water-soluble monomer) must penetrate the membrane and penetrate the oil phase to react with TDI (the organ soluble monomer). This makes the polyurea shell denser and more durable. As a result, the formation of a polymer shell (consisting of a polyuria network) is observed on the interface of the emulsifier and the oil phase, due to the reaction between amine and TDI. The temperature that the reaction takes place is the ambient temperature [4]. The series of experiments on the composition of polyuria shell micro-containers were carried out in order to find the most suitable experimental conditions for an efficient result. Thus, the experiments are related to each other by following each time the most promising result. Thus, in the first series SDS is used as an emulsifier and EDA and DETA are used as amines. Between the two amines, the other studies used DETA, as it showed the best micro-pot formation. The second study showed that PEG and Triton x lead to the formation of micro-containers. Below is the SEM image of the sample where DETA has been used as an amine while PEG has been used as an emulsifier. **Figure 13** shows SEM of the sample of polyurea shell containers from the second series of experiments, emulsifier PEG.

As well, the study of specific microcontainers is indicative that an average microsphere size has not been found as in all previous microsphere studies. The above image is not representative for the entire sample. An indicative study of the composition of microspheres with polyuria shell was carried out in this section. Polyuria has very good strength and is insoluble in most solvents. That is why it was chosen to study such a system. Experiments were conducted changing parameters of the experiment. Promising results give those cases where PEG and Triton x with amine DETA are used as emulsifiers and as an oil phase 1-octadecene.

Figure 13.
SEM image of the sample of polyurea shell micro-containers from the second series of experiments, emulsifier PEG.

3.6 Microbial precipitation $CaCO_3$

As has already been mentioned, the application of bacteria in concrete for the purpose of self-healing in case of cracks is not impossible. There have been many studies that prove their effectiveness in healing cracks. Bacteria that can withstand the extreme environment of concrete (high voltages and intense alkalinity) are those of the genus Bacillus (Gram-positive bacteria). Species of this genus have the potential to form endospores, forms resistant to non-bacterium-friendly environments. When the endospores are found in appropriate conditions, they are activated (a process also known as eblastosis) and begin to develop. Bio-precipitation of a mineral by a microorganism occurs when it is found in an environment with suitable nutrient material. In the case of cement self-healing, the mineral chosen is $CaCO_3$. Although the most suitable bacterium for this work is some species from the genus Bacillus, this study looked at two other types of bacteria, namely *Escherichia coli* (Gram negative) and *Staphylococcus aureus* (Gram positive). Thus, the possibility of submersion of $CaCO_3$ salt from a Gram-negative and a Gram-positive bacterium was studied, with the main emphasis on *Staphylococcus aureus*. The reason this was done is because *St. aureus* has a cell structure similar to that of Bacillus, although it does not have the ability to form endospores. It should be stressed that the bacteria studied in this work were pathogenic, and all safety rules were followed during their use. As mentioned in the above section, $CaCO_3$ precipitation is a two-way chemical process controlled mainly by four factors, first calcium concentration, second concentration of dissolved inorganic carbonate ions (DIC), third the pH, and fourth availability of nucleation centers.

There are three mechanisms associated with bio-precipitation, and in this work, the mechanism of urea breakdown (hydrolysis of urea, HU) was studied through the enzyme urease, a course easily manageable and controlled. The general reaction is as follows:

$$CO_3^{\ 2-} + Ca^{2+} \rightarrow CaCO_3 \tag{2}$$

The bacterium plays the role of the nuclearization center with the following mechanism:

$$Ca^{2+} + Cell \rightarrow Cell - Ca^{2+} \tag{3}$$

$$Cell - Ca^{2+} + CO_3^{\ 2-} \rightarrow Cell - CaCO_3 \downarrow \tag{4}$$

The exponential growth phase is the ideal phase to study any cellular function.

3.6.1 Precipitation $CaCO_3$ from bacteria

3.6.1.1 $CaCO_3$ precipitation rating study by Staphylococcus aureus

In order to investigate the possibility of precipitation of $CaCO_3$ from *St. Aureus*, four microorganism solutions with a different amount of $CaCl_2.2H_2O$ and urea were prepared. In this way, the ability of this bacterium to precipitate salt at different nutrient concentrations has been qualitatively studied. **Table 6** gives the composition of *St. aureus* solutions with different concentration of $CaCl_2.2H_2O$ and urea.

For the preparation of the above solutions, a quantity of microorganism solution was isolated and the corresponding quantity of $CaCl_2.2H_2O$ and urea was added

Materials	1° Solution	2° Solution	3° Solution	4° Solution	5° Solution
$CaCl_2.2H_2O$	—	0.05 g	0.1 g	0.25 g	0.5 g
Urea	—	0.05 g	0.1 g	0.25 g	0.5 g
Cultivation St. aureus	20 ml	20 ml	20 ml	20 ml	20 ml

Table 6.
Composition of St. aureus solutions with different concentrations of $CaCl_2.2H_2O$ and urea.

each time. The solutions were incubated at 37°C at 100 rpm. After 72 h, in these conditions, the solutions were removed and compared. The solution of the microorganism containing the most $CaCl_2.2H_2O$ and urea showed the greatest turbidity of the other solutions with apparent precipitation of $CaCO_3$. There is a marked change in the turbidity of the solutions, as distinguished from the images above. The change in clarity between the nutrient solution, LB and the solution after the development of the microorganism, certifies that the microorganism has been properly incubated to study any of its metabolic processes. An apparent precipitation of sediment is observed between the 5th solution and the microorganism solution (**Figure 14C**). *St. aureus* solution becomes cloudier while $CaCO_3$ submersion certifies its ability to break down urea and lead to precipitation of the mineral. In order to characterize more fully the precipitation of $CaCO_3$, a kinetic study is carried out by the microorganism in solutions of (a) in nutrient material LB in the presence of $CaCl_2.2H_2O$ and urea and (b) in agar medium (Petri dish) in the presence of $CaCl_2.2H_2O$ and urea. Taking samples at regular intervals, and through infrared spectroscopy (FT-IR), the existence of the mineral is studied in two different environments. The following describes the composition of microorganism solutions in LB nutrient solution and

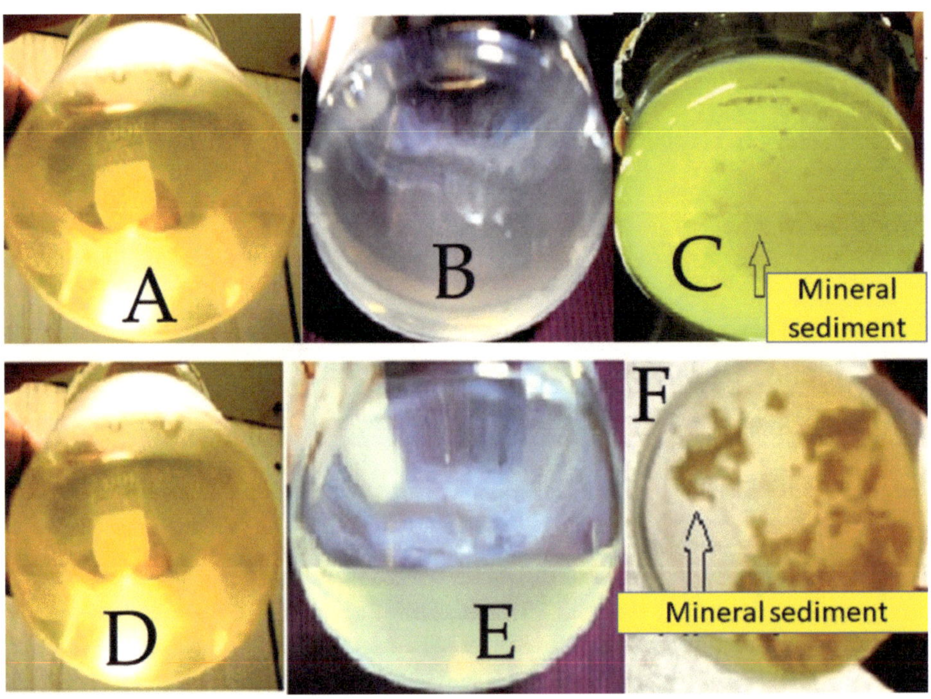

Figure 14.
Change in clarity of St. aureus solution. (A) LB solution, (B) microorganism solution after incubation in LB, and (C) precipitation of $CaCO_3$ from the 5th solution of the bacterium. Change in the turbidity of the E-coli crop: (D) LB solution, (E) microorganism solution after incubation in LB, and (F) precipitation of $CaCO_3$.

agar nutrient medium for the $CaCO_3$ precipitation kinetic study. Maximum concentrations of $CaCl_2.2H_2O$ and urea were selected from $CaCO_3$ precipitation quality control, as in these quantities the mineral was more evident.

In this particular case, however, *St. Aureus*, which develops, is placed in the stirring incubator for approximately 24 days in certain conditions (37 °C, 100 rpm). At regular intervals, a sample is isolated, centrifuged, and sterilized to measure the FT-IR spectrum for the purpose of certifying the presence of $CaCO_3$.

In this particular case, however, o *St. aureus* is left in the incubation oven for about 21 days under certain conditions. (37 °C). At regular intervals a sample is isolated, centrifuged, sterilized and measured the FT-IR spectrum for the purpose of certifying the presence of $CaCO_3$. The environment in which the mineral develops is quite complex, with many different compounds, which come from the nutrient and microorganism. These compounds show peaks in IR spectra. Therefore, in order to verify the existence of $CaCO_3$ in the cases where the study is carried out, a comparison of the range of $CaCO_3$ with the spectrum $CaCO_3$ LB nutrient solution (**Figure 15A**), in order to identify the tops of the mineral in the nutrient microorganism. Then, after determining the carbonic ion peaks in the IR spectrum, each kinetic study spectrum is included in addition to the spectra of the isolated samples, the $CaCO_3$ LB and the spectrum of the microorganism. In this way, it is possible to identify the peaks corresponding to the $CaCO_3$ in any case but also the peaks corresponding to the microorganism. The results of this study are listed below.

Since $CaCO_3$ is a crystalline salt, the vibrations that occur in an IR spectrum correspond to the bonds of CO_3^{2-}. The carbonic ion peaks in the IR spectrum are as follows: a strong wide peak in the 1530–1320 cm^{-1} range, medium intensity peaks at 1160 cm^{-1} and in the 890–800 cm^{-1} range, and a peak in the 745–670 cm^{-1} range [5]. Based on the above and the IR spectrum of **Figure 15A**, we observe that $CaCO_3$ in LB solution shows peaks at 712, 873, and 1409 cm^{-1}. The top at 1638 cm^{-1} can be attributed to any of the components of the nutrient medium LB. Finally, the peak shown in the range 3000–3500 cm^{-1} is attributed to the –OH of the aqueous solution. An indicative range of IR bacterium is shown in **Figure 15B**. Based on

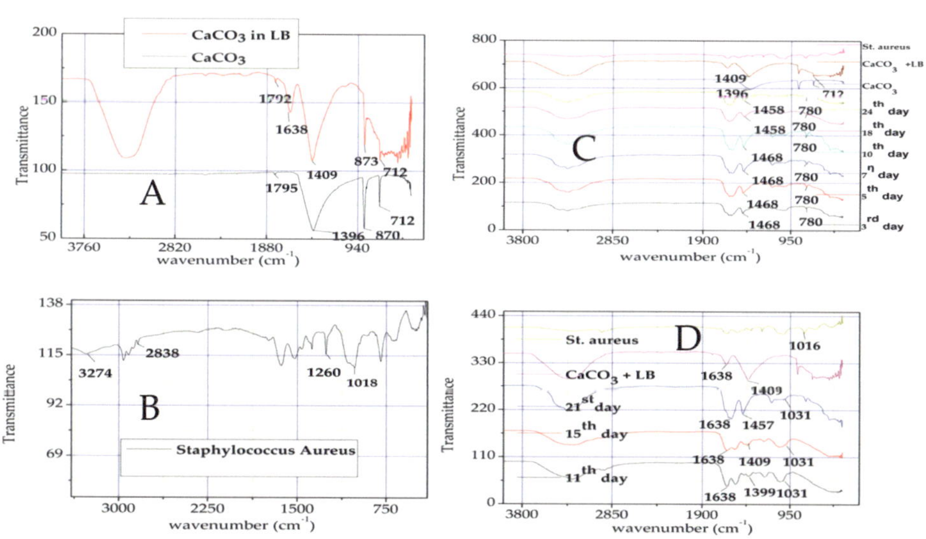

Figure 15.
(A) IR spectrum of the crystalline $CaCO_3$ and $CaCO_3$ in the nutrient medium LB, (B) spectrum IR St. aureus in nutrient medium LB, (C) IR spectrum of $CaCO_3$ precipitation kinetic study from St. aureus to nutrient LB, and (D) IR spectrum of $CaCO_3$ kinetic precipitation study from St. aureus in the nutrient agar medium in a dish petri.

the range of the bacterium, we can assign some peaks to the various components that make up the microorganism. The peak at 1018 cm^{-1} is attributed to polysaccharides compounds, the peak at 1260 cm^{-1} at the asymmetric vibration of the PO_2 cm^{-1} bond, at 3274 cm^{-1} we observe the vibration of N-H, while at 2838 cm^{-1} we observe the symmetrical vibration of the CH_2 bond [5]. The IR spectrum of the $CaCO_3$ kinetic precipitation study from *St. aureus* to the nutrient medium LB is given below. By comparing the spectra of the samples isolated each time, the spectra of the microorganism and $CaCO_3$ in LB draw the following conclusions. From the 3rd day of incubation of the microorganism with $CaCl_2.2H_2O$ and urea, precipitation $CaCO_3$ is observed. This is certified from the top at 1468 cm^{-1}, which as mentioned above belongs to the display area 1530–1320 cm^{-1} for carbonate ions. The new peak at 780 cm^{-1} is also attributed to carbonate ions. Although this peak does not correspond to any area of appearance of carbonate ions, it can be attributed to the range 745–670 cm^{-1} but as shifted. The presence of carbonate ions is observed in the kinetic study for all periods of time when the sample is isolated. The remaining peaks showing the sample spectra can be attributed to compounds of the microorganism and nutrient material. Finally, with this comparative study of the various IR spectroscopy, it appears that *St. aureus* is capable of precipitating $CaCO_3$ in LB nutrient solution when dissolved in sufficient $CaCl_2.2H_2O$ and urea. Due to the ureolytic action of the microorganism, carbonate ions are released, and in a calcium-rich environment there is precipitation of the mineral. Below is the IR spectrum of the kinetic precipitation study $CaCO_3$ from *St. aureus* to the nutrient in Petri dish. The comparative precipitation study of $CaCO_3$ by St. aureus in the nutrient medium, in a Petri dish, observed the carbonic peaks corresponding to carbonate ions in the isolated samples. Thus, according to the above range, the IR peak at 1638 cm^{-1} is attributed to nutrients and the peak to 1031 cm^{-1} to the polysaccharides of the microorganism. The peak attributed to carbonate ions is in the range 1399–1457 cm^{-1}, as it is included in the range 1530–1320 cm^{-1} where these ions appear. We can conclude, for example, that *St. aureus* has the ability to precipitate $CaCO_3$ in a Petri dish (**Tables 7** and **8**).

3.6.1.2 Precipitation characterization study of $CaCO_3$ in microorganism solution

In this way, the ability of this bacterium to precipitate salt has been qualitatively studied. **Table 9** shows the composition of E-coli solutions with different concentrations of $CaCl_2.2H_2O$ and urea.

For the preparation of the above solutions, a quantity of microorganism solution has been isolated, and the corresponding amount of $CaCl_2$ and urea every time. These solutions were incubated at 37°C and 100 rpm. The solution of the microorganism that contained the most amount of $CaCl_2.2H_2O$ and urea showed the greatest turbidity of the other solutions with apparent precipitation of $CaCO_3$. After 72 h in these conditions, the solutions were removed and compared. Below are

Materials	Quantities
LB	49.5 ml
$CaCl_2.2H_2O$	2.5 g
Urea	2.5 g
Solution St. aureus	0.5 ml

Table 7.
Quantities of materials used for the kinetic precipitation study $CaCO_3$ from microorganism solution to LB nutrient material.

Materials	Quantities
Nutrient agar medium	25 ml in a dish Petri
$CaCl_2.2H_2O$	0.5 g
Ουρία	0.5 g
Solution St. aureus	Capable of its growth in the dish

Table 8.
Quantities of materials used for the kinetic precipitation study of $CaCO_3$ from microorganism solution to a nutrient agar medium in a petri dish.

Materials	1° Solution	2° Solution	3° Solution	4° Solution	5° Solution
$CaCl_2$	—	0.05 g	0.1 g	0.25 g	0.5 g
Urea	—	0.05 g	0.1 g	0.25 g	0.5 g
Cultivation *E-coli*	20 ml	20 ml	20 ml	20 ml	20 ml

Table 9.
Composition of E-coli solutions with different concentrations of $CaCl_2.2H_2O$ and urea.

the photos of the nutrient solution, the solution of the bacterium after incubation, and the 5th solution. As shown above, during the development of the microorganism *E-coli,* a comparative study is carried out, and there is a marked change in the turbidity of the solutions. The change in clarity between the nutrient solution, LB and the solution after the development of the microorganism, certifies that the microorganism has been properly incubated to study any of its metabolic processes. **Figure 13** C has a marked turbidity in the solution of the microorganism that contained the most $CaCl_2.2H_2O$ and urea, in relation to the incubation solution of the microorganism. The sediment submerged can be attributed to the creation of $CaCO_3$ from the *E-coli*. The visual characterization of calcium carbonate precipitation shows that the *E-coli* through the ureolytic breakdown sink $CaCO_3$. Two types of bacteria have been studied in this paper, *E-coli* and *St. aureus*, for their ability to break down urea and precipitate $CaCO_3$ in an environment rich in Ca^{2+}. Through the quality tests carried out, it was observed that both bacteria precipitate the desired mineral. Further characterization was carried out with the *St. aureus,* where a kinetic study was carried out on the existence of $CaCO_3$, in two different nutrient environments of the microorganism. In both cases, precipitation of the mineral. These bacteria in force can be used in the phenomenon of self-feeding of cement, since it is evident that they can precipitate $CaCO_3$ under certain circumstances.

3.7 Interaction between microbeads and microorganisms

In this chapter, bacteria behavior is studied in the presence of microspheres. Specifically, the interaction between microspheres and bacteria in vitro is studied. With the ultimate aim of these micro-containers being introduced into cement as carriers of bacteria, it is necessary to study whether polymer spheres are toxic to humans and bacteria. Of the five types of polymeric microspheres compositional (a) water traps, (b) coated water traps with SiO_2, (c) SiO_2 microspheres, and (d) PMMA microspheres for their interaction with bacteria were studied. The fifth type of polymeric micro-cell, polyuria shell micro-containers, needs further optimization before any coupling with a microorganism. Following this work, it is studied whether microns affect the ureolytic action of bacteria for $CaCO_3$ submersion. In other words, it is being studied whether the bacteria are still capable of submerging the mineral after contact with the microcontainers (**Figure 16**).

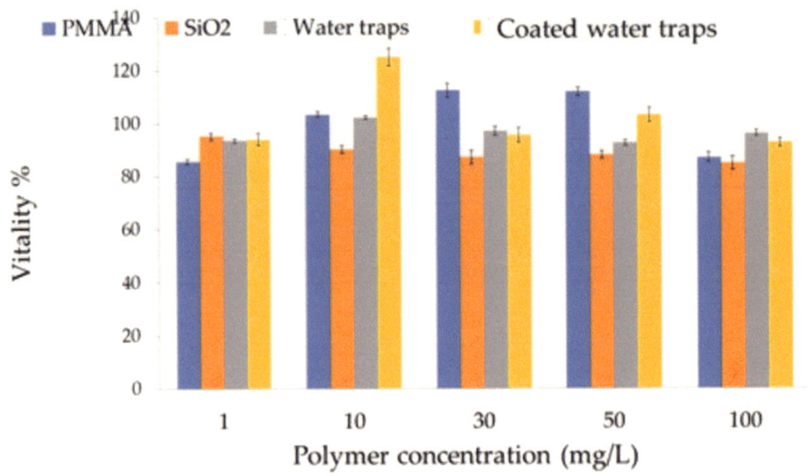

Figure 16.
HEK-293 cell vitality rates, presence of microspheres at various concentrations.

3.7.1 Toxicity study

3.7.1.1 Study of the cytotoxic action of organic-inorganic microspheres

The investigation of cell vitality is a characteristic indicator and a necessary condition in cases of cytotoxic activity of polymeric inorganic-organic microspheres where they will be used in materials that will come into direct contact with humans, such as cement. The MTT method studies the cytotoxic activity of the microspheres that were compositional in the HEK-293 cells. Studied, the cytotoxic action of microspheres at various concentrations. The rates of cell vitality after incubation with these materials are shown in the **Figure 14**. The diagram shows that the microspheres studied are not toxic to humans. In general, we conclude that regardless of polymer concentration the metabolic function of cells is not affected in the presence of micro-transfer systems.

3.7.1.2 Toxicity study of organic-inorganic microbeads to St. aureus

After it has been shown that microspheres are not toxic to humans, it was then investigated whether microspheres affect the development of the microorganism. In other words, quality control is carried out on the toxicity of micro-transport systems in *St. aureus*. Quality control is done by vaccinating the microorganism in Petri plates, in the presence of microspheres. In this case, however, before vaccinating the microorganism in the dish, four *St. aureus* solutions were prepared to which a quantity of each case of microspheres was added to each. The resulting sediment, consisting of microorganisms and microspheres, was inoculated in a Petri dish. Thus, the development of *St. aureus* was observed in an environment containing micro-transport systems. **Figure 17** shows photos of the Petri plates of *St. aureus* for each case of microcephaly studied.

According to **Figure 17**, it is certified that microspheres do not affect the development of *St. aureus*. The colonies formed by the microorganism are visible in all cases of plates. These colonies prove that *St. aureus* remains active metabolically even in the presence of microspheres. This quality control finally shows that the four species of micropesties are not toxic to the microorganism and do not affect it in its development.

The next check concerns whether the presence of microspheres is affected by precipitation $CaCO_3$ by *St. aureus*.

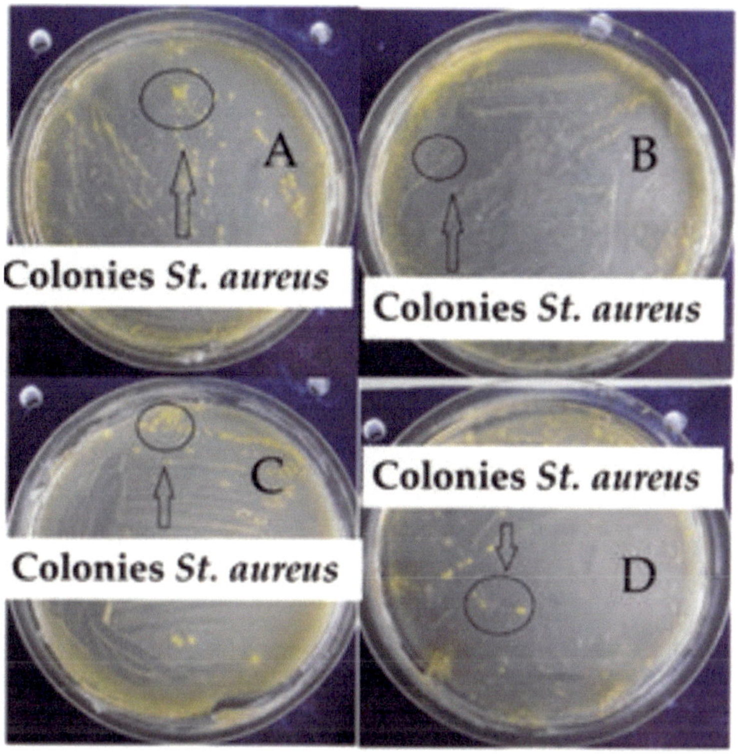

Figure 17.
Development of St. aureus in petri plates in the presence of (A) PMMA microspheres, (B) SiO_2 microspheres, (C) water traps, and (D) SiO_2 coated water microtraps.

3.7.1.3 Precipitation characterization study $CaCO_3$ by Staphylococcus aureus in the presence of microspheres

The effect of microspheres on the precipitation of the $CaCO_3$, from the microorganism, a kinetic medium, was studied in *St. aureus* solutions in the presence of $CaCl_2.2H_2O$, urea, and microspheres. Taking samples at regular intervals, and by characterization of infrared spectroscopy (FT-IR), the existence of the mineral is studied. The microspheres used for this study are water traps.

Table 10 describes the composition of the solutions of the microorganism in a nutrient solution LB for the kinetic precipitation study of $CaCO_3$.

The development of the microorganism in LB nutrient material with the above quantities of materials. In this particular case, however, *St. Aureus*, which develops, is placed in the stirring incubator for about 24 days in certain conditions (37°C, 100 rpm). At regular

Materials	Quantities
LB	49.5 ml
$CaCl_2.2H_2O$	2.5 g
Urea	2.5 g
Solution St. aureus	0.5 ml
Micro-water traps	1 g

Table 10.
Quantities of materials used for the kinetic precipitation study of $CaCO_3$ from microorganism solution to LB nutrient material in the presence of micro-water traps.

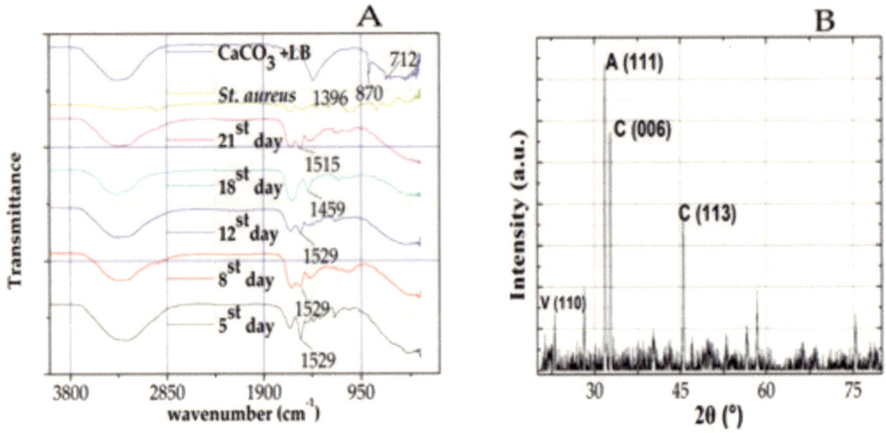

Figure 18.
(A) FT-IR spectrum of $CaCO_3$ kinetic precipitation study by St. aureus in LB nutrient material in the presence of water traps. (B) X-ray radiance spectrum for St. aureus sample in the presence of water microtraps after precipitation of $CaCO_3$.

intervals, a sample is isolated, centrifuged, sterilized, and eventually characterized via FT-IR, for development certification of $CaCO_3$. In other words, in order to identify the carbonic ion peaks in the FT-IR spectra of the isolated samples, a comparative study is carried out between the sample spectra, the microorganism, and the $CaCO_3$ in LB nutrient. It has emerged that $CaCO_3$, in LB solution, has peaks of 712, 873, and 1409 cm^{-1} (**Figure 18A**), while the peaks corresponding to the microorganism are as follows: the peak at 1018 cm^{-1} is attributed to polysaccharides compounds, the peak at 1260 cm^{-1} in the asymmetric vibration of the bond PO_2, at 3274 cm^{-1} we observe the vibration of N-H, while at 2838 cm^{-1}, we observe the symmetrical vibration of the CH_2 bond.

Based on the FT-IR spectrum of **Figure 18**, it appears that $CaCO_3$ is precipitated by *St. aureus* in the LB nutrient material in the presence of water traps. This is demonstrated by the appearance of a peak at 1529 cm^{-1} where it belongs to the carbonate display area 1530–1320 cm^{-1}. The remaining peaks of the sample spectrum are attributed to components of the microorganism itself and the nutrient LB. Further characterization of $CaCO_3$ precipitation by *St. aureus* in the presence of micro water traps was carried out using the XRD method (**Figure 18B**). According to the X-ray spectrum of the sample, the existence of $CaCO_3$ is certified by the coexistence of all three of its crystalline forms: (a) calcite, C with crystalline levels (006) and (113) and (b) baterite, V (110) and aragonite, A (111).

4. Conclusions

The purpose of this work is the synthesis and characterization of an economical and easily manageable material, which will have an impact on microorganisms by attributing to specific conditions of $CaCO_3$, a mineral that allows the healing of cracks in building materials. More specifically, the composition of microspheres of various types such as PMMA has been studied and their interaction with microorganisms, Gram-positive and Gram-negative, which under appropriate conditions cause $CaCO_3$ precipitation. According to the literature, the most appropriate bacterium in the study of such a case is bacillus (Gram positive, alkalophilic, spore-like bacterium). Due to the inability to find bacillus bacterium, the study was carried out on two types of bacteria, *Escherichia coli* (Gram negative) and *Staphylococcus aureus* (Gram positive). More specifically, however, St. aureus was investigated as it is a Gram-positive bacterium. The first system studied is water traps. Microspheres of 1 μm size were made, with a uniform size distribution, and from the water absorption study, it was shown that they can respond to aqueous stimuli by increasing their size due to swelling. Such a system could be used to absorb the seeds of the microorganism. The water traps were then coated with SiO_2 in order to give the water traps a durable shell, so that in the future when added to the concrete, it protects the adsorbed bacterium from the extreme environment of cement. In the second microsphere synthesis system, the development of an organic (PMMA) and an inorganic (SiO_2) transport microsystem is observed. Both types of microspheres were synthesized in aqueous environment, which is chosen by industry for the production of a material, as an economical composition. This contrasts with the composition of the water microtraps and their coatings as the composition was carried out in organic solvent (acetonitrile), which is economically unprofitable for production. The composition of the PMMA and SiO_2 microspheres led to production of approximately 1 μm in size, with a uniform size distribution.

The third system concerns the synthesis of microspheres with polyurea shells for the purpose of encasing a substance inside them. Successful synthesis of urea shell capsules was observed when PEG and Triton-x were used as an emulsifier. This shell has useful properties for the industry, as it has great durability and is stable. In the second part of the work, the ability of *Escherichia coli* and *Staphylococcus aureus* to precipitate $CaCO_3$ was studied. The submersion of the mineral is certified by visual characterization as well as by FT-IR spectroscopy. After it has been shown that microspheres are toxic not only to humans (average MTT method) but also to bacteria (visual reculture control after vaccination in a petri dish), it was studied whether they affect the microorganism's submersion of $CaCO_3$. As observed from the data of this study, microorganism remains metabolically active, which leads to precipitation of the mineral, in the presence of microspheres under specific incubation conditions.

Although the bacterium-sphere interaction mechanism has been fully studied, this study shows that the synthesis of micro-bacteria transport systems for the purpose of self-feeding building materials is a promising way to protect the bacterium within cement. Bibliographically, it has been observed that the viability of seeds of the genus Bacillus within cement decreases over time. Thus, the composition of a material where it can protect the bacterium in such an environment, is promising in the field of self-healing. In this way, the performance of the bacterium in precipitation increases, and more effective healing of the crack will be observed. More objective for the composition of such a material is the resulting product to be as manageable and easy to implement as possible, in order to become industrially competitive.

Author details

George Kordas
Sol-Gel Laboratory, INN, NCSR Demokritos, Aghia Paraskevi Attikis, Greece

*Address all correspondence to: gckordas@gmail.com

References

[1] Jonkers HM. Self healing concrete: A biological approach. In: Zwaag SVD, editor. Self Healing Materials. Netherlands: Springer

[2] Standars A-IPWL. Available from: http://www.lgcstandards-atcc.org/products/all/CRL-1573.aspx?geo_country=gr#357C3571006A4259B64650D34DF19048

[3] Ltd., MI. Zetasizer Nano Series, User Manual. MAN0485 Issue 1.1; 2004

[4] Zhang H, Wang X. Synthesis and properties of microencapsulated n-octadecane with polyurea shells containing different soft segments for heat energy storage and thermal regulation. Solar Energy Materials and Solar Cells. 2009;**93**(8):1366-1376

[5] Socrates G. Infrared and Raman Characteristic Group Frequencies-Tables And Charts. 3rd ed; 2001

Chapter 6

New Bismuth Sodium Titanate Based Ceramics and Their Applications

Hengchang Nie, Genshui Wang and Xianlin Dong

Abstract

Ferroelectric materials are widely investigated due to their excellent properties and versatile applications. At present, the dominant materials are lead-containing materials, such as Pb $(Zr,Ti)O_3$ solid solutions. However, the use of lead gives rise to environmental concerns, which is the driving force for the development of alternative lead-free ferroelectric materials. $(Bi_{0.5}Na_{0.5})TiO_3$-based ceramics are considered to be one of the most promising lead-free materials to replace lead-containing ferroelectric ceramics due to their excellent ferroelectric properties, relaxation characteristics, and high Curie point. After decades of efforts, great progress has been made in the phase structure characterization and properties improvement of BNT based ceramics. However, most of the studies on BNT system mainly focuses on its piezoelectric properties and application of piezoelectric sensors and strain actuators, little attention is paid to its ferroelectric properties and related applications. In this chapter, new BNT-based ceramics via composition modification and special focuses on the ferroelectric properties, phase transition behaviors under external fields and related applications, such as application in energy storage, pulsed power supply and pyroelectric detection were proposed.

Keywords: bismuth sodium titanate, ferroelectric properties, energy storage, pulsed power supply, energy storage, pyroelectric effect

1. Introduction

$(Bi_{0.5}Na_{0.5})TiO_3$ (BNT) was first reported by Smolenskii et al. in 1960 [1]. BNT ceramic is a kind of ABO3 type ferroelectrics which is replaced by Na^+ and Bi^{3+} complex ions at A-site. The A-site ions of BNT ceramics are located at the eight corner positions of octahedron, and the B-site ions are at the body center of octahedral structure [2]. Well sintered BNT ceramics have been obtained by hot pressing sintering method with d_{33} of 94–98 pC/N [3]. BNT ceramics exhibit a high Curie temperature (~320°C) and high polarization of 38 μC/cm^2, which is considered to be one of the most promising environment-friendly ceramic system to replace lead-based ceramics [4].

Pure BNT ceramics exhibits some problems such as high conductivity, and large coercive field, consequently giving problems in the poling process [4], which seriously hinder its practical application. Studies show that the comprehensive properties of BNT system can be significantly improved by doping or by incorporation

IntechOpen

BNT-based piezoelectric ceramics	d_{33} (pC/N)	Ref.
$(Bi_{1/2}Na_{1/2})TiO_3$	94–98	[3]
$(Bi_{1/2}K_{1/2})TiO_3$	69	[10]
(1-x) $(Bi_{1/2}Na_{1/2})TiO_3$-$x$$(Bi_{1/2}K_{1/2})TiO_3$	140–192	[11]
$(Na_{1-x}K_x)_{0.5}Bi_{0.5}TiO_3$	192	[12]
$85(Bi_{1/2}Na_{1/2})TiO_3$-$12(Bi_{1/2}K_{1/2})TiO_3$-$3BaTiO_3$	158	[13]
$(Bi_{0.5}Na_{0.5})TiO_3$-Ba(Zr, Ti)O_3 + 2 mol% CuO	156	[14]
$BiNaTiO_3$-$BiKTiO_3$-$BiLiTiO_3$	230	[15]
$[(Bi_{0.98}La_{0.02}Na_{1-x}Li_x)_{0.5}]_{0.94}Ba_{0.06}TiO_3$	212	[16]
$(Bi_{1/2}Na_{1/2})TiO_3$-$(Bi_{1/2}K_{1/2})TiO_3$-$(Bi_{1/2}Li_{1/2})TiO_3$-$BaTiO_3$	163	[17]
$Bi_{0.5}Na_{0.5}TiO_3$-$Bi_{0.5}K_{0.5}TiO_3$-$Bi_{0.5}Li_{0.5}TiO_3$	147–231	[18]

Table 1.
Piezoelectric properties (d_{33}) of BNT-based ceramics [9].

with other components to form solid solutions. In recent years, investigation of BNT based ceramics mainly focuses on two aspects. On one hand, the phase transformation and structure change of BNT under external field (electric field, temperature and stress field). On the other hand, the properties enhancement of BNT-based were obtained by doping or by incorporation with other components to form solid solutions, such as $Bi_{0.5}Na_{0.5}TiO_3$-$BaTiO_3$(BNT-BT) [5], $Bi_{0.5}Na_{0.5}TiO_3$-$Bi_{0.5}K_{0.5}TiO_3$(BNT-BKT) [6] and other systems [7, 8] to promote its application in sensors and actuators.

At present, BNT system is mainly about its piezoelectric properties. The piezoelectric properties of well-known BNT-based system were reported in **Table 1** [9]. However, because of its strong ferroelectricity, BNT-based ceramics also exhibit great potential in the fields of energy storage, energy conversion and pyroelectric applications. In this chapter, new bismuth sodium titanate ceramics were synthesized and characterized, the ferroelectric properties, phase transition behaviors under external fields and related applications, such as application in energy storage, energy conversion and pyroelectric detection were proposed.

2. New BNT-based ceramics for energy storage applications

BNT-based materials possess a superior potential for energy storage due to their high saturation polarization which originates from hybridization between the Bi *6p* and O *2p* orbitals. However, the pure BNT materials at room temperature own a ferroelectric perovskite structure with the polar *R3c* space group, usually exhibiting a saturated polarization loop with high remnant polarization, which is very unfavorable to obtain good energy storage performance [19]. Fortunately, the BNT materials can show an antiferroelectric-like behavior at around 200–320°C, which opens a door to the energy storage application of BNT-based materials, and the 200°C is identified as the depolarization temperature (T_d) of the BNT materials, which correspond with a peak in the temperature-dependent dielectric loss curve. The structure at this temperature range is still under debate. Zvirgzds et al. [20] proposed a rhombohedral (*R3c*)-tetragonal (non-polar *P4bm*) phase transition over the broad temperature range (255–400°C). Moreover, Schmitt et al. [21] suggested the phase transformation from non-polar *P4bm* phase to polar *R3c* phase under applied electric field accounted for the antiferroelectric-like characteristic, but this could not reasonably explain a large temperature hysteresis of different physical properties about the phase transition between 200 and 320°C. Dorcet et al. [22] revealed a modulated phase at 200–300°C through in-situ Transmission electron microscope (TEM) characterization, it was formed of *Pnma* orthorhombic sheets which are locally analogous to an antiferroelectric phase, and these sheets are twin boundaries between *R3c* ferroelectric domains. The phase structure evolution disclosed by Zvirgzds et al. [21] well matches the macroscopic physical properties of BNT materials during the heating process.

In 1947, Sakata et al. reported an antiferroelectric-like behavior in the 0.85BNT-0.15$SrTiO_3$ ceramics [23]. Later, Zhang et al. introduced (K, Na)NbO_3 (KNN) into BNT-$BaTiO_3$ (BT) ceramics to low the phase transition temperature and achieved the antiferroelectric-like behavior in BNT-BT-KNN ceramics with slanted polarization hysteresis loops at room temperature [24]. In 2011, Gao et al. [25] first investigated the energy storage properties of the BNT-BT-KNN system, the 0.89BNT-0.06BT-0.05KNN ceramics was chosen as the object, **Figure 1(a)** is the temperature-dependent dielectric properties of the 0.89BNT-0.06BT-0.05KNN ceramics, it can be seen that these ceramics showed much lower T_d compared with pure BNT materials, indicating the antiferroelectric-like behavior at a lower temperature. **Figure 1(b, c)**show the temperature dependence of polarization hysteresis loops of the 0.89BNT-0.06BT-0.05KNN ceramics under different electric fields. At 20°C, the polarization hysteresis loop was more of ferroelectric featured with coercive field E_c = 0.9 kV/mm and remnant polarization P_r = 6.2 μC/cm^2 under 6 kV/mm. At 110°C, the polarization hysteresis loop was more of an antiferroelectric-like feature with a pronounced shrinkage in both E_c and P_r compared with those at 20°C. The energy density as a function of the temperature of the 0.89BNT-0.06BT-0.05KNN ceramics are displayed in **Figure 1(d)**. An energy density of

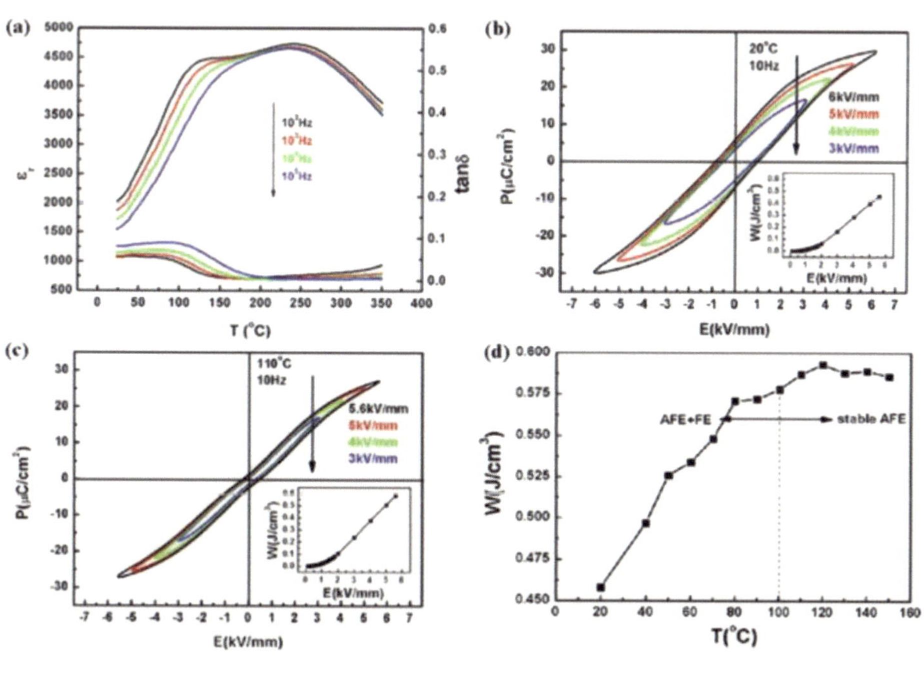

Figure 1.
Properties of 0.89BNT-0.06BT-0.05KNN ceramics: (a) the temperature-dependence of dielectric properties, (b) the polarization-electric field (P-E) loops at 20°C, (c) the P-E loops at 110°C, (d) the energy density as function of temperature [25].

around 0.59 J/cm^3 under 5.6 kV/mm at 10 Hz was obtained in 0.89BNT-0.06BT-0.05KNN ceramics from 100 °C to 150 °C, indicating high stability of temperature in the antiferroelectric-like region. Although the obtained energy density was very small and only existed above 100°C, this work is still meaningful because it inspires the further way for studying energy-storage in BNT-based materials. After, researches about the energy storage properties in BNT-based ceramics have been extensively reported.

Ren et al. [26] reported that the introduction of KNN would decrease the T_d of BNT-$BiAlO_3$ (BA) ceramics and the KNN content exerts a significant influence on the polarization hysteresis loops of BNT-BA-KNN materials as shown in **Figure 2b**. For 0.93 (0.96BNT-0.04BA)-0.07KNN ceramics, the T_d was below the room temperature as depicted in **Figure 2a** and these ceramics were more of antiferroelectric-like behavior. Ren et al. [26] also investigated the energy storage properties of 0.93 (0.96BNT-0.04BA)-0.07KNN ceramics, an energy storage density of 0.65 J/cm^3 was obtained under 8 kV/mm at room temperature, and these ceramics exhibited good stability of energy density as a function of temperature and frequency at 7 kV/mm, which can be seen from **Figure 2c,d**.

Due to the high energy loss of the antiferroelectric-like BNT-based materials, the BNT-based relaxor ferroelectrics have attracted more and more attention for energy storage and usually can show superior energy storage performance. In fact, by modifying composition and temperature in BNT-based systems, a normal or square P-E loop can transform into a slim P-E loop due to the occurring of an ergodic relaxor phase, which can be contributed to the energy storage properties. Wu et al. [27] focused on the energy storage characteristics of BNT-based relaxor ferroelectric ceramics and introduced $Sr_{0.85}Bi_{0.1}\square_{0.05}TiO_3$ (□ represents the A site vacancy) and $NaNbO_3$ into the BNT matrix as illustrated in **Figure 3**. The introduced A site

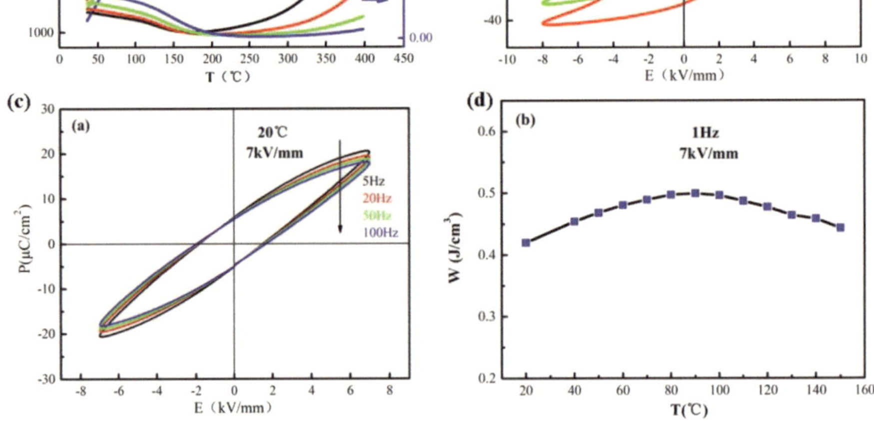

Figure 2.
(a) The temperature dependence dielectric properties of the 0.93 (0.96BNT-0.04BA)-0.07KNN ceramics. (b) The P-E loops of the (1-x)(0.96BNT-0.04BA)-xKNN ceramics. (c) The P-E loops of the 0.93 (0.96BNT-0.04BA)-0.07KNN ceramics under different frequencies. (d) The energy density of 0.93 (0.96BNT-0.04BA)-0.07KNN ceramics at different temperatures [26].

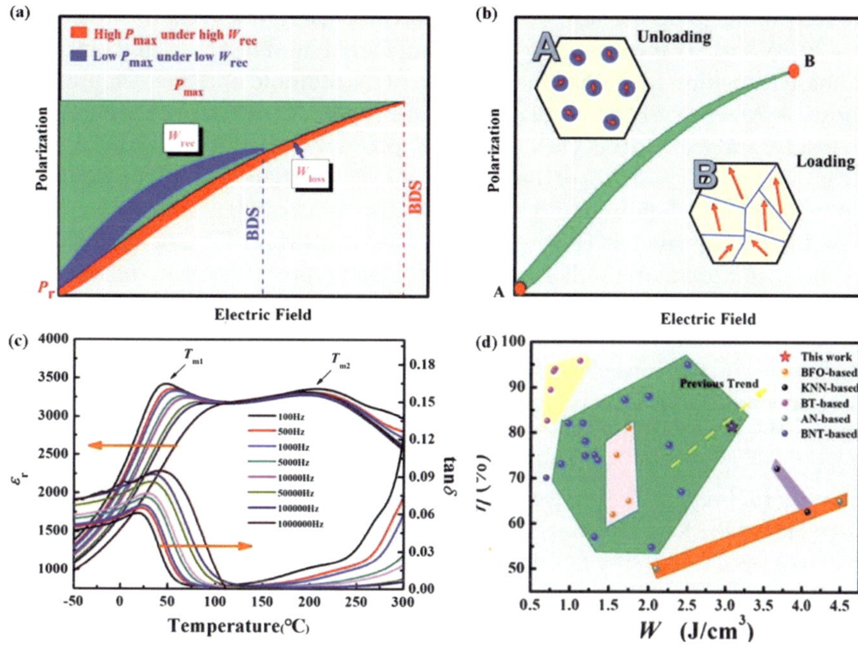

Figure 3.
(a) Schematic image showing energy storage properties under different electric fields. (b) Schematic image showing polar structure in relaxor ferroelectrics under loading and unloading electric fields. (c) The dielectric permittivity and loss as a function of temperature, measured at different frequencies from 0.1 kHz to 1 MHz for the 0.96 ($0.65Bi_{0.5}Na_{0.5}TiO_3$–$0.35Sr_{0.85}Bi_{0.1}TiO_3$)–$0.04NaNbO_3$ ceramics. (d) Energy efficiency versus recoverable energy density value for the 0.96 ($0.65Bi_{0.5}Na_{0.5}TiO_3$–$0.35Sr_{0.85}Bi_{0.1}TiO_3$)–$0.04NaNbO_3$ ceramics compared to other lead-free systems [27].

vacancy and Sr^{2+}, Nb^{5+} ions replaced the A- and B- sites ions respectively, which led to the stress mismatch and charge imbalance. These effects acted together to effectively form a local random field, which broke the long-range ordered structure of the dipole in the matrix and formed a weakly coupled polar nanodomain. Under the applied electric field, the modified ceramics exhibited a small hysteresis and a small remnant polarization, achieving high energy storage density (3.08/cm^3) and high energy storage efficiency (81.4%). To evaluate the practicability of the modi fied ceramic, energy storage performance test in a wide range of temperature and frequency found that the variations of its energy storage performance at RT ~ 100°C and 1 Hz ~ 100 Hz was less than 10%. The modified ceramics with excellent application prospects are excellent candidate materials for dielectric energy storage capacitors.

3. New BNT-based ceramics for pulse power supply application

Ferroelectric materials have an important application in pulse power supply due to their shock compression induced depolarization behavior [28]. At present, the main material systems studied are PZT52/48 piezoelectric ceramics [28], PZT95/5 ceramics [28, 29] and PIN-PMN-PT single crystals [30]. However, due to the toxicity of Lead, it is urgent to develop lead-free materials for high ferroelectric pulse power supply.

$Bi_{0.5}Na_{0.5}TiO_3$ (BNT) is explored as an alternative lead-free candidate for pulse power supply, in view of its high P_r, high breakdown strength E_b, low bulk density, and relatively high Curie temperature (T_c). Gao et al. [31] reported that the BNT can be fully depolarized by shock compression and generate a giant power output (3.04×10^8 W/kg). This power output is mainly attributed to a two-step polar-non-polar phase transition from rhombohedral to orthorhombic under shock pressure.

Figure 4 shows that BNT is polar phase and rhombohedral (space group R3c) at low pressure, and transforms via a first-order phase transition to a nonpolar phase (space group Pnma), which is orthorhombic and centrosymmetric. The electrical output of BNT from depoling under shock compression can be attributed to the ferroelectric-to-paraelectric (R3c – Pnma) phase transition. The energy output under shock compression in BNT is larger than that reported in other ferroelectric materials, mainly due to a first-order R-O phase transition under high dynamic pressure. This phase transition undergoes two steps, which correspond to the unit-cell shrinkage and O^{2-} ions chain rearrangement, respectively, as shown in **Figure 5**. These results will extend the potential application of the pressure induced depolarization effects and guide the application and development of BNT ferroelectric materials.

Liu et al. [32] report the pressure driven depolarization behavior in 0.97[(1-x) $Bi_{0.5}Na_{0.5}TiO_3$-x $BiAlO_3$]-0.03$K_{0.5}Na_{0.5}NbO_3$ (BNT-xBA-0.03KNN) ceramics. Particularly, with increasing hydrostatic pressure from 0 MPa to 495 MPa, the polarization of BNT-0.04 decreases from 30.7 μC/cm^2 to 8.2 μC/cm^2, decreasing ~73%. The observed depolarization effect is associated with the pressure induced polar ferroelectric -nonpolar relaxor phase transition. The results revealed that BNT-xBA-0.03KNN ceramics as promising lead-free candidates for energy conversion applications based on the pressure driven depolarization effect.

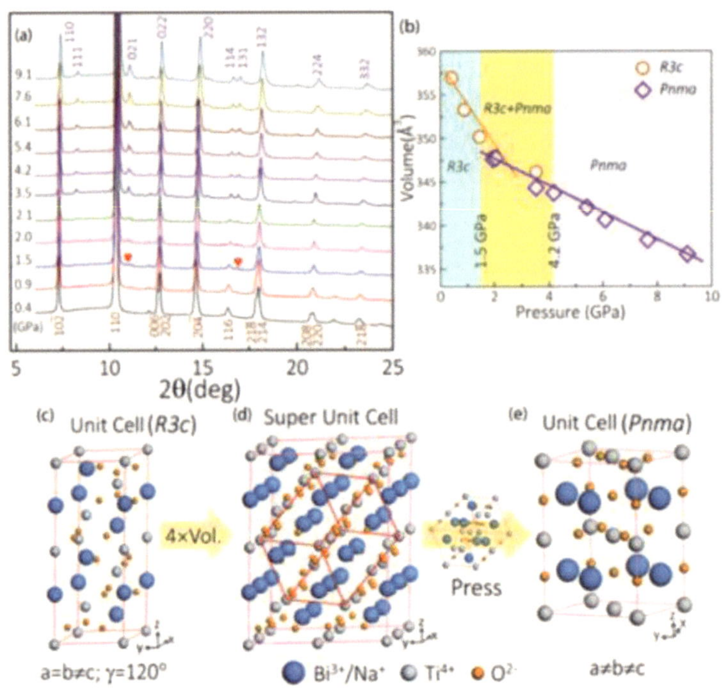

Figure 4.
Pressure dependence of the phase transition in BNT has been studied by the in situ synchrotron x-ray diffraction [31]. (a) The x-ray diffraction spectra of NBT ferroelectric materials at selected pressures. (b) The XRD peaks of the phase are marked by the red spades. The NBT is rhombohedral (R3c) structure at low pressure, and it changes into orthorhombic structure (Pnma) at high pressure. The normalized P-V curve of NBT according to the Z = 6, and (c)–(e) the schematic diagram of the structure phase transition during the phase transition.

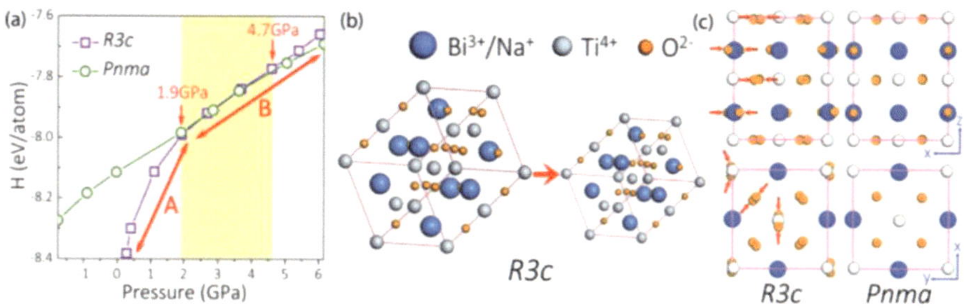

Figure 5.
First-principles calculations of the R3c and Pnma phases as a function of pressure [31]. (a) The enthalpy (H) calculated by first-principles simulation for R3c and Pnma phases at different pressures, respectively. The enthalpy change of R3c phase could be divided into two regions (A and B). When the pressure is below 1.9 GPa (region A), the enthalpy of R3c increases sharply due to the volume decreasing as shown in (b). When the pressure is above 1.9 GPa (region B), the enthalpy of R3c phase increases gently, which is mainly due to the O2-ions displacing following the red arrows in (c).

Figures 6 and 7 show the effect of hydrostatic pressure on the ferroelectric properties of BNT-0.01BA-0.03KNN and BNT-0.04BA-0.03KNN, respectively. It is clear that the P_r and P_m decrease monotonically with increasing pressures, which further confirms the increasing the instability of the long-range FE order and the energy barrier for the formation of FE domains under hydrostatic pressure conditions. In addition, the response of BNT-0.04BA-0.03KNN under pressure is more sensitive than that of BNT-0.01BA-0.03KNN. And the thermally induced depolarization is also stronger for BNT-0.04BA-0.03KNN. These phenomena should be related to their different depolarized temperature values. The ER phase exhibits smaller volume than the FE phase. Therefore, applying compressive pressure favors

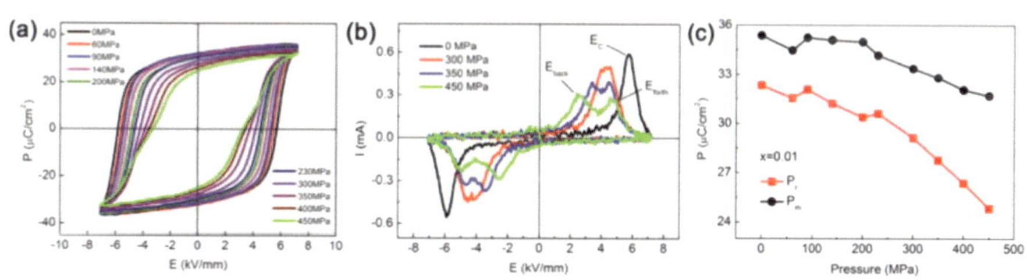

Figure 6.
(a) P-E loops and (b) I-E curves of BNT-0.01BA-0.03KNN ceramics under different hydrostatic pressures; and (c) the pressure dependence of maximum polarization and remanent polarization of BNT-0.01BA-0.03KNN ceramics [32].

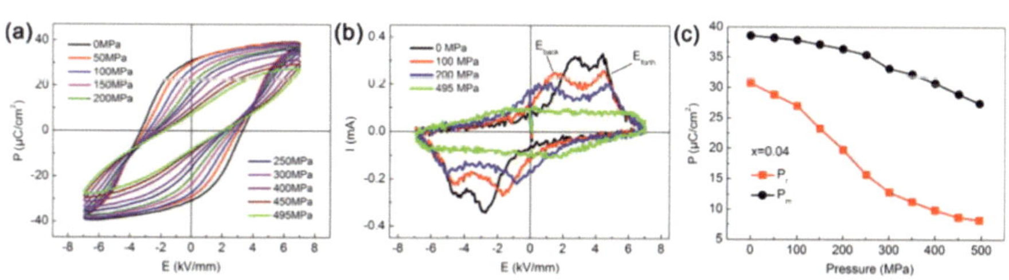

Figure 7.
(a) P-E loops and (b) I-E curves of BNT-0.04BA-0.03KNN ceramics under different hydrostatic pressures; (c) the pressure dependence of maximum polarization and remanent polarization of BNT-0.04BA-0.03KNN ceramics [32].

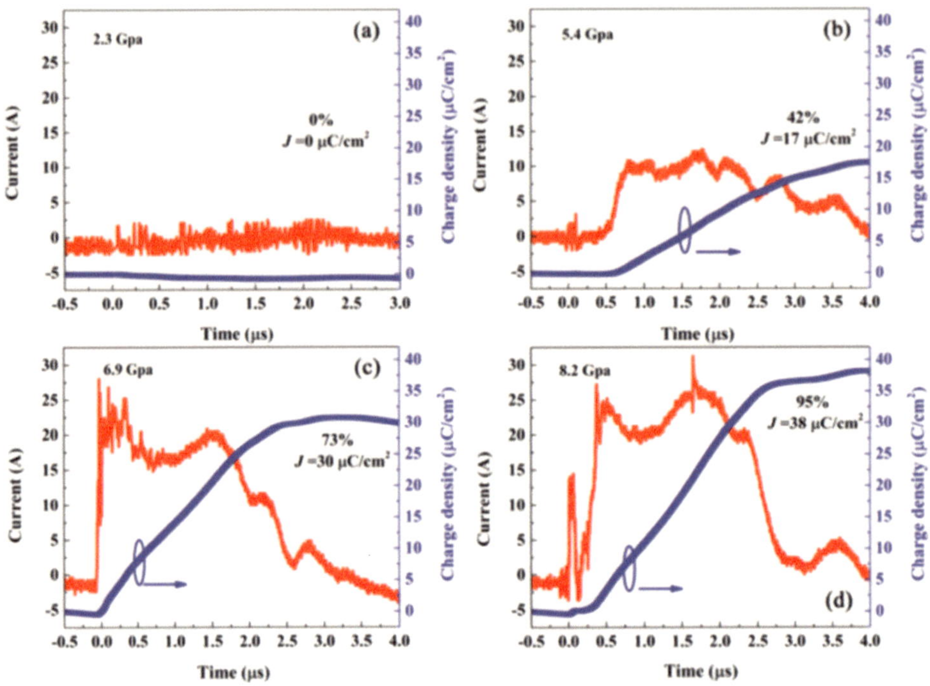

Figure 8.
Dynamic response behaviors of BNT-BA-0.01NN ceramics in a short-circuit mode under different shock pressures [34]. (a) 2.3 GPa, (b) 5.4 GPa, (c) 6.9 GPa, (d) 8.2 GPa.

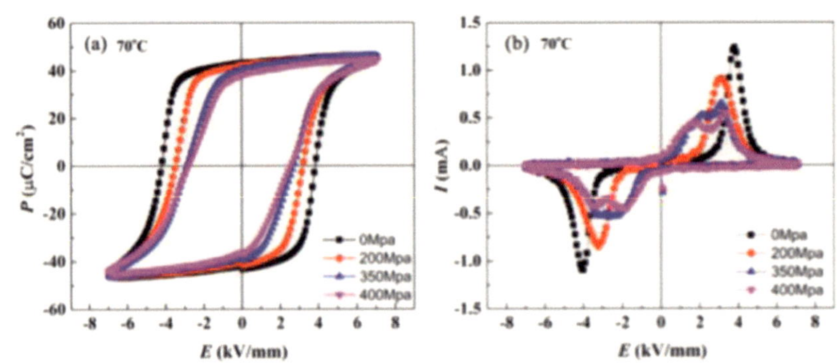

Figure 9.
Pressure-dependent (a) P-E and (b) I-E loops of unpoled BNT-BA-0.01NN ceramics at 70°C [34].

the FE-ER phase transition. This is quite similar to the case of Nb doped PZT95/5, in which pressure can drive the larger volume FE phase to transform into the smaller volume AFE phase.

Peng et al. [33, 34] report the depolarization behavior of lead-free ternary $0.99[0.98\ (Bi_{0.5}Na_{0.5})(Ti_{0.995}Mn_{0.005})\ O_3–0.02BiAlO_3]$-$0.01NaNbO_3$ (BNT-BA-0.01NN) ferroelectric ceramics under shock wave compression. Particularly, approximately complete depolarization under shock compression was observed in the poled BNT-BA-0.01NN ceramics, releasing a high discharge density J of 38 μC/cm^2. The released J was 96% of thermal-induced discharge density (~40 μC/cm^2). This discharge density J was 18% higher than that of PZT95/5 ceramics [29]. The shock-induced depolarization mechanism can be attributed to the ferroelectric-ergodic relaxor phase transition. These results reveal the BNT-based ceramics as promising candidates for pulsed power applications.

Figure 8 shows the BNT-based ceramics were almost completely depolarized, similar to PZT95/5 ceramics [29] and PIN-PMN-PT crystals [30], which indicate a similar depolarization mechanism, that is, a stress-induced phase transition. Although the released J in BNT-based ceramics is 26% lower than that obtained in PIN-PMN-PT crystals, the simple preparation methods together with environmental friendliness will be a benefit to their applications in the future. **Figure 9** unveils the possible shock-induced depolarization mechanism of BNT-BA-0.01NN ceramics. The pinched P-E loops gradually emerge and the sharp current peak splits into four peaks, indicating a pressure-induced FE-ER phase transition. It is suggested that applying compressive pressure favors the formation of the ER phase for its smaller volume.

4. New BNT-based ceramics for pyroelectric applications

At present, the most widely used intrinsic pyroelectric materials are perovskite-type lead-containing materials, such as $Pb(Zr, Ti)O_3$ (PZT), $PbTiO_3$ doped with Ca (PCT), $xPb(Mg_{1/3}Nb_{2/3})O_3$-(1-x)$PbTiO_3$ (PMN-PT) [35–40]. Over the last few decades, continuous efforts have been devoted to the development of lead-free pyroelectric ceramics. Many lead-free ceramics such as $Sr_{0.3}Ba_{0.7}Nb_2O_6$-based, (Ba, Ca)TiO_3-based, $CaBi_4Ti_4O_{15}$-based bismuth layer-structured and $Bi_{0.5}Na_{0.5}TiO_3$-based ceramics have been investigated [41–44]. Among them, BNT-based ceramics have been regarded as one of most promising alternative lead-free ceramics due to its high pyroelectric coefficient (p), high remnant polarization P_r (around 38μC/cm^2), high Curie temperature T_c (around 320°C), low-cost, and simple synthesis process. In recent decades, pyroelectric properties of BNT-based materials, including pyroelectric coefficient and detection rate, have been greatly improved. The pyroelectric coefficient of BNT-based lead-free pyroelectric materials has been comparable to commercial PZT [45–47]. However, the enhanced pyroelectric property is usually at the cost of degraded depolarization temperature (<150°C) and thermal stability, which are the hurdles to application. The BNT-based pyroelectric ceramics with low Td will depolarize partially or completely during the heat treatment (typically >100°C) processes, causing degradation of pyroelectric performance. Therefore, from the viewpoint of practical application, it is urgent for BNT-based materials to optimize their depolarization temperature, thermal stability and pyroelectric performance, thus further to promote their applications in infrared detection [48, 49].

4.1 BNT- BNN pyroelectric ceramics

(1-x)$(Bi_{0.5}Na_{0.5})TiO_3$-xBa$(Ni_{0.5}Nb_{0.5})O_3$ lead-free pyroelectric ceramics (abbreviated as (1-x)BNT-xBNN) were synthesized by a conventional solid-state reaction method [50], and the thermal stability and depolarization temperature is enhanced at the same time as the excellent pyroelectric performance is maintained. BNN is a compound with a mixed valence state at the b position, which can be solid-solved with BNT and expand a wide range of composition adjustment. The (1-x)BNT-xBNN take into account the advantages of b-position acceptor substitution and donor substitution. The effect of BNN content on phase structure, electrical properties and thermal stability was systematically studied. After the solid-state reaction of BNN, (1-x)BNT-xBNN exhibits enhanced pyroelectric performance with a high depolarization temperature. In addition, it can be exposed to temperature up to ~145°C with negligible deterioration of pyroelectric properties, showing excellent thermal stability.

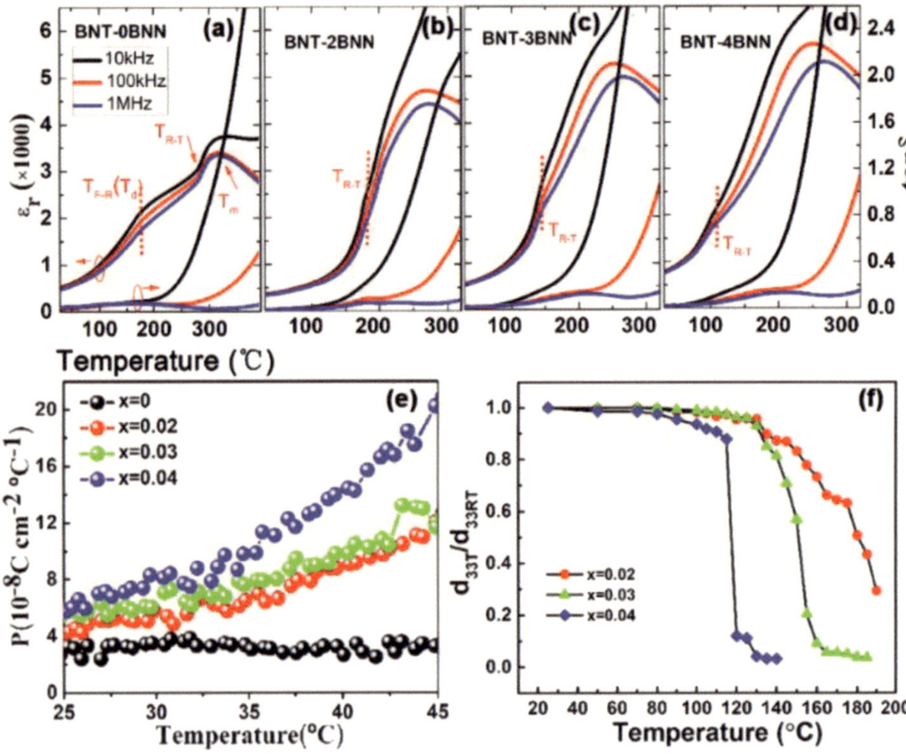

Figure 10.
Temperature dependence of dielectric constant (εr) and dielectric loss (tan δ) for poled (1-x)BNT-xBNN (a) x = 0, (b) x = 0.02, (c) x = 0.03, (d) x = 0.04 in frequency range between 10 kHz and 1 MHz; (e) Pyroelectric coefficient (p) of poled (1-x)BNT-xBNN ceramics as a function of temperature; (f) d33RT/d33T at room temperature after annealing at T_a [50].

The temperature-dependent properties of poled (1 – x)BNT-xBNN ceramics are displayed in **Figure 10a–d**. With the increasing BNN content, the Curie temperature Tc indicated by the maximum dielectric constant decreases and dielectric constant and dielectric loss of BNN decrease first and then increase. The minimum value of dielectric constant and dielectric loss occurs when the BNN content is 2%, which further improve the pyroelectric detection rate figure of merit. The depolarization temperature T_d can be characterized by the first anomal point of temperature dependent dielectric properties, and the content of 2% has the highest depolarization temperature. As shown in **Figure 10e**, after the increase of BNN, the room temperature p values rise from $3.01 \times 10^{-8}C/cm^2K$ of pure BNT to $5.94 \times 10^{-8}C/cm^2K$ of 0.96BNT-0.04BNN with the increasing addition of BNN, which gains advantage compared with many other lead-free ceramics. The *p* value of (1-x)BNT-xBNN ceramics increases sharply, which indicates that the (1-x)BNT-xBNN sample is sensitive to ambient temperature. In addition, it can be seen that the p value increases with the increasing temperature, which indicates that the (1 – x)BNT-xBNN samples are sensitive to the surrounding temperature. Besides, 0.98BNT-0.02BNN ceramics have the best thermal stability and it can withstand heat treatment at 145°C without depolarization (**Figure 10f**), which is attributed to the domain switching and phase transition.

4.2 BNT-BT pyroelectric ceramics

BNT-BT possesses a rich phase structure, which can be easily adjusted by varying the BT content. Because of the low tripartite-tetragonal transition barrier, the morphotropic phase boundary (MPB) of BNT-BT, located at where the BT content is approximately 6%, exhibits the best pyroelectric properties and has received much attention. But it is not advisable to blindly pursue a high pyroelectric coefficient. The improvement of pyroelectric performance is often at the cost of low depolarization temperature, which is not helpful to practical applications. However, it is found that the sample with high BT content is in the tetragonal phase, which brings a higher T_d than that of the tripartite, but there is no relevant report on the pyroelectric performance of high BT content.

Based on the above ideas, the tetragonal phase 0.8BNT-0.2BT lead-free pyroelectric material with high BT content was successfully prepared, and the microstructure, dielectric properties, pyroelectric properties, and thermal stability were studied [51]. Owing to its high T_d, this composition can endure high-temperature environment (180°C) for half hour with the value of p at room temperature remains ~90% of its initial value, demonstrating that the 0.8BNT-0.2BT samples show excellent thermal stability. Moreover, the T_d of the samples is up to ~209°C, which is far higher than that of the reported BNT-based, pyroelectric materials, and it is also comparable to the commercial PZT materials.

The pyroelectric properties of 0.8BNT-0.2BT pyroelectric ceramics between 25 and 70°C are investigated. With the increase of temperature, the pyroelectric performance shows an increasing trend, indicating that the material has good pyroelectric performance in a wide temperature range. Meanwhile, because the 0.8BNT-0.2BT sample has a low dielectric constant and dielectric loss, it will show a larger detection merit (**Figure 11a**). In order to study the depolarization temperature of the material, the dielectric thermo diagram of the sample was shown in **Figure 11b**. When the temperature rises to about 209°C, the dielectric constant of the sample suddenly increases with a dielectric loss peak appearing, indicating that this temperature is the

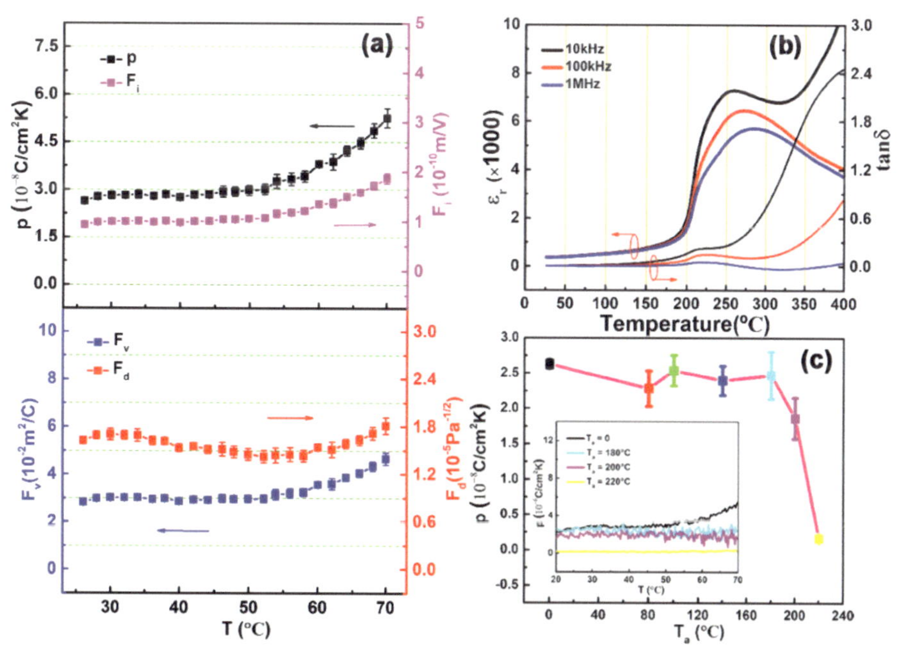

Figure 11.
(a) Pyroelectric coefficient (p) and figures of merit (F_i, F_v, F_d) as a function of temperature on heating during the range of RT to 70°C. The figures of merit are determined based on the values of p, C_V, ε_r, and tanδ; (b) temperature-dependent dielectric constant (ε_r) and dielectric loss (tanδ) of poled 0.8BNT–0.2BT ceramics; (c) pyroelectric coefficient at room temperature after annealing at Ta. The inset shows the temperature-dependent pyroelectric coefficient on heating after annealing at Ta [51].

depolarization temperature T_d. Notably, the depolarization temperature of reported BNT-based pyroelectric materials is generally lower than 180°C. The materials with high Td (209°C) and high pyroelectric coefficient discovered lay the foundation for the further development of lead-free pyroelectric materials. Moreover, it can be observed from **Figure 11c** that the room temperature pyroelectric coefficient of 0.8BNT-0.2BT maintains about 90% of the original data after being treated at 180°C, indicating that the material has good temperature stability and can withstand high temperature treatment up to180°C without pyroelectric performance loss.

4.3 BNT-BA-NN pyroelectric ceramics

A new ternary system 0.98BNT-0.02BA-xNN ceramic was obtained by solid solution of $NaNbO_3$ (NN) in the BNT-BA system and Mn element substitution

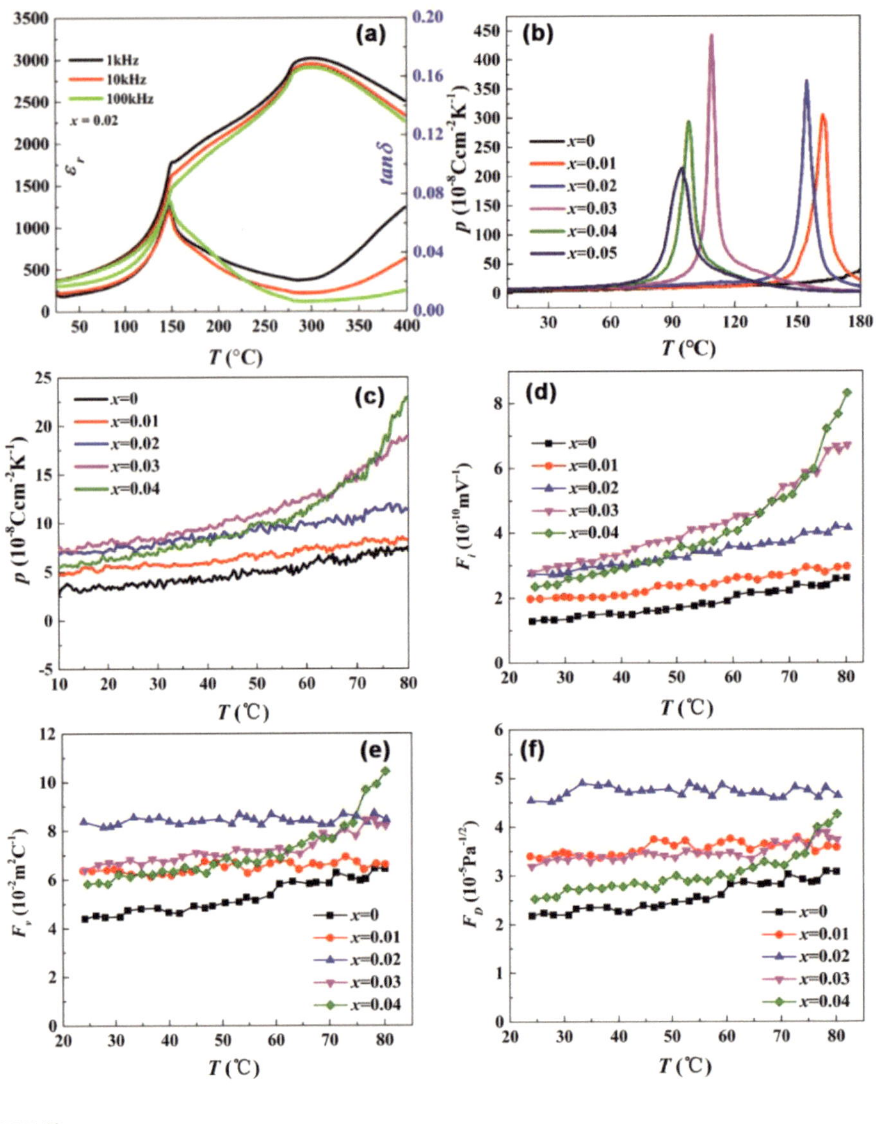

Figure 12.
(a) Temperature dependent dielectric constant (ε_r) and tangent loss (tanδ) of 0.98BNT-0.02BA-xNN ceramics; (b) temperature-dependent pyroelectric coefficient of 0.98BNT-0.02BA-xNN ceramics; (c) the temperature-dependent pyroelectric coefficient p of 0.98BNT-0.02BA-xNN ceramics in the temperature range from 10–80°C; merit (d) F_i, (e) F_v and (f) F_d of 0.98BNT-0.02BA-xNN ceramics measured at 1 kHz over the range of 20–80°C [52].

modification [52]. The NN solution significantly affect the microstructure, phase transition and pyroelectric properties of 0.98BNT-0.02BA-*x*NN ceramics. It was found that NN addition tends to reduce the rhombohedral phase while favoring the formation of the tetragonal phase. The compositions exhibit excellent pyroelectric performance. All components exhibit excellent ferroelectric properties at room temperature, and the Pr values are all higher than 35 μC/cm^2, of which the Pr of the x = 0.03 component is the largest, reaching 45 μC/cm^2.

Furthermore, the influence of NN solid solution on the relaxation characteristics and phase transition of BNT-BA-based ceramics was analyzed by testing the temperature-changing dielectric properties in **Figure 12a**. **Figure 12b** shows the change curve of the pyroelectric coefficient of 0.98BNT-0.02BA-xNN after polarization with temperature changing. The FE-RE phase transition occurs at T_d, corresponding to the sudden drop in the polarization intensity P_r. the largest peak appears at the composition x = 0.03, reaching 441.0 × 10^{-8} C/cm^{2}K, which is much larger than other BNT-based ceramics reported. As the NN content increases, the T_d continuously decreases. Notably, t the T_d of the x = 0.02 component is still as high as 155°C. It can be observed from **Figure 12c** that the introduction of NN significantly improves the room temperature pyroelectric coefficient. With the increase of NN content, the p under room temperature (25°C) first increases and then decreases, and the maximum value is obtained at x = 0.03 (p = 8.45 × 10^{-8} C/cm^{2}K), which improved about 54% compared to the matrix (x = 0, p = 3.87 × 10^{-8} C/cm^{2}K). Moreover, the optimal figure of merit (FOMs) at room temperature were obtained at x = 0.02 with F_i = 2.66 × 10^{-10} m/V, F_v = 8.07 × 10^{-2} m^2/C, and F_d = 4.22 × 10^{-5} Pa$^{-1/2}$ (**Figure 12d–f**). Furthermore, the compositions with $x \leq 0.02$ possess relatively high depolarization temperature (≥155°C). Those results unveil the potential of 0.98BNT-0.02BA-*x*NN ceramics for infrared detector applications.

5. Conclusion

Due to its strong ferroelectric properties, BNT-based ceramics exhibit great potential in the fields of energy storage, pulsed power supply and pyroelectric applications. In this chapter, new bismuth sodium titanate ceramics were synthesized and characterized via composition modifications, the ferroelectric properties, phase transition behaviors under external fields and related applications were proposed in this chapter. To detail, BNT-BT-KNN, BNT-BA-KNN, and BNT-SBT-NN ceramics for energy storage application, BNT, BNT-BA-KNN, and BNT-BA-NN ceramics for pulsed power supply, as well as BNT-BNN, BNT-BT, and BNT-BA-NN for pyroelectric detection application were presented.

Acknowledgements

The authors would like to thank the financial support by Youth Innovation Promotion Association, CAS (Grant No. 2017296), National Natural Science Foundation of China (NSFC) (Grant No. 51872312), and Natural Science Foundation of Shanghai (Grant NO.18ZR1444900).

Conflict of interest

The authors declare no conflict of interest.

Author details

Hengchang Nie[1], Genshui Wang[1,2,3] and Xianlin Dong[1,2,3]*

1 Key Laboratory of Inorganic Functional Materials and Devices, Shanghai Institute of Ceramics, Chinese Academy of Sciences, Shanghai, China

2 Hangzhou Institute for Advanced Study, UCAS, Hangzhou, China

3 Center of Materials Science and Optoelectronics Engineering, University of Chinese Academy of Sciences, Beijing, China

*Address all correspondence to: xldong@mail.sic.ac.cn

References

[1] GA Smolenskii, VA Isupov, AI Agranovskaya, NN Krainik. Ferroelectrics with diffuse phase transition. Soviet Physics Solid State. 1961;2:2651.(doi is not available)

[2] YM Li, W Chen, Q Xu, J Zhou, HJ Sun, R Xu. Dielectric and piezoelectric properties of lead-free $(Na_{0.5}Bi_{0.5})TiO_3$-$NaNbO_3$ ceramics. Materials Science & Engineering B. 2004;112:5-9. DOI:10.1016/j.mseb.2004.04.019.

[3] H Nagata, T Shinya, Y Hiruma, T Takenaka. Developments in Dielectric Materials and Electronic Devices, Ceramic Transactions. 2004;167:213-221.

[4] PK Panda. Review: environmental friendly lead-free piezoelectric materials, Journal of Materials Science. 2009;44:5049-5062. DOI: 10.1007/s10853-009-3643-0.

[5] T Takenaka, K Maruyama, K Sakata. $(Bi_{1/2}Na_{1/2})TiO_3$-$BaTiO_3$ system for lead-free piezoelectric ceramics. Japanese Journal of Applied Physics. Part 1, 1991;30: 2236-2239. DOI: 10.1143/JJAP.30.2236.

[6] J Rödel, W Jo, K Seifert, EM Anton, T Granzow, D Damjanovic. Perspective on the development of lead-free piezoceramics. Journal of American Ceramic Society. 2009;92:1153-1177. DOI: 10.1111/j.1551-2916.2009.03061.x.

[7] CR Zhou, XY Liu. Dielectric and piezoelectric properties of bismuth-containing complex perovskite solid solution of $Bi_{1/2}Na_{1/2}TiO_3$-$Bi(Mg_{2/3}Nb_{1/3})O_3$. Journal of Materials Science. 2008;43:1016. DOI:10.1007/s10853-007-2246-x.

[8] S Mahboob, G Prasad, GS Kumar. Impedance spectroscopy and conductivity studies on B site modified $(Na_{0.5}Bi_{0.5})(Nd_xTi_{1-2x}Nb_x)O_3$ ceramics. Journal of Materials Science. 2007; 42:10275. DOI:10.1007/s10853-006-1122-4.

[9] P. K. Panda & B. Sahoo, PZT to Lead Free Piezo Ceramics: A Review, Ferroelectrics, 2015; 474:1, 128-143, DOI: 10.1080/00150193.2015.997146.

[10] Y Hiruma, R Aoyagi, H Nagata, T Takenaka, Ferroelectric and Piezoelectric Properties of $(Bi_{1/2}K_{1/2})TiO_3$ Ceramics. Japanese Journal of Applied Physics. Part 1, 2005; 44: 5040-5044. DOI: 10.1143/JJAP.44.5040.

[11] Z Yang, B Liu, L Wei, Y Hou. Structure and Electrical Properties of $(1-x)Bi_{0.5}Na_{0.5}TiO_3$–$xBi_{0.5}K_{0.5}TiO_3$ Ceramics Near Morphotropic Phase Boundary. Materials Research Bulletin. 2008;43:81-89. DOI: 10.1016/j.materresbull.2007.02.016.

[12] YR Zhang, JF Li, BP Zhang, Enhancing Electrical Properties in NBT–KBT Lead-Free Piezoelectric Ceramics by Optimizing Sintering Temperature, Journal of the American Ceramic Society. 2008;91:2716-2719. DOI: 10.1111/j.1551-2916.2008.02469.x.

[13] H Hu, M Zhu, F Xie, N Lei, J Chen, Y Hou, H Yan. Effect of Co_2O_3 Additive on Structure and Electrical Properties of $85(Bi_{1/2}Na_{1/2})TiO_3$-$12(Bi_{1/2}K_{1/2})TiO_3$-$3BaTiO_3$ Lead-Free Piezoceramics. Journal of the American Ceramic Society. 2009;92:2039-2045. DOI: 10.1111/j.1551-2916.2009.03183.x.

[14] HY Tian, KW Kwok, HLW Chan, C E Buckley. The effects of CuO-doping on dielectric and piezoelectric properties of $Bi_{0.5}Na_{0.5}TiO_3$–$Ba(Zr,Ti)O_3$ lead-free ceramics. Journal of Materials Science. 2007;42:9750-9755. DOI: 10.1007/s10853-007-2005-z.

[15] DQ Xiao, DM Lin, JG Zhu, P Yu. Investigation on the design and

synthesis of new systems of BNT-based lead-free piezoelectric ceramics, Journal of Electronceramics. 2006; 16: 271-275. DOI: 10.1007/s10832-006-9863-7.

[16] D Lin, K W. Kwok. Ferroelectric and piezoelectric properties of $[(Bi_{0.98}La_{0.02}Na_{1-x}Li_x)_{0.5}]_{0.94}Ba_{0.06}TiO_3$ lead-free ceramics. Journal of Materials Science. 2009;44:4953-4958.

[17] SH Choy, XX Wang, HLW Chan, CL Choy. Electromechanical and ferroelectric properties of $(Bi_{1/2}Na_{1/2})TiO_3$-$(Bi_{1/2}K_{1/2})TiO_3$-$(Bi_{1/2}Li_{1/2})TiO_3$-$BaTiO_3$ lead-free piezoelectric ceramics for accelerometer application. Applied Physics A. 2007; 89: 775-781. DOI: 10.1007/s00339-007-4170-y.

[18] D Lin, Q Zheng, C Xu, K W. Kwok. Structure, electrical properties and temperature characteristics of $Bi_{0.5}Na_{0.5}TiO_3$–$Bi_{0.5}K_{0.5}TiO_3$–Bi0.5$Li_{0.5}TiO_3$ lead-free piezoelectric ceramics. Applied Physics A, 2008; 93: 549-558. DOI: 10.1007/s00339-008-4667-z.

[19] SB Vakhrushev, BG Ivanitskij, BE Kvyatkovskij, AN Majstrenko, RS Malysheva, NM Okuneva, NM Parfenova: Neutron scattering studies of the structure of sodium bismuth titanate. Soviet Physics Solid State, 1983; 25: 1504. (doi is not available)

[20] J. A. Zvirgzds, P. P. Kapostin, J. V. Zvirgzde, T. V. Kruzina. X-ray study of phase transitions in efrroelectric $Na_{0.5}Bi_{0.5}TiO_3$. Ferroelectrics, 1982; 40(1): 75-77. DOI: 10.1080/00150198208210600.

[21] Ljubomira Ana Schmitt, Hans-Joachim Kleebe: Single grains hosting two space groups—a transmission electron microscopy study of a lead-free ferroelectric. Functional Materials Letters. 2010;3:55-58. DOI: 10.1142/S1793604710000920.

[22] V. Dorcet, G. Trolliard, P. Boullay : Reinvestigation of phase transitions in Na0. 5$Bi_{0.5}TiO_3$ by TEM. Part I: First order rhombohedral to orthorhombic phase transition. Chemistry of Materials. 2008;20:5061-5073. DOI: 10.1021/cm8004634.

[23] Koichiro Sakata, Yoichiro Masuda : Ferroelectric and antiferroelectric properties of $(Na_{0.5}Bi_{0.5})TiO_3$-$SrTiO_3$ solid solution ceramics. Ferroelectrics, 1974; 7(1): 347-349. DOI: 10.1080/00150197408238042

[24] ST Zhang, Alain Brice Kounga, Emil Aulbach : Giant strain in lead-free piezoceramics $Bi_{0.5}Na_{0.5}TiO_3$–$BaTiO_3$–$K_{0.5}Na_{0.5}NbO_3$ system. Applied Physics Letters. 2007;91:112906. DOI: 10.1063/1.2783200.

[25] F Gao, X Dong, C Mao, W Liu, H Zhang, L Yang, F Cao, and G Wang. Energy-storage properties of $0.89Bi_{0.5}Na_{0.5}TiO_3$-$0.06BaTiO_3$-$0.05K_{0.5}Na_{0.5}NbO_3$ lead-free anti-ferroelectric ceramics. Journal of the American Ceramic Society, 2011;94: 4382-4386. DOI: 10.1111/j.1551-2916.2011.04731.x.

[26] WJ Ren, Phase transition behaviors and energy-storage properties of BNT-BA-KNN ceramics, [thesis]. University of Chinese Academy of Sciences, 2016.

[27] YC Wu, YZ Fan, NT Liu, P Peng, MX Zhou, SG Yan, F Cao, XL Dong, GS Wang. Enhanced energy storage properties in sodium bismuth titanate-based ceramics for dielectric capacitor applications. Journal of Materials Chemistry C. 2019;7:6222-6230. DOI: 10.1039/C9TC01239G.

[28] S I Shkuratov, J Baird, V G Antipov, E F Talantsev, H R Jo, J C Valadez, C S Lynch. Depolarization mechanisms of $PbZr_{0.52}Ti_{0.48}O_3$ and $PbZr_{0.95}Ti_{0.05}O_3$ poled ferroelectrics under high strain rate loading. Applied Physics Letters. 2014;104: 212901. DOI: 10.1063/1.4879545.

[29] R E Setchell. Shock wave compression of the ferroelectric ceramic $Pb_{0.99}(Zr_{0.95}Ti_{0.05})_{0.98}Nb_{0.02}O_3$: depoling currents. Journal of Applied Physics, 2005;97:013507. DOI: 10.1063/1.1828215.

[30] SI Shkuratov, J Baird, V G Antipov, E F Talantsev, J B Chase, W Hackenberger, J Luo, H R Jo, C S Lynch. Ultrahigh energy density harvested from domain-engineered relaxor ferroelectric single crystals under high strain rate loading. Scientific Reports. 2017;7:46758. DOI: 10.1038/srep46758.

[31] ZP Gao, W Peng, B Chen, Simon A.T.Redfern, K Wang, BJ Chu, Q He, Y Sun, XF Chen, HC Nie, W Deng, LK Zhang, HL He, GS Wang, XL Dong. Giant power output in lead-free ferroelectrics by shock-induced phase transition. Physical Review Materials. 2019;3:2475-9953. DOI:10.1103/physrevmaterials.3.035401.

[32] Z Liu, W Ren, H Nie, P Peng, Y Liu, X Dong, F Cao, G Wang, Pressure driven depolarization behavior of $Bi_{0.5}Na_{0.5}TiO_3$ based lead-free ceramics. Applied Physics Letters, 2017;110: 212901. DOI:10.1063/1.4984088.

[33] P Peng, H Nie, W Guo, F Cao, G Wang, X Dong. Pressure-induced ferroelectric-relaxor phase transition in $(Bi_{0.5}Na_{0.5})TiO_3$-based ceramics. Journal of the American Ceramic Society. 2018.102:2569-2577. DOI:10.1111/jace.16069.

[34] P Peng, H Nie, G Wang, Z Liu, F Cao, X Dong. Shock-driven depolarization behavior in BNT-based lead-free ceramics. Applied Physics Letters. 2018. 113, 082901. DOI:10.1063/1.5045392.

[35] F Zhuo, Q Li, J Gao, Y Wang, Q Yan, Z Xia, Y Zhang, X Chu. Structural phase transition, depolarization and enhanced pyroelectric properties of $(Pb_{1-1.5x}La_x)(Zr_{0.66}Sn_{0.23}Ti_{0.11})O_3$ solid solution. Journal of Materials Chemistry C. 2016; 29:7110-7118. DOI: 10.1039/c6tc01326k.

[36] YX Tang, XY Zhao, XQ Feng, WQ Jin, HS Luo. Pyroelectric properties of 111-oriented $Pb(Mg_{1/3}Nb_{2/3})O_3$-$PbTiO_3$ crystals. Applied Physics Letters. 2005; 86:082901. DOI: 10.1063/1.1865337.

[37] YX Tang, LH Luo, YM Jia, HS Luo, XY Zhao, HQ Xu, D Lin, JL Sun, XJ Meng, JH Zhu, M Es-Souni. Mn-doped $_{0.71}Pb(Mg_{1/3}Nb_{2/3})O_3$-$0.29PbTiO_3$ pyroelectric crystals for uncooled infrared focal plane arrays applications. Applied Physics Letters. 2006;89: 162906. DOI: 10.1063/1.2363149.

[38] S Zhang, F Li. High performance ferroelectric relaxor-$PbTiO_3$ single crystals: Status and perspective. Journal of Applied Physics. 2012;111:031341. DOI: 10.1063/1.3679521.

[39] Y Li, Y Tang, J Chen, X Zhao, L Yang, F Wang, Z Zeng, H Luo. Enhanced pyroelectric properties and thermal stability of Mn-doped $0.29Pb(In_{1/2}Nb_{1/2})O_3$-$0.29Pb(Mg_{1/3}Nb_{2/3})O_3$-$0.42PbTiO_3$ single crystals. Applied Physics Letters. 2018; 112: 172901. DOI: 10.1063/1.5024286.

[40] M Kobune, A Mineshige, S Fujii, Y Maeda. Preparation and pyroelectric properties of Mn-modified (Pb, La)(Zr, Ti)O_3 (PLZT) ceramics. Japanese Journal of Applied Physics Part 1-Regular Papers Short Notes & Review Papers. 1997:36:5976-5980. DOI: 10.1143/JJAP.36.5976.

[41] H Chen, S Guo, X Dong, F Cao, C Mao, G Wang, $Ca_xSr_{0.3-x}Ba_{0.7}Nb_2O_6$ lead-free pyroelectric ceramics with high depoling temperature. Journal of Alloys and Compounds. 2017; 695: 2723-2729. DOI: 10.1016/j.jallcom.2016.11.192.

[42] KS Srikanth, S Patel, S Steiner, Vaish R. Pyroelectric signals in (Ba,Ca)TiO_3-xBa(Sn,Ti)O_3 ceramics: A viable alternative for lead-based ceramics.

Scripta Materialia. 2018;146:146-149. DOI: 10.1016/j.scriptamat.2017.11.027.

[43] Y Tang, Z Shen, S Zhang, T R. Shrout, Improved Pyroelectric Properties of CaBi4Ti4O15 Ferroelectrics Ceramics by Nb/Mn Co-Doping for Pyrosensors. Journal of the American Ceramic Society. 2016;99:1294-1298. DOI: 10.1111/jace.14075.

[44] F Guo, B Yang, S Zhang, F Wu, D Liu, P Hu, Y Sun, D Wang, W Cao. Enhanced pyroelectric property in (1-x)$(Bi_{0.5}Na_{0.5})$ $TiO_{3\text{-}x}Ba(Zr_{0.055}Ti_{0.945})O_3$: Role of morphotropic phase boundary and ferroelectric-antiferroelectric phase transition. Applied Physics Letters. 2013; 103: 182906. DOI: 10.1063/1.4828675.

[45] AM Balakt, CP Shaw and Q Zhang. Giant pyroelectric properties in La and Ta co-doped lead-free $0.94Na_{0.5}Bi_{0.5}TiO_3$-$0.06BaTiO_3$ ceramics. Journal of Alloys and Compounds. 2017;709: 82-91. DOI: 10.1016/j.jallcom.2017.03.143

[46] AM Balakt, CP Shaw and Q Zhang. The decrease of depolarization temperature and the improvement of pyroelectric properties by doping Ta in lead-free $0.94Na_{0.5}Bi_{0.5}TiO_3$-$0.06BaTiO_3$ ceramics. Ceramics International. 2017;43: 3726-33. DOI: 10.1016/j.ceramint.2016.12.004.

[47] AM Balakt, CP Shaw and Q Zhang. Enhancement of pyroelectric properties of lead-free 0.94 Na0.5Bi0.5TiO3-0.06 $BaTiO_3$ ceramics by La doping. Journal of the European Ceramic Society. 2017;37:1459-66. DOI: 10.1016/j.jeurceramsoc.2016.12.021.

[48] M Kobune, A Mineshige, S Fujii and Y Maeda. Preparation and pyroelectric properties of Mn-modified (Pb, La) $(Zr, Ti)O_3$ (PLZT) ceramics. Japanese Journal of Applied Physics Part 1-Regular Papers Short Notes & Review Papers. 1997;36:5976-80. DOI: 10.1143/JJAP.36.5976.

[49] Y Li, Y Tang, J Chen, X Zhao, L Yang, F Wang, Z Zeng and H Luo. Enhanced pyroelectric properties and thermal stability of Mndoped $0.29Pb(In_{1/2}Nb_{1/2})O_3$-$0.29Pb(Mg_{1/3}Nb_{2/3})O_3$-$0.42PbTiO_3$ single crystals. Applied Physics Letters. 2018; 112:172901. DOI: 10.1063/1.5024286.

[50] J Jia, S Guo, F Cao, S Yan, C Yao, X Dong, G Wang. The improved pyroelectric properties, high depolarization temperature and excellent thermal stability in lead-free $(1-x)(Bi_{0.5}Na_{0.5})TiO_3$-$xBa(Ni_{0.5}Nb_{0.5})O_3$ ceramics. Materials Research Express. 2019;6:046308. DOI: 10.1088/2053-1591/aafcd0.

[51] J Jia, S Guo, S Yan, F Cao, C Yao, X Dong, G Wang. Simultaneous large pyroelectric response and high depolarization temperature in sodium bismuth titanate-based perovskites. Applied Physics Letters. 2019;114:032902. DOI: 10.1063/1.5063318.

[52] P Peng, H Nie, Z Liu, F Cao, G Wang, X Dong. Enhanced pyroelectric properties in $(Bi_{0.5}Na_{0.5})TiO_3$–$BiAlO_3$–$NaNbO_3$ ternary system lead-free ceramics. Journal of the American Ceramic Society. 2018;101:4044-4052. DOI: 10.1111/jace.15568.

Section 4

Properties Studies

Chapter 7

Investigation of Structural, Magnetic and Electrical Properties of Chromium Substituted Nickel Ceramic Nanopowders

Rapolu Sridhar, D. Ravinder, J. Laxman Naik, K. Vijaya Kumar, N. Maramu and S. Katlakunta

Abstract

Nano-ceramic of $NiCr_xFe_{2-x}O4$ ($0.1 \leq x \leq 1.0$) ferrites were synthesized by citrate-gel auto combustion method. The structural parameter such as lattice parameter, X-ray density, bulk density and porosity variations with Cr doping were studied. The average crystallite size is in the range 8.5–10.5 nm. The surface morphology and elemental analysis was studied with SEM (EDAX) spectrum and the structural information analyzed with FTIR spectra. Magnetic properties were discussed with Cr^{3+}ion concentration. Electrical parameters like dc resistivity and drift mobility were reported with function of temperature and dopent concentration from room temperature to well beyond Curie temperature and explained with hopping mechanism between $Fe^{2+} \leftrightarrow Fe^{3+}$ ions. The activation energies in ferri and para magnetic regions were investigated. Dielectric parameters like dielectric constant, dielectric loss and ac conductivity were investigated variation with frequency and composition.

Keywords: nano-ceramics, citrate-gel auto combustion technique, structural parameters, magnetic properties, electrical properties, dielectric properties

1. Introduction

In the field of science and technology nanoscale ceramics are play an very important role because the nanostructure ceramic material are show evidence of novel properties and all other properties together are different than that of their bulk ceramic materials [1, 2]. From the past few years researchers are focus to result new materials used in cooling and energy conversion system [3]. Hence, nano-sized ceramic materials have the most important because nano-sized material structure, magnetic and electrical properties studies and their interrelation is still incomplete.

Among magnetic ceramic materials, spinal structure ceramics are most significant materials for research in fundamental electronic components due to their tremendous magnetic and electrical properties [4, 5]. The spinal structure ceramics are quite stable and they are important in a wide range of technological applications like magnetic recording, sensors, magnetic resonance imaging, transformer etc. [6, 7]. These materials are used as high frequency magnetic materials, microwave applications and data storage devices due to its high electrical resistivity values [8]. Nano-sized spinel ceramics are important dielectric materials in high frequency applications due to their high resistivity, low magnetic and dielectric losses [9, 10]. These materials are posses with high dielectric constant at low frequencies (10^2-10^5 Hz) with low conductivity. Hence these are use in microwave applications and several devices like high frequency transformer cores, resonators, switches and phase shifter etc. [11, 12].

Presently, many researchers are focused on preparation the nanoscale dimension ceramics because they have incredible changes in their properties, therefore these ceramics are largely synthesized in nanoscale for new and improved properties such as low saturation magnetization, enhanced coercivity etc. [13, 14]. Since last few years, numbers of wet chemical methods are developed to synthesis nano-sized spinel ceramic materials and these methods advantage over the physical methods to prepare homogeneity materials. Several chemical methods are used to prepare the material samples like electro-deposition [15], co-precipitation [16], micro emulsion method [17], glyoxylate precursor technique [18], hydrothermal technique [19], reverse micelle method [20], Solid state reaction [21], Sol–gel method [22] and Citrate-gel technique [23]. In most recent research work, Citrate-gel auto combustion technique attained immense significance, since it provides the pure and homogeneous nanoparticles and cost is low as compared to other chemical processes [24].

Among the spinel ceramics, Ni nano ceramic is a soft ferrimagnetic material due to its nanocrystalline nature and it useful for novel applications like gas sensing [25] and drug delivery [26]. In the recent years nano-sized Ni ceramic and substituted Ni nano ceramic extensively studied due to having high electrical resistance, low cost, high mechanical hardness, and eddy current losses low etc. [27, 28]. These material scientifically interest because of its promising and interesting applications in microwave devices, color imaging, magnetic refrigerators and high density recording devices [29]. Particularly trivalent like Cr is substituted in it, it's fascinating effect on electromagnetic and dielectric properties of nickel nano ceramics. From a review of earlier work it is evident that very less attention has been paid to study of structural, magnetic, electrical and dielectric properties in systematic manner.

2. Experimental technique

Nano ceramics chemical composition $NiCr_XFe_{2-X}O_4$ (X = 0.1, 0.3, 0.5, 0.7, 0.9 and 1.0) have been prepared with Nickel nitrate, Chromium nitrate, Ferric nitrate, Citric acid and Ammonia as raw materials at low temperature by citrate-gel auto combustion technique.

$$\begin{gathered}Ni\,(NO_3)_2 6H_2O + 2Fe\,(NO_3)_2 9H_2O + C_6H_8O_7.H_2O \\ \rightarrow NiFe_2O_4 + 20H_2O + 3\,N_2 + 6CO_2 + 2.5O_2\end{gathered}$$

Calculated quantities of the molar quantity AR grade of metal nitrates described in the starting materials by sensitive digital balance and they were dissolved in deionized double distilled water in separate in 100 ml beakers. The metal nitrates solutions are mixed together in beaker and added the citric acid in 1:3 ration of nitrate to citric acid; it was systematically stirred placed on magnetic hot plate stirrer. To this nitrate-citrate mixture added the ammonia to maintain the p^H to 7. The homogeneous solution was heated at about 80°C and to attain a thick gel. Further heated the gel maintained at a temperature of 180–200°C. Finally, water molecules were removed from the mixture, the viscous gel then began frothing.

The gel starts auto combustion reaction with flameless express in the most up-to-date regions of the beaker and it spread from the bottom to the top, the reaction was finished in a couple of minutes produced the structure with branched tree in an dark gray loose-fitting product. Finally the as-burnt ferrite powders were grained by using Agate Mortar and Pestle then calcined in programmable furnace through "Eurotherm" programmer-cum-controller at 700°C for 5 hr. The calcined ferrite powders were again grained by using Agate Mortar and Pestle to obtain a better crystallization and homogeneous distribution in the spinel. The step by step procedure for the synthesis of Nano ceramics is shown in the form of flow chart in **Figure 1**. For electrical and dielectric measurement pellet was prepared with KBr hydraulic press (*Model: M-15*) in 2–3 mm thickness and 10 mm diameter size.

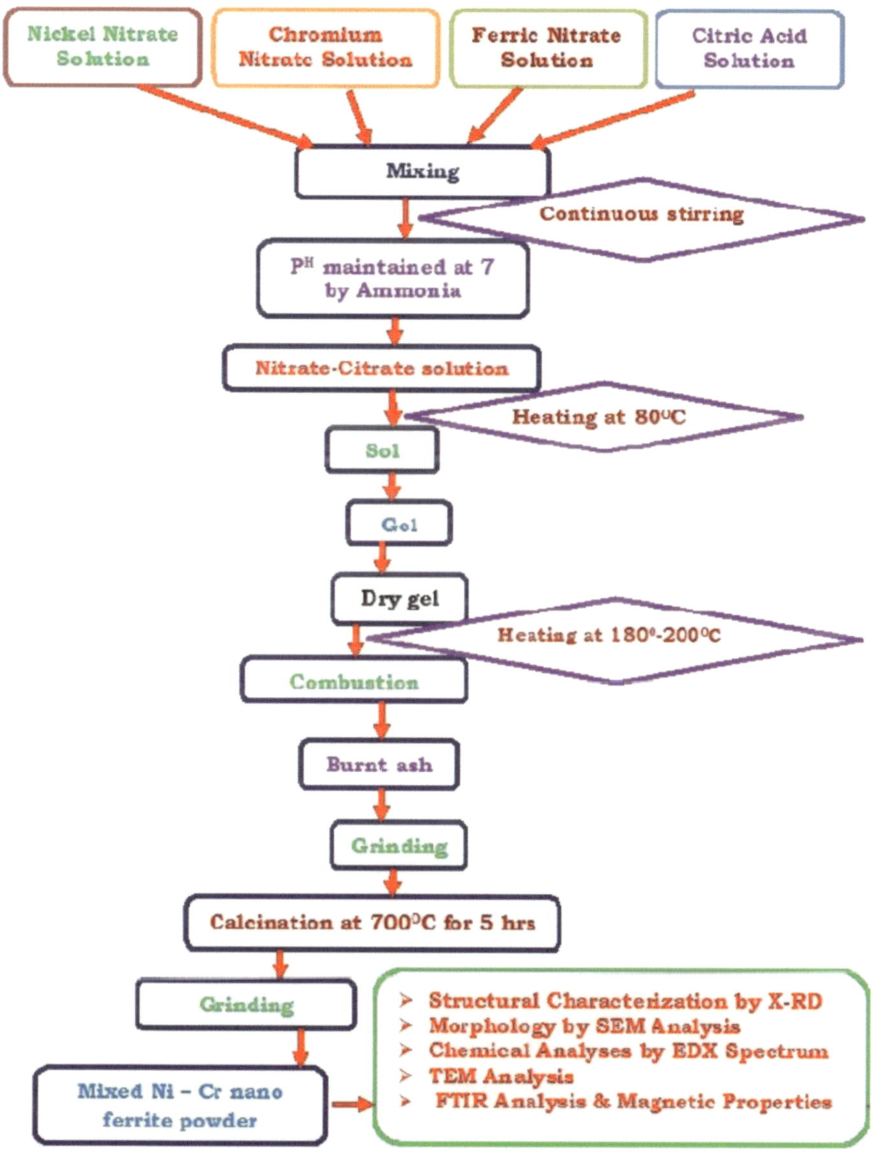

Figure 1.
Flow chart for the synthesis of Ni-Cr nano ferrites using Citrate-gel auto combustion technique.

The X-ray diffraction analysis was studied by Diffractomerter Bruker (Karlsruhe, German) D8 advanced system with Cu Kα radiation (λ = 1.5405Å) between phase angle 20° to 80° by step 0.04^O/sec and scanning speed of 1.5 sec/step.

The prepared samples crystallite sizes were measured using the Scherer's Equation [30]

$$D_{hkl} = \frac{0.91\,\lambda}{\beta\,\mathrm{Cos}\theta} \tag{1}$$

Where D_{hkl} is the crystalline size perpendicular to (h k l) plane, λ is the incident X-ray wave length, β is the full width half maxima (FWHM) of (311) peak, and θ is the peak position (Bragg's angle at (311) peak).

The lattice parameter value is measured with the following equation:

$$a = \frac{d_{hkl}}{\sqrt{h^2 + k^2 + l^2}} \tag{2}$$

Hopping length at Tetrahedral (A-site) and Octahedral (B-site) were measured by following relations.

$$\text{A site (Tetrahedral)hopping length } d_A = 0.25a\sqrt{3} \tag{3}$$

$$\text{B site (Octahedral)hopping length } d_B = 0.25a\sqrt{2} \tag{4}$$

The X-ray density (dx) measured with the following relation:

$$d_x = \frac{8M}{Na^3}\left(\frac{g}{cm^3}\right) \tag{5}$$

Where 8 is the number of molecules in a unit cell, M is the composition molecular weight, and N is the Avogadro's number.

The bulk density dm was determined using formula:

$$d_m = \frac{m}{\pi r^2 h} \tag{6}$$

Where m is the sample mass, r is the sample radius, and h is the sample thickness The porosity (P) of the ferrite was determined using formula:

$$P = 1 - \frac{d_m}{d_x} \tag{7}$$

Where d_m is the bulk density and d_x is the X-ray density.

The surface morphology was performed by using SEM technique (Scanning Electron Microscope). Elemental analysis was analyzed by energy dispersive X-ray spectroscopy (EDS). The calcined powders microstructure and average crystallite size were characterized by TEM (Tecnai-12, FEI, Netherlands) technique. FTIR is gives the absorption band positions and it analyzed to get structural information about the prepared ferrite systems.

The magnetic properties were carryout at room temperature with obtained M-H loops by using VSM (GMW Magnet System, model 3473) From M-H loops saturation magnetization (M_S) and coercivity (H_C) are directly extracted. From the above measurements calculated the following parameters.

The anisotropy constant (K) was calculated with the following relation [31]:

$$H_c = \frac{0.98K}{M_s} \tag{8}$$

The magnetic moment (μB) was calculated with the following relation [32]:

$$\mu_B = \frac{M_w X M_s}{5585} \tag{9}$$

Where M_W is the composition molecular weight, M_S is the saturation magnetization.
The Yefet-Kittel (Y-K) angles are calculated with the following relation [33]:

$$\mu_B = (6 + x)\, cos\, \alpha_{Y-K} - 5(1 - x) \tag{10}$$

Where x is the Cr^{3+} concentration.

Temperature and composition dependent dc electrical properties were measured with two probe method [34].

The resistivity (ρ) and temperature (T) Kelvin relationship may be expressed as Arrhenius relation [35].

$$\rho = \rho_o e^{\Delta E / K_B T} \tag{11}$$

Where ρ_o is the resistivity at room temperature, K_B is the Boltzmann constant (8.617×10^{-5} eV K^{-1}), and ΔE is the activation energy;

The activation energies were measured with the following relation:

$$\Delta E = 2.303 X K_B X 10^3 X slope \ \ (eV) \tag{12}$$

The drift mobility (μ_d) of charge carriers were measured with the following relation [36].

$$\mu_d = \frac{1}{\eta e \rho} \tag{13}$$

Where η is the number of charge carriers, e is the electron charge, and ρ is the resistivity at a given temperature.

The charge carrier concentration measured with the following relation [37]:

$$\eta = \frac{N_A d_B P_{Fe}}{M} \tag{14}$$

Where M is the sample molecular weight, N_A is the Avogadro number, d_B is the bulk density, and P_{Fe} is the number of iron atoms in ferrite composition.

Dielectric properties of the prepared pellets measured in between 20 to 2 MHz frequency at room temperature with LCR meter. Dielectric properties like dielectric constant loss tangent and ac conductivity were determined by the following formulae.

The dielectric constant was measured with the following relation [38]:***

$$\varepsilon' = \frac{\text{Cd}}{\varepsilon_o \text{A}} \tag{15}$$

Where C is the material capacitance, ε_O is the permittivity of the air ($8.854 x 10^{-12} Fm^{-1}$)

The AC conductivity was measured with the following relation [39]:

$$\sigma_{ac} = \omega \varepsilon_O \varepsilon' tan\delta \quad (16)$$

Where ε_O is the permittivity of free space (8.854×10^{-12}F/m), ε' is the dielectric constant, and tanδ is the loss tangent.

3. Results and discussion

3.1 Structural characterization

3.1.1 XRD analysis

X-ray diffraction pattern of Ni-Cr nano ceramic particles is depicted in **Figure 2**. It shows the crystalline phases were identified with standard data PDF# 862267 from the ICDD data. It was observed that X-ray diffraction pattern can be well indexed with peaks corresponding to cubic spinel structure such as (111), (220), (311), (400), (511), (440), and (533). The highest reflection comes from (311) peak that shows spinel structure and all samples represents formation of cubic spinel structure in single phase without other evident additional impurity phases or secondary phases for chromium substituted nickel nano ceramic [40]. The crystallite size was computed and is given in **Table 1**. It shows that the prepared samples crystallite size is in the nanometer scale between 8.55 nm–10.36 nm.

Figure 3 shows the slightly decreases the lattice parameter with dopent Cr ion increases in mixed Ni-Cr ceramics, that means it obeys Vegard's law [41]. It is because of the large ionic radii of Fe^{3+} (0.67Å) is replaced by low ionic radius Cr^{3+} (0.64Å) in B site [42]. Similar behavior was reported in Ni-Cr nanoceramic system [43].

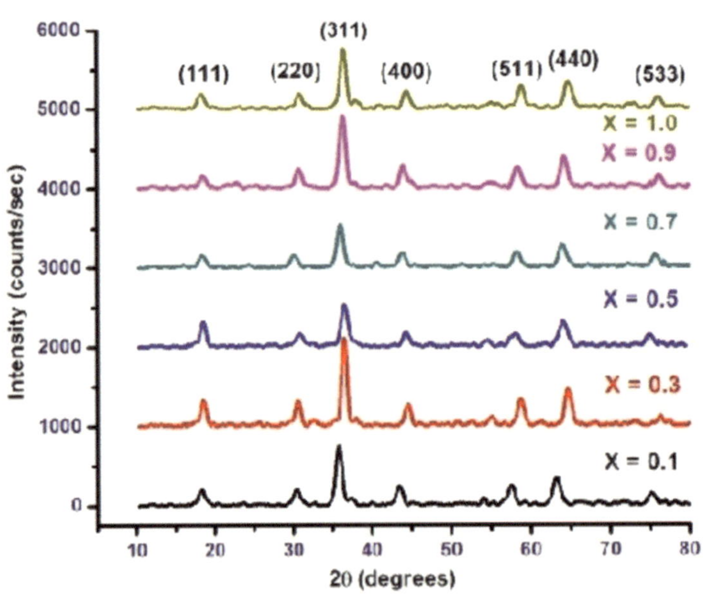

Figure 2.
X-ray diffraction pattren of mixed $NiCr_XFe_{2-X}O_4$ nano ferrites.

Sl. No.	Composition	Crystallite size (nm)	Lattice parameter (a) (Å)	Hopping length (A-site) (d_A) (Å)	Hopping length (B-site) (d_B) (Å)	X-ray density (gram/cc)	Bulk Density (gram/cc)	Porosity (P %)
1	$NiCr_{0.1}Fe_{1.9}O_4$	8.96	8.356	3.618	2.954	5.326	5.218	2.01937
2	$NiCr_{0.3}Fe_{1.7}O_4$	10.36	8.342	3.612	2.949	5.334	5.192	2.67538
3	$NiCr_{0.5}Fe_{1.5}O_4$	7.95	8.329	3.606	2.944	5.34	5.07	5.0628
4	$NiCr_{0.7}Fe_{1.3}O_4$	8.55	8.316	3.601	2.940	5.35	5.012	6.31649
5	$NiCr_{0.9}Fe_{1.1}O_4$	8.84	8.301	3.594	2.934	5.36	4.941	7.82673
6	$NiCrFeO_4$	9.26	8.293	3.591	2.932	5.368	4.813	10.3369

Table 1.
Structural analysis of synthesized and heat treated $NiCr_XFe_{2-X}O_4$ nano ferrites.

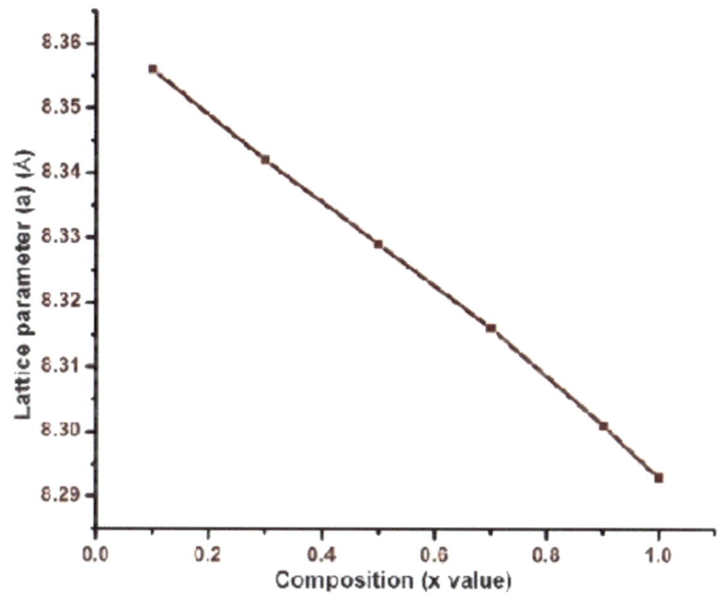

Figure 3.
Lattice parameter variation with Cr concentration.

The measured hopping lengths are given in **Table 1**. It shows the decreases the distance between magnetic ions with increase Cr ion concentration in Ni nano ceramic, which makes decreasing the hopping length. It may be due to that Cr^{3+} ion (0.63Å) has smaller radius than Fe^{3+} ion (0.67Å). Similar trend was studied for the Ni-Cr ceramic system prepared by impregnation technique [44].

The density measurements were illustrated in **Table 1**. It was found that **Figure 4** shows the increases of the X-ray density (dx) from 5.326 to 5.368 gram/cc and the bulk density (dm) decreases from 5.218 to 4.813 gram/cc with increases Cr^{3+} ions concentration in nickel nanoferrite. It may be because of larger atomic weight and density of Fe (55.847gm/mole, 7.874gm/cm^3) compare with atomic weight and density of Cr (51.996gm/mole, 7.14gm/cm^3). The X-ray density is more than the apparent density due to the existence of pores which depends on the preparation state. Porosity increase with increase Cr ion concentration and it shows similar behavior of X-ray density. Similar behavior observed for the Cr substitution nano ceramic system with other researcher reports [45, 46].

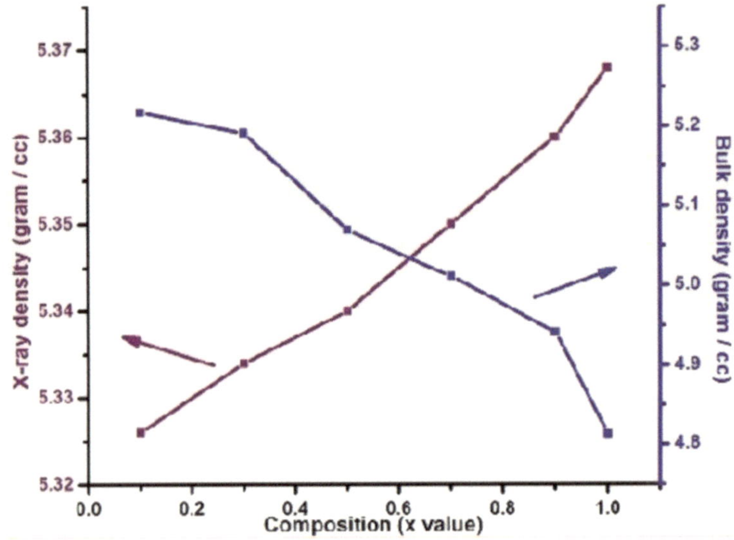

Figure 4.
X-ray density and bulk density variation with Cr concentration.

3.1.2 Morphological studies

The SEM representative micrographs of the prepared Ni-Cr nanoceramic system, with various Cr concentration, are shown in **Figure 5(a-f)**. It shows that the morphology is similar and they are in nanoscale with almost inhomogeneous. The obtained patterns energy dispersive X-ray spectroscopy of various composition of Ni-Cr nanoceramic systems are shown in **Figure 5(a-f)**. The corresponding elemental and atomic percentage of various chromium concentrations were illustrated in **Table 2**. It reveals that the compositions representing the elements Ni, Cr, Fe, and O without precipitating cations

TEM micrographs of $NiCr_{0.3}Fe_{1.7}O_4$ and $NiCr_{0.7}Fe_{1.3}O_4$ nanoceramics are represented in **Figure 6(a)** and **6(b)**. The crystallite size is in nanometer scale and also be in agreement well with crystallite size estimated from XRD analysis.

3.1.3 FTIR analysis

The FTIR spectra of the mixed $NiCr_XFe_{2-X}O_4$ nanoceramic, recorded in the range of 400–4000 cm^{-1}, is shown in **Figure 7** and absorption band results are reported in **Table 3**. It shows the two absorption bands ν_1 and ν_2 at around 600 cm^{-1} and 400 cm^{-1} respectively. The high frequency band (ν_1) corresponds to Fe^{3+}-O^{2-} vibrations at tetrahedral (A site) and low frequency band (ν_2) corresponds to M^{+2}-O^{-2} vibrations at octahedral sites (B site) [47] and these are representing the spinel ceramic in single phase [48]. The bands around 3400 cm^{-1}, 2400 cm^{-1} and 1600 cm^{-1} are the contribution of the stretching vibration of free and absorbed water, indicated the removal of the -OH, -CO and -NO groups. Similar trend have been observed for Ni-Cr nano ceramic system prepared with impregnation technique by others [49, 50].

3.2 Magnetic properties

The obtained magnetic hysteresis loops are illustrated in **Figure 8**. It shows that the loop area is very narrow therefore the samples present soft ferrite nature with

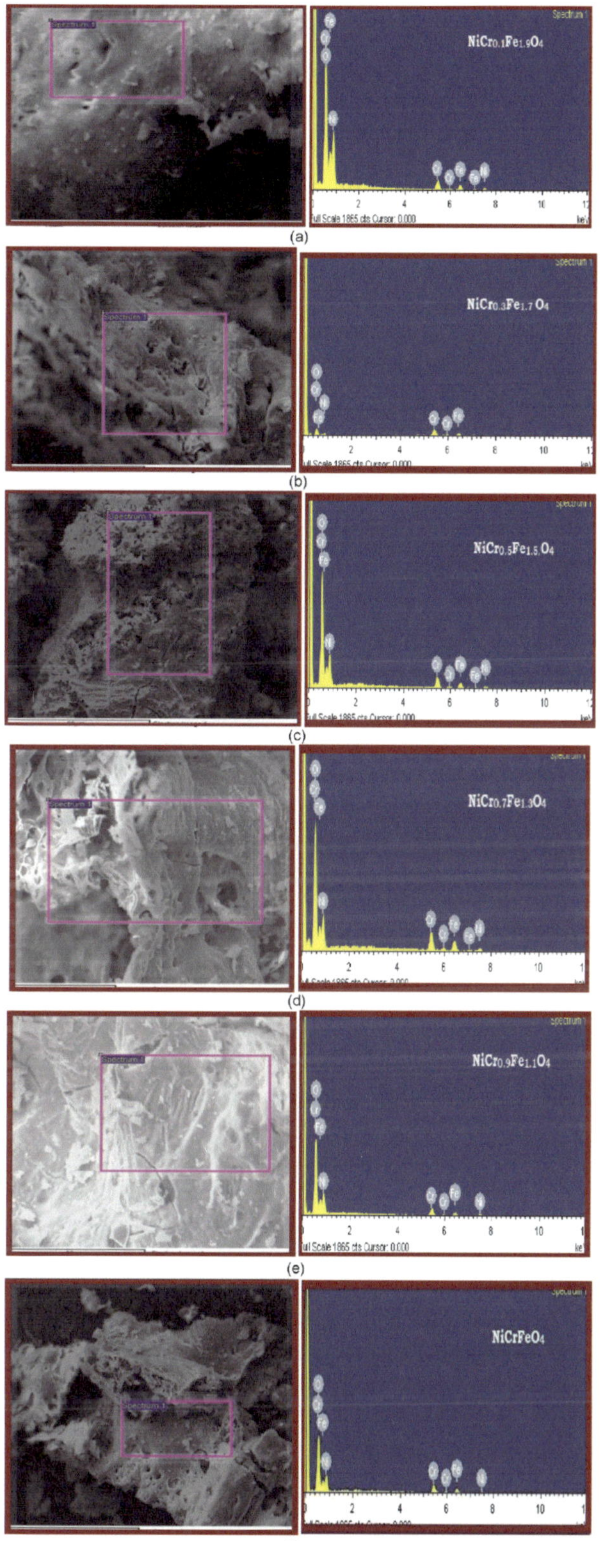

Figure 5.
(a-f) SEM images and EDS images of $NiCr_XFe_{2-X}O_4$ nano ferrites. (a) $NiCr_{0.1}Fe_{1.9}O_4$ (X = 0.1), (b) $NiCr_{0.3}Fe_{1.7}O_4$ (X = 0.3), (c) $NiCr_{0.5}Fe_{1.5}O_4$ (X = 0.5), (d) $NiCr_{0.7}Fe_{1.3}O_4$ (X = 0.7), (e) $NiCr_{0.9}Fe_{1.1}O_4$ (X = 0.9), and (f) $NiCrFeO_4$ (X = 1.0).

Element	O		Fe		Ni		Cr	
Ferrite composition	Element %	Atomic %	Element %	Atomic %	Element %	Atomic %	Element %	Atomic %
$NiCr_{0.1}Fe_{1.9}O_4$	17.01	41.45	29.07	21.80	27.72	19.35	26.20	17.40
$NiCr_{0.3}Fe_{1.7}O_4$	19.50	45.20	29.47	21.99	25.02	16.67	26.01	16.14
$NiCr_{0.5}Fe_{1.5}O_4$	19.16	44.87	33.03	23.81	24.28	16.30	23.53	15.02
$NiCr_{0.7}Fe_{1.3}O_4$	19.79	45.86	30.76	21.96	27.30	18.14	22.15	14.04
$NiCr_{0.9}Fe_{1.1}O_4$	17.55	42.24	31.56	24.11	27.89	19.23	23.00	14.42
$NiCrFeO_4$	20.41	46.97	27.28	19.32	23.00	14.42	24.72	15.51

Table 2.
Elements of each sample composition analyzed by EDS pattern.

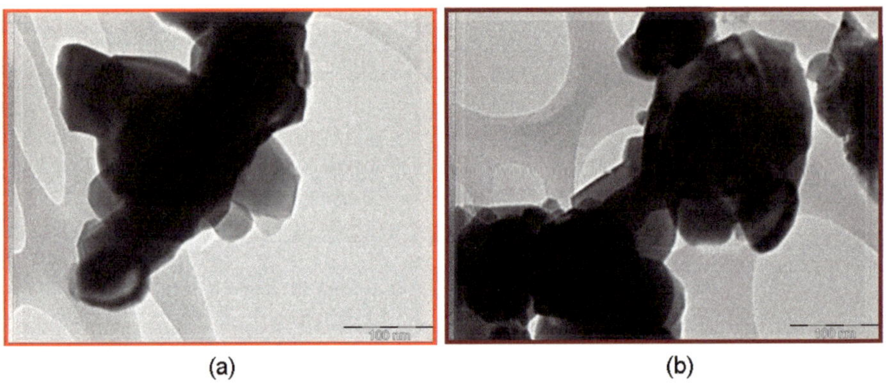

Figure 6.
(a) TEM image of $NiCr_{0.3}Fe_{1.7}O_4$. nano ferrite. (b) TEM images of Ni $Cr_{0.7}Fe_{1.3}O_4$ nano ferrite.

less coercivity [51]. The measured magnetic properties of Ni-Cr nano ceramics at room temperature were reported in **Table 4**.

It is observed that the Ni-Cr nanoceramics have less saturation magnetization and less coercivity due to the smaller grain size, as illustrated in **Table 4**. That means the grain size is small, the saturation magnetization is less. The saturation magnetization decreases from 4.49 to 2.97 emu/gr with increases of the Cr^{3+} concentrations in Ni nanoceramics at room temperature, as evident in **Figure 9**. Because of that less magnetic moment Cr^{3+} ions ($\sim 3\mu_B$) are substituted in the place of higher magnetic moment Fe^{3+} ions ($\sim 5\mu_B$) at octahedral sublattice. As increases Cr^{3+} ion concentration, decreases the iron ions ratio between octahedral and tetrahedral sites. As a result, super exchange interaction between A-site & B-site decreases. Which lead to decrease in saturation magnetization. It is attributing to the weak magnetic interactions in Ni-Cr ceramics. Therefore, material becomes converted into soft magnetic material. Similar report was observed by Bhukal et al. [52].

The measured Coercitive field values are reported in **Table 4**. It shows that this parameter decreases from 136.19 to 63.03 Oe (minimum for x = 0.7 composition) and thereafter it increases to 106.08 Oe with increases the Cr^{+3} concentration in nickel nanoceramic. It shows that decreases in coercive field with increase Cr^{+3} ion concentration because of anisotropy field decreases which in order decreases the domain wall energy [53]. Increases in coercive field with composition X = 0.9 and 1.0 due to anisotropy field increases which in order domain wall energy increases [54, 55].

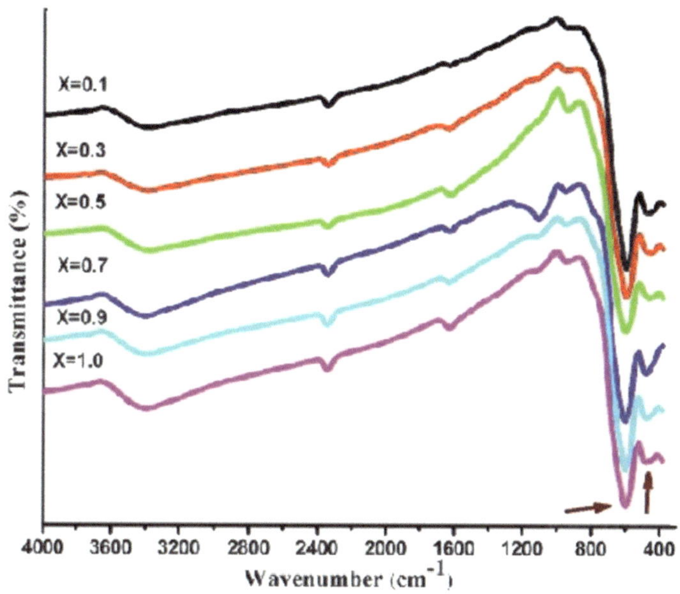

Figure 7.
FT-IR patterns of mixed $NiCr_XFe_{2-X}O_4$ nano ferrites.

Sl.No.	Composition	ν_1 (cm^{-1})	ν_2(cm^{-2})
1	$NiCr_{0.1}Fe_{1.9}O_4$	599.962	485.354
2	$NiCr_{0.3}Fe_{1.7}O_4$	607.036	477.055
3	$NiCr_{0.5}Fe_{1.5}O_4$	602.453	479.694
4	$NiCr_{0.7}Fe_{1.3}O_4$	591.435	481.109
5	$NiCr_{0.9}Fe_{1.1}O_4$	594.864	492.428
6	$NiCrFeO_4$	597.132	472.619

Table 3.
FT-IR parameters of mixed Ni-Cr nano ferrites.

From **Table 4**, it shows that the magnetic moment values are decreases from 0.188μ_B to 0.122μ_B with increases Cr^{3+} ion concentrations in Ni nano ceramic. The decrease in magnetic moment is credited to greater tenancy of Cr^{3+} at B sites. Therefore the materials are getting changed into soft ferrite materials.

Magnetic moment values are low due to the Cr^{3+} ions substituted in nickel nano ceramic. It is explained based on the non-collinear spin arrangement [56, 57]. The B–O–B coupling interactions at the B sublattice become stronger than that of A–O–B coupling between magnetic ions at the A and B sublattice due to the presence of a small canting of the B site moment with respect to the direction of the A site moment. The B–O–B coupling leads to the random existence of the small canted structure at the B site and forms triangular configuration in the ferrite system. As a result, the magnetic moments of the Fe ions at the B site are shifted from the collinear parallel to nonparallel arrangements. Therefore, the saturated magnetization is being decreased corresponding to the magnetic moment which is also decreased.

The decrease of magnetization has been proposed by Yafet and Kittel(Y–K) by triangular arrangement of spins [58]. The Y-K angles of Ni-Cr nano ceramic system are reported here in the **Table 4**. It is clear that increases the Y-K angles with

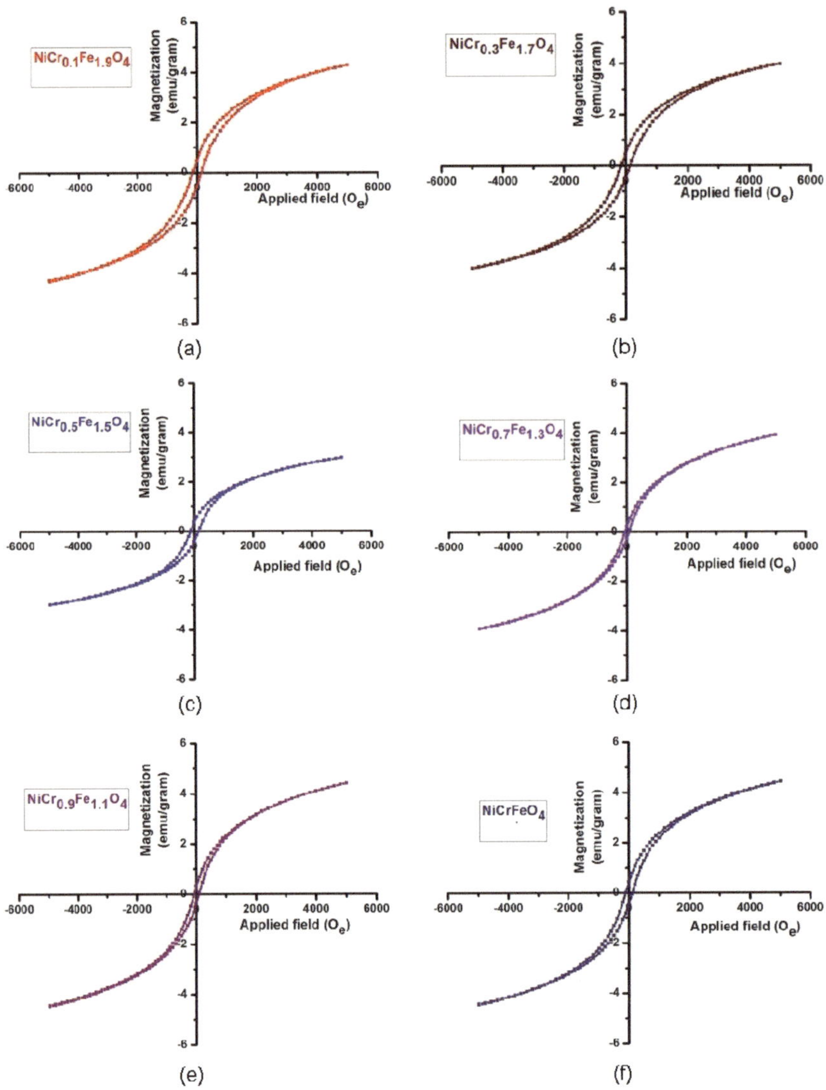

Figure 8.
(a-f) Magnetic hysteresis loops for $NiCr_xFe_{2-x}O_4$ nano ferrites. (a) X = 0.1, (b) X = 0.3, (c) X = 0.5, (d) X = 0.7, (e) X = 0.9, and (f) X = 1.0.

Sl. No.	Composition	Saturation magnetization (M_S)(emu/gr)	Coercive field (H_C) (O_e)	Magnetic moment (μ_B) (Bohr magneton)	Y-K angle (θ)
1	$NiCr_{0.1}Fe_{1.9}O_4$	4.49	136.19	0.18823	39.7751
2	$NiCr_{0.3}Fe_{1.7}O_4$	4.29	125.95	0.17911	54.2687
3	$NiCr_{0.5}Fe_{1.5}O_4$	4.15	116.58	0.17273	65.7204
4	$NiCr_{0.7}Fe_{1.3}O_4$	3.99	63.02	0.16529	75.6083
5	$NiCr_{0.9}Fe_{1.1}O_4$	3.67	79.19	0.15154	84.5817
6	$NiCrFeO_4$	2.97	106.07	0.12272	88.9955

Table 4.
Magnetic parameters from hysteresis loops of mixed Ni-Cr nano ferrites.

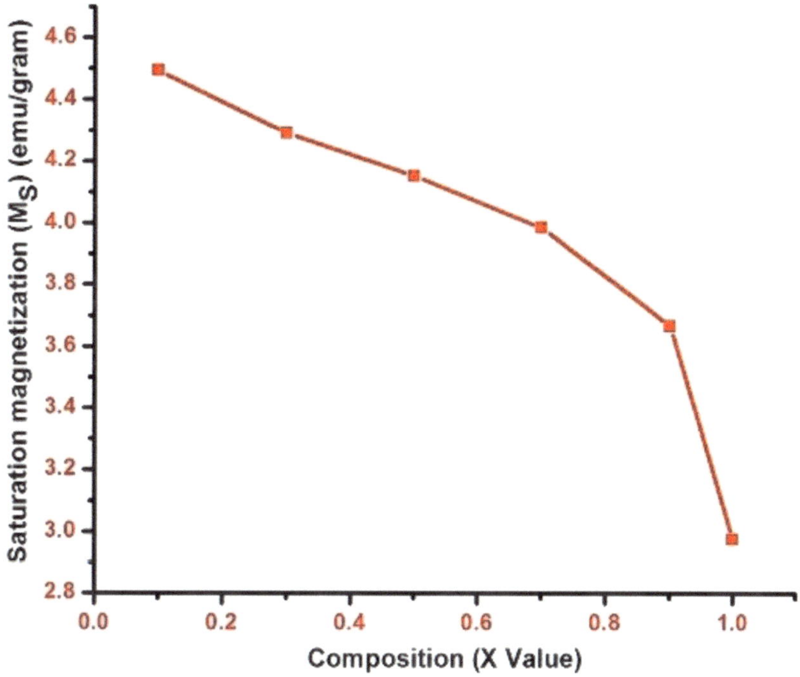

Figure 9.
Variation of saturation magnetization with Cr concentration.

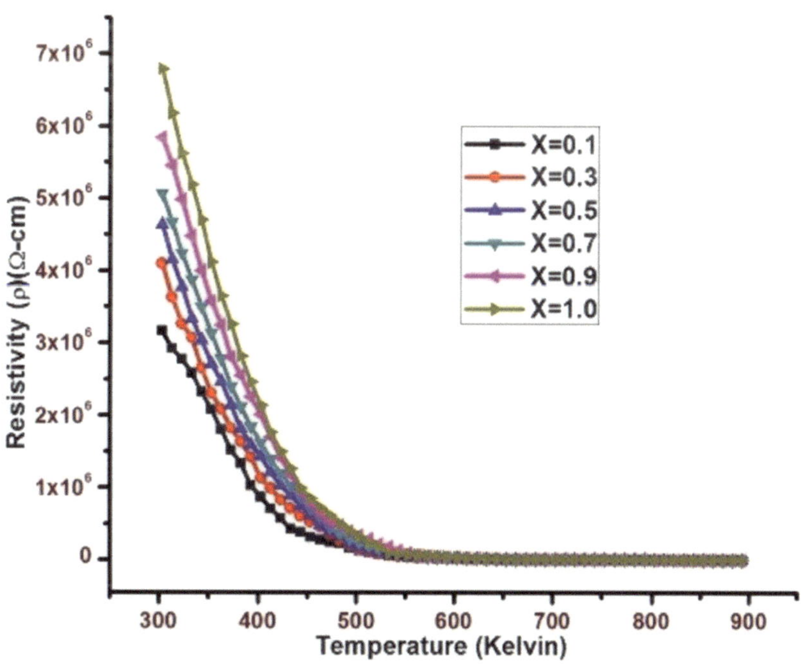

Figure 10.
DC resistivity variation with temperature of $NiCr_XFe_{2-X}O_4$ nano ferrites.

increase Cr^{3+} ion concentration in Ni nano ferrite. It indicates that the spin canting takes place significantly at higher concentration of Cr content. Therefore, increases the spin arrangement at B-site. As a result, decrease A-B exchange interaction consequent decreases in magnetization. From obtained hysteresis loops it is proved that the prepared samples are enhanced soft magnetic performance. Hence these materials are desirable for transformers and these are useful for low inductance cores and coils [59].

3.3 Electrical properties

DC resistivity variation with temperature of Ni-Cr nano ceramics are illustrated in **Figure 10**. It shows the normal behavior for semiconductors of prepared samples [60]. As temperature in increasing dc resistivity decreases because hopping rate is increases due to which activation of electrons jumps from Fe^{3+} to Fe^{2+} ions at B-sites. Similar trend reported by Iqbal et al. [61]. The calculated dc electrical values such as resistivity, conductivity, and drift mobility of Ni-Cr nano ceramic compositions at room temperature were reported in **Table 5**.

From **Table 5** the DC resistivity increases from $3.17.10^6$ to $6.79.10^6$ Ω-cm and the conductivity decreases from $3.15.10^{-7}$ to $1.47.10^{-7}$ Ω-cm, as given in **Figure 11**. It shows that the resistivity has increased and conductivity has decreased with Cr^{3+} ion concentration increase in Ni nano ceramic. Because of that Fe (9.7×10^{-6} Ω-cm) has smaller value of resistivity as compared with Cr (1.3×10^{-5} Ω-cm). Resistivity of Ni-Cr nano ferrites indicate that have high resistivity which show the way to low eddy current losses and they are popular in electronic inductors, transformers and electromagnets. Similar result was observed by Sagar E. Shirsath et al. [62].

Figure 12 shows the Arrhenius plots ln(ρT) versus 10^3/T of Ni-Cr nano ceramic system, it observed that a change at a point, it indicates a change of magnetic ordering. The curve divided into two regions as ferrimagentic region and paramagnetic region. Activation energy values of prepared samples are calculated and they are reported in **Table 6**. It shows that the activation energy values in paramagnetic region (E_P) are higher than those in ferrimagnetic region (E_F). Because of charge carriers need more energy for the conduction in paramagnetic region as compared with ferrimagnetic region. As a result, the conduction process is affected by the change in magnetic ordering. Similar results are reported by other researchers [63, 64].

It may be due to the replacement of Fe^{3+} ions with Cr^{3+} ions at B-sites [65]. Cr^{3+} ions are strongly preferred at octahedral site (B-site). When Cr ion substituted in Ni ceramic, Fe^{3+} ions are partially replaced with Cr^{3+} ions at octahedral site (B-site). Hence decrease the number of Fe^{3+}/Fe^{2+} ion pairs at octahedral site (B-site). Therefore decrease in hopping of ions consequently an increase in resistivity and decrease

Sl. No.	Composition	At room temperature			Curie temperature (T_C) (Kelvin)	
		DC Conductivity (σ) (Ω^{-1}-cm^{-1})	Resistivity (ρ)(Ω-cm)	Drift mobility (μ_d)(cm^2/Vs)	DC Resistivity	Loria-Sinha technique
1	$NiCr_{0.1}Fe_{1.9}O_4$	3.15X10^{-07}	3.17 X10^{06}	7.73 X10^{-11}	775.39	789
2	$NiCr_{0.3}Fe_{1.7}O_4$	2.44 X10^{-07}	4.10 X10^{06}	7.55 X10^{-11}	747.58	762
3	$NiCr_{0.5}Fe_{1.5}O_4$	2.16 X10^{-07}	4.64 X10^{06}	7.42 X10^{-11}	705.32	729
4	$NiCr_{0.7}Fe_{1.3}O_4$	1.98 X10^{-07}	5.06 X10^{06}	7.29 X10^{-11}	690.71	697
5	$NiCr_{0.9}Fe_{1.1}O_4$	1.71 X10^{-07}	5.84 X10^{06}	6.84 X10^{-11}	659.53	678
6	$NiCrFeO_4$	1.47 X10^{-07}	6.79 X10^{06}	6.69 X10^{-11}	635.34	642

Table 5.
DC electrical values of mixed Ni-Cr nano ferrites.

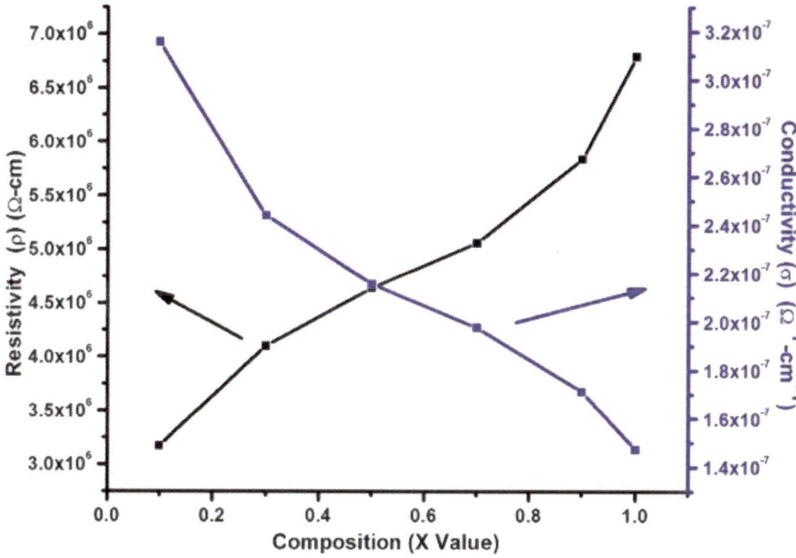

Figure 11.
DC resistivity and conductivity variation with Cr^{3+} *concentration at room temperature.*

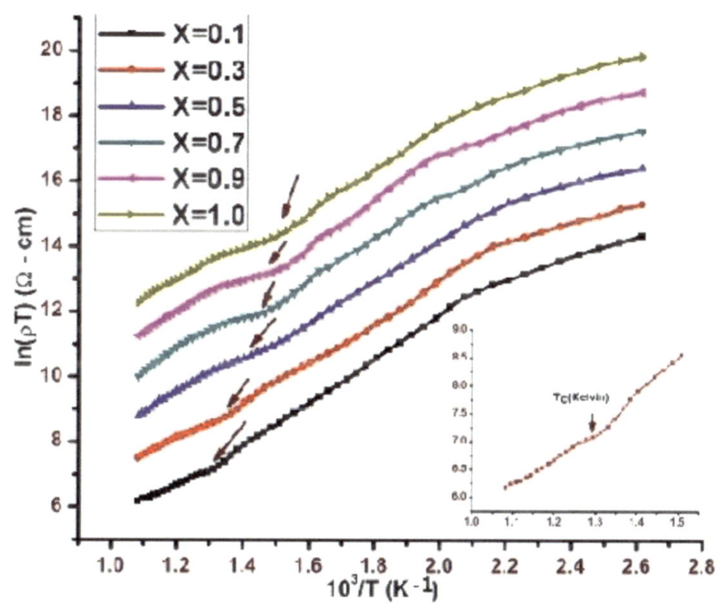

Figure 12.
DC resistivity variation with inverse temperature of $NiCr_XFe_{2-X}O_4$ *nano ferrites.*

in conductivity with increases Cr^{3+} ion in Ni nano ceramic system. A similar behavior was reported by Khan et al. [66]. Hence, the activation energy increases from 0.167 to 0.341 eV with increase Cr3+ ions concentration in Ni nanoceramic systems (see **Figure 13**). Due to resistivity increases with increase in Cr^{3+} ion concentration. Similar result reported by others [67, 68].

The activation energy increases with increase in Cr^{3+} content (x) recommend that the Cr^{3+} ions are toward the inside into octahedral sites barricade the electron hopping between $Fe^{2+}\leftrightarrow Fe^{3+}$ ions for electrical conduction. This show the way to an

Sl.No.	Composition	Para region (E_P) ev	Ferri region (E_F) ev	Activation energy (ΔE) ev
1	$NiCr_{0.1}Fe_{1.9}O_4$	0.412	0.244	0.167
2	$NiCr_{0.3}Fe_{1.7}O_4$	0.522	0.324	0.197
3	$NiCr_{0.5}Fe_{1.5}O_4$	0.635	0.394	0.241
4	$NiCr_{0.7}Fe_{1.3}O_4$	0.726	0.44	0.286
5	$NiCr_{0.9}Fe_{1.1}O_4$	0.794	0.489	0.305
6	$NiCrFeO_4$	0.852	0.511	0.341

Table 6.
Activation energy values of mixed Ni-Cr nano ferrites.

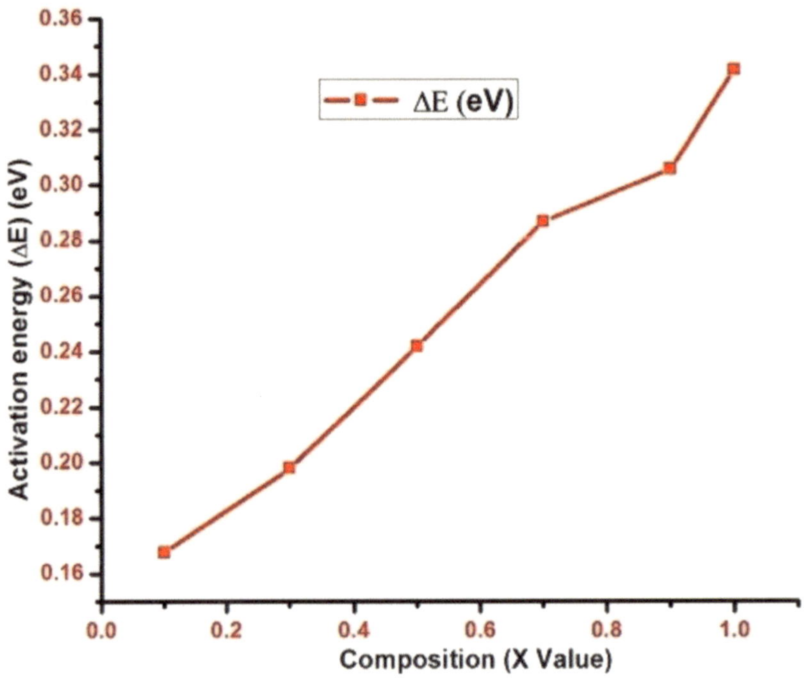

Figure 13.
Activation energy (ΔE) variation with Cr^{3+} concentration.

decrease in the conduction loss and increase in resistivity, as a result material with higher resistivity has higher values of activation energies and vice versa [69].

The drift mobility variation with temperature of Ni-Cr nano ceramic system reported in **Figure 14**. It shows that drift mobility increases with increase in temperature. Because the change in charge carrier mobility rather than the change in carrier concentration. Hence, that charge carriers begin hopping from one site to another site as increasing the temperature, therefore drift mobility increases. Similar result was observed by Bhukal et al. [70].

From **Table 5**, it shows that the drift mobility decreases from $7.73X10^{-11}$ cm^2/Vs to $6.69X10^{-11}$ cm^2/Vs with increases Cr^{3+} concentration in Ni nano ceramics. It is due to the material with higher resistivity have lower mobility and vice versa. The decreasing trend in drift mobility with the composition shown materials is good choice for high frequency applications. Similar results were also reported by Ashiq et al. [71].

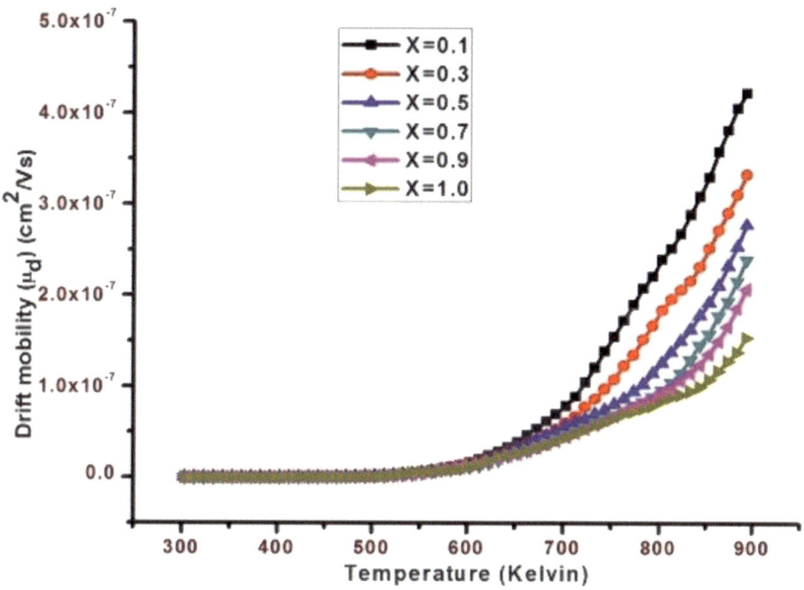

Figure 14.
Drift mobility variation with inverse temperature of $NiCr_XFe_{2-X}O_4$ nano ferrites.

From Arrhenius plot of ln(ρT) with 10^3/T as shown in **Figure 12**, straight line must take place a change at a particular temperature, there the ferrimagnetism will change into paramagnetism is known as Curie temperature. From dc resistivity measurements values of the Curie temperature with composition of the prepared samples and it has been determined by using another method (Loria-Sinha technique). The values were tabulated in **Table 5** as evident in **Figure 15**.

In the present study, from Loria-Sinha method, the Curie temperature is observed as decreases from 789 to 642 K and from DC resistivity experimental observation it was found to decreases from 775.39 to 635.34 K on increasing the Cr^{3+} ions concentration in Ni nanoceramics, due to the Fe3+ ions, which have been placed by paramagnetic Cr^{3+} ion concentration [72].

When the Fe^{3+} ion replacement with Cr^{3+} ion concentration increases, the magnetization decreases in B-sublattice without disturbing the A-sublattice, therefore decrease in A-B interaction hence, decreases in Curie temperature. Similar behavior was observed in the trivalent substitution nanoceramic system [73]. The Curie temperature results from Loria-Sinha method (gravity method) were good in agreement with determined transition temperature values by dc resistivity measurements.

3.4 Dielectric properties

The frequency dependence dielectric constant (ε') of Ni-Cr nano ceramic system represent in **Figure 16**. It is observed that at lower frequency dielectric constant is high and it decreases sharply and then decreases slowly with the increase in frequency and shows almost frequency independent behavior at high frequency range [74]. The variation of dielectric constant with frequency is explained according to Maxwell–Wagner theory [75, 76], which is in good agreement with Koop's phenomenological theory [77]. Similar kind of trend for dielectric constant with frequency was observed by others [78, 79].

The dielectric loss tangent (tan δ) variation with frequency is representing in **Figure 17**. It depicts the value of tanδ increases initially and exhibits the loss factor,

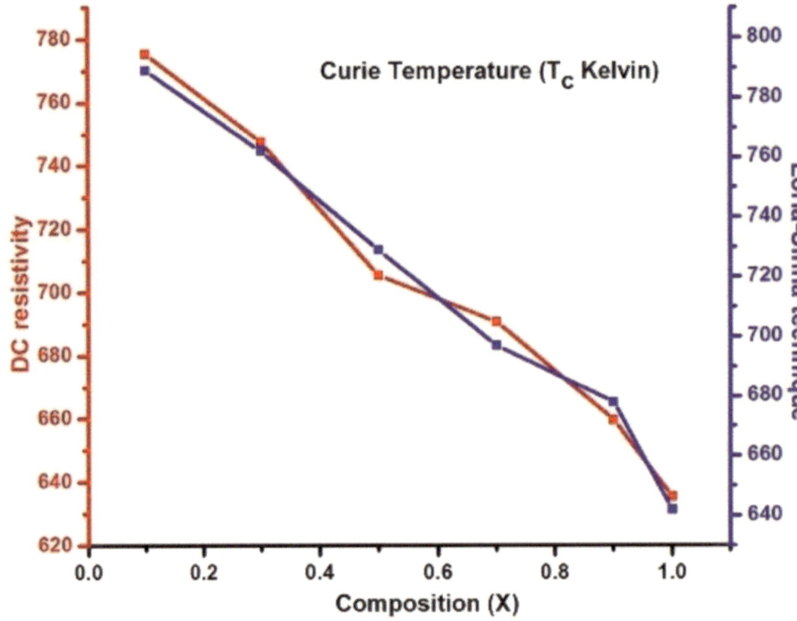

Figure 15.
Curie temperature variation with Cr^{3+} concentration.

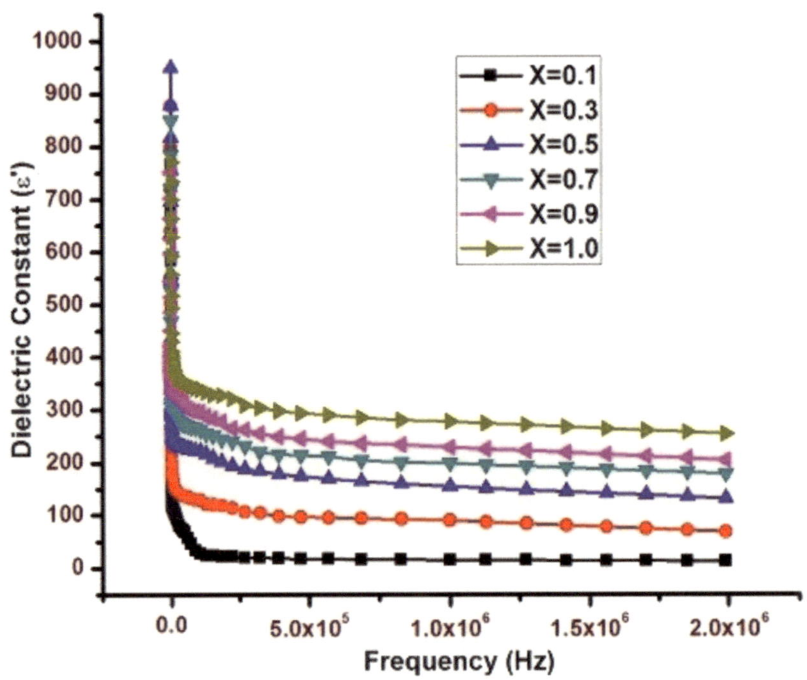

Figure 16.
Dielectric constant (ε') variation of Ni-Cr nano ferrites with frequency.

which is maximum between 1.103 to 4.104 Hz, and further decreases by increasing the frequency. It shows the Debye-type relaxation and this type of peaking behavior is observed when the jumping frequency between $Fe^{2+} \leftrightarrow Fe^{3+}$ ions is exactly equal to

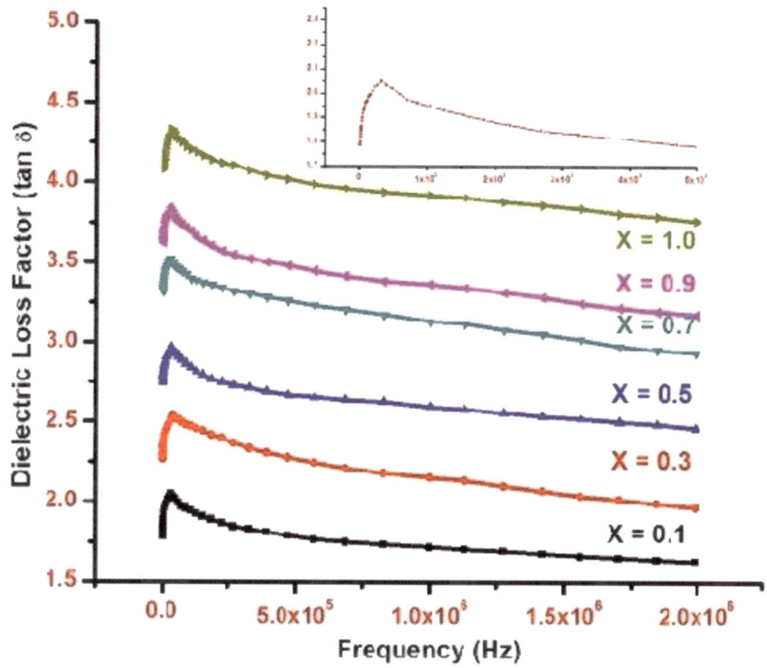

Figure 17.
Dielectric loss tangent (tanδ) variation of Ni-Cr nano ferrites with frequency.

the frequency of the applied field [80]. Similar type of variation is reported by other researchers [81, 82]. It is clear that at low frequency region dielectric loss decreases sharp and at high frequency region the rate of decrease is slow and it almost frequency independent behavior. A similar trend was observed in case of mixed Ni-Cu nano ceramic system for X = 0.4, 0.5, 0.8 and 1.0 under presents investigation. The decrease in tan δ with an increase in frequency may be explained on the basis of Koop's phenomenological model.

The AC conductivity variation with frequency is represented in **Figure 18**. It is clear that with increasing frequency ac conductivity increases at low frequency and at high frequency almost independent behavior. This behavior is like to Maxwell-Wagner type [75, 76] in good agreement with Koop's phenomenological theory [77]. According to Koop's phenomenological theory, at lower frequencies region majority are grain boundaries and they act as interference for mobility of charge carriers, hence the hopping electrons between Fe^{2+} and Fe^{3+} ions is less, so conductivity is less. At higher frequency region the conductive grains which become more active and promote the hopping electrons between Fe^{2+} and Fe^{3+} ions, hence the conductivity is more and they are take part for creating charge carriers, these charge carriers are responsible for increasing the ac conductivity. It is in good agreement with the other reports [83, 84].

Dielectric parameter variation with Cr^{+3} ion composition at frequencies 2 MHz are reported in **Table 7**. The dielectric parameters are increases with increase Cr^{3+} ion concentration. The dielectric constant and the AC conductivity reach maximum values at X = 0.5 and the loss factor maximum at X = 0.7 further increase in Cr^{3+} ion concentration these parameters are decreases. It is due to the formation of Fe^{3+} ions at octahedral sites and it explained with dielectric polarization which is similar to the conduction mechanism in ferrites. Cr ions prefer the octahedral site until the Cr concentration becomes greater than 50%, thereafter Cr ions may increase in tetrahedral sites causing migration of equal number of ions to the octahedral sites [85]

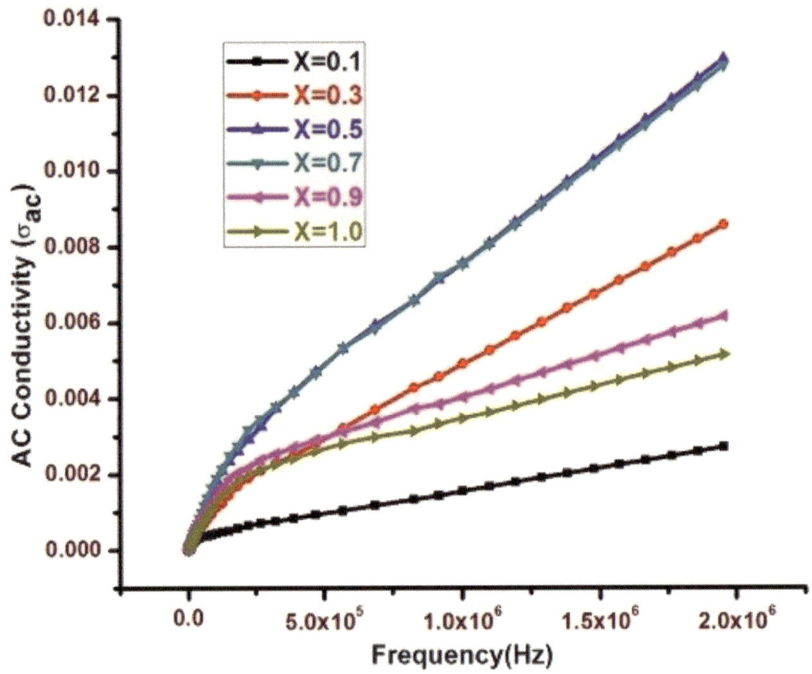

Figure 18.
AC conductivity (σ_{ac}) variation of Ni-Cr nano ferrite system with frequency at different frequency.

Sr.No.	Cr content (x)	Dielectric parameters at frequency of 2 MHz		
		ε'	tan δ	σ_{ac}
1	0.1	11.850	1.537	2.70 X10^{-03}
2	0.3	17.861	1.608	8.56 X10^{-03}
3	0.5	31.243	1.812	1.29 X10^{-02}
4	0.7	28.650	2.237	1.28 X10^{-02}
5	0.9	4.643	2.044	6.15 X10^{-03}
6	1.0	4.166	1.714	5.14 X10^{-03}

Table 7.
Dielectric parameters of mixed Ni-Cr nano ferrite system.

and decreasing the Fe^{3+}ions at octahedral site there is a least possibility of electronic exchange interaction between Fe^{2+} and Fe^{3+} ions, hence decrease in dielectric parameters with increase in Cr concentration. Similar behavior reported by Raghasudha et al. [86].

4. Conclusions

- The Citrate-gel auto combustion technique is convenient, more efficient and economical for synthesis of investigated nano-cermics.
- X-ray diffraction analysis confirms the single phase cubic spinel structure without impurity peak. From X-ray diffraction analysis data found that the

crystallite size was in range of 8.5–10.5 nm, which indicates the nanocrystalline form.

- In mixed Ni-Cr nano ceramic system the lattice parameter and the hopping length at tetrahedral and octahedral sites are slightly decreases with Cr^{3+} ion dopant. It is due to difference between ionic radii of Fe^{3+} (0.067Å) and Cr^{3+} (0.64Å) ions and it indicating that the systems are obeys Vegard's law.

- In mixed Ni-Cr nano ceramic system increases in X-ray density, porosity and decrease in bulk density with increases in Cr^{3+} ion dopent in Ni nano ceramic. This is due to the atomic weight and density of Fe^{3+}(55.847gram/mole, 7.874gram/cc) more than that Cr^{3+} (51.996gram/mole, 7.14gram/cc).

- Microstructure of mixed Ni-Cr nano ceramic system the existing of Ni, Cr, Fe and O only. The TEM pictures confirm the crystallite formation in nano size.

- FT-IR absorption spectra of the mixed Ni-Cr nano ceramic system revealed the presence of two significant absorption bands around at 600 cm^{-1} and 400 cm^{-1} and it confirms spinel structure in single phase with two sub lattices.

- The saturation magnetization and magnetic moment decreases and the Y–K angles increase with increase Cr^{3+} ion concentrations in Ni nano ceramic. Since the fact that less magnetic moment Cr^{3+} ions ($\sim 3\mu_B$) are substituted in the place of higher magnetic moment Fe^{3+} ions ($\sim 5\mu_B$) at octahedral sublattice site.

- Temperature dependent of dc resistivity shows decreasing trend that suggest the mixed Ni-Cr nano ceramic systems are shown semiconductor behavior. At room temperature resistivity is low and it increases and conductivity decreases with the increase of Cr^{3+} ion concentration. It explains with conduction mechanism between $Fe^{2+} \leftrightarrow Fe^{3+}$ ions.

- Prepared samples resistivity is high, which show the way to low eddy current losses are desirable as core materials in electronic applications such as electronic inductors, power transformers, electromagnets and in telecommunication applications.

- The activation energy increases with Cr3+ ion concentration increases in Ni nano ceramic. It shows low conductivity of the ceramics goes hand with high activation energy, vice versa.

- Drift mobility of mixed Ni-Cr nano ceramic system has increases with increase in temperature. At room temperature the drift mobility decreases with increases Cr3+ concentration. It explain that with higher resistivity have lower mobility and vice versa. It shows the materials are good choice for high frequency applications.

- The Curie temperature (Tc) decreases from 775.39–635.34 K with increase in Cr3+ concentration. The Curie temperature (Tc) is good agreement which is measured form Loria-Sinha technique.

- Frequency dependent dielectric measurements such as dielectric constant, loss tangent and ac conductivity results are shows normal dispersion with frequency in Ni-Cr nano ceramic system.

- For all prepared samples the value of tanδ exhibits maximum between 1X103-4X104Hz and further decrease by increasing the frequency and ac conductivity increase with frequency dependent.
- In prepared ceramic systems dielectric loss is low at higher frequencies and conductivity value is low. Hence these ceramic compositions are shows the potential applications like high frequency micro wave devices and in microwave applications.

Author details

Rapolu Sridhar[1], D. Ravinder[2]*, J. Laxman Naik[2], K. Vijaya Kumar[3], N. Maramu[4] and S. Katlakunta[5]

1 Department of BS&H, Vignan Institute of Technology and Science, T.S. State, India

2 Department of Physics, Osmania University, Hyderabad, Telangana State, India

3 Department of Physics, JNTUH College of Engineering Sultanpur, Telangana State, India

4 Department of Physics, Kakatiya Institute of Technology and Science, Telangana, India

5 Department of Physics, University College of Science, Saifabad, Osmania University, Hyderabad, Telangana, India

*Address all correspondence to: ravindergupta28@rediffmail.com

References

[1] Z. Gao, B. Xu, M. Ma, A. Feng, Y. Zhang, X. Liu, Z. Jia and G. Wu. Electrostatic self-assembly synthesis of ZnFe2O4 quantum dots (ZnFe2O4@C) and electromagnetic microwave absorption Composites. Part B 2019; 179 107417. https://doi.org/10.1016/j.compositesb.2019.107417

[2] Noppakum Sanpo, James Wang and Christopher C. Berndt. Influence of Chelating Agents on the Microstructure and Antibacterial Property of Cobalt Ferrite Nanopowders. *Journal of Austr. Cer. Soc.,* 2013; 49 84-91. www.austceram.com/ACS-Journal

[3] Ch.Venkateshwarlu, Ch.Ashok, B. AppaRao, D.Ravinder, B.S.Boyanov. Electrical conductivity of Co–Zr substituted hexagonal barium ferrites. *Journal of Alloys and Compounds* 2006; 426(4) 1-3. https://doi.org/10.1016/j.jallcom.2006.02.001

[4] R. M. Borade, S. B. Somvanshi, S. B. Kale, R. P. Pawar and K. Jadhav. Influence of trivalent Cr ion substitution on the physicochemical, optical, electrical, and dielectric properties of sprayed $NiFe_2O_4$ spinel-magnetic thin films. *Mater. Res. Express 2020;* 7 016116. https://doi.org/10.1039/D0RA04319B

[5] Kharat, P.B., More, S.D., Somvanshi, S.B. *et al.* Exploration of thermo-acoustics behavior of water based nickel ferrite nanofluids by ultrasonic velocity method. *J Mater Sci: Mater Electron* 2019; 30 6564–6574. https://doi.org/10.1007/s10854-019-00963-4

[6] A. Abou-Hassan, S. Neveu, V. Dupuis and V. Cabuil. Synthesis Of Cobalt Ferrite Nanoparticles In Continuous-Flow Microreactors. *RSC Adv.,* 2012; 2 11263–11266. https://doi.org/10.1039/C2RA21799F

[7] M. Babrekar and K. Jadhav. Synthesis and Characterization of Spray Deposited Lithium Ferrite Thin Film, *Int. Res. J. Sci. Eng.,* 2017; 73–76. http://www.irjse.in

[8] Manik Gupta, B.S. Randhawa. Microstructural, magnetic and electric properties of mixed Cs–Zn ferrites prepared by solution combustion method, *Solid State Sciences.* 2012; 14 849-856. https://doi.org/10.1016/j.solidstatesciences.2012.04.010

[9] H. J. Kardile, S. B. Somvanshi, A. R. Chavan, A. A. Pandit and K. M. Jadhav. Effect of Cd^{2+} doping on structural, morphological, optical, magnetic and wettability properties of nickel ferrite thin films. *Optik,* 2010; 207 164462. https://doi.org/10.1016/j.ijleo.2020.164462

[10] Q. Yue, C. Liu, Y. Wan, X. Wu, X. Zhang and P. Du. Defect engineering of mesoporous nickel ferrite and its application for highly enhanced water oxidation catalysis. *J. Catal.,* 2018; 358 1–7. https://doi.org/10.1016/j.jcat.2017.10.027

[11] *R. Zhang, M. Liu, L. Lu, S.-B. Mi and H. Wang. Ultra-low temperature epitaxial growth of lithium ferrite thin films by high-pressure sputtering.* Cryst Eng comm., 2015; 17 8256–8263. https://doi.org/10.1039/C5CE01477H

[12] V. Sepel_ak, K. Baabe, K. Mienert, K. Schultze, F. Krumeich, F.J. Litterst, K.D. Becker. Mater Evolution Of Structure And Magnetic Properties With Annealing Temperature In Nanoscale High-Energy-Milled Nickel Ferrite. *J. Magn. Magn.,* 2003; 257 377-386. https://doi.org/10.1016/S0304-8853(02)01279-9

[13] Pavlović, M., Jovalekić, Č., Nikolić, A.S. *et al.* Mechanochemical synthesis of stoichiometric $MgFe_2O_4$ spinel. *J Mater Sci: Mater Electron,* 2009; 20 782–787. https://doi.org/10.1007/s10854-008-9802-2

[14] K.M. Batoo, S. Kumar, C.G. Lee, Alimuddin. Study of ac impedance spectroscopy of Al doped $MnFe_{2-2X}Al_{2x}O_4$. *J. Alloys Comp.* 2009; 480 596-602. https://doi.org/10.1016/j.jallcom.2009.01.137

[15] U. Erb. Electrodeposited Nanocrystals: Synthesis, properties and industrial applications. *Nanostructured Materials* 1995; 6(5) 533-538. https://doi.org/10.1016/0965-9773(95)00114-X

[16] M. Pal and D. Chakravorty. Nanocrystalline magnetic alloys and ceramics, *Sadhana. Indian Academy of Sciences* 2003; 28(1-2) 283-297.https://doi.org/10.1007/BF02717138

[17] J. Chargles, N. O'Connor, E. Kolesnichenko, C. Carpenter, S. W. Zheu, A. Kumbhar, S. Jessica and A. Fabrice. *Synthetic Metals,* 2001; 122(3) 547-557. https://doi.org/10.1016/S0379-6779(01)00328-9

[18] C. Caizer and M. Stefanescu. Magnetic characterization of nanocrystalline Ni–Zn ferrite powder prepared by the glyoxylate precursor method. *Journal of Physics D: Applied Physic,* 2002; 35(23) 3035-3040. https://doi.org/10.1088/0022-3727/35/23/301

[19] A. Dias. Microstructural evolution of fast-fired nickel–zinc ferrites from hydrothermal nanopowders. *Materials Research Bulletin* 2000; 35(9) 1439-1446. https://doi.org/10.1016/S0025-5408(00)00337-8

[20] S. Gubbala, H. Nathani, K. Koizol and R. D. K. Misra. Magnetic properties of nanocrystalline Ni–Zn, Zn–Mn, and Ni–Mn ferrites synthesized by reverse micelle technique. *Physica B Condenced Matter* 2004; 348(1-4) 317- 328.https://doi.org/10.1016/j.physb.2003.12.017

[21] Liu. Y., Liu. Z., Yang. Y., Yang. H., Shen. G., Yu. R. Sens. synthesis of $MgFe_2O_4$ Nanoparticles As Gas Sensing Materials. *Actuators. B Simple* 2005; 107 600-604. https://doi.org/10.1016/j.snb.2004.11.026

[22] A. Chatterjee, D. Das, S. K. Pradhan, D. Chakravarty. Synthesis of nanocrystalline Ni-Zn ferrite by the sol-gel method. *Journal of Magnetism and Magnetic Materials* 1993; 127(1-2) 214-218. https://doi.org/10.1016/0304-8853(93)90217-P

[23] A. Pradeep, C. Thangasamy, G. Chandrasekaran. Synthesis and structural studies on $Ni_{0.5+x}$ $Zn_{0.5}Cu$ x Fe_{2-2x} O_4, *J. Mater.* 2004; 15 797-802. https://doi.org/10.1023/B:JMSE.0000045302.52854.0b

[24] B. P. Jacob, S. Thankachan, S. Xavier, E. M. Mohammed. Dielectric behavior and AC conductivity of Tb^{3+} doped $Ni_{0.4}Zn_{0.6}Fe_2O_4$ nanoparticles. *Journal of Alloy of Compounds* 2012; 541 29-35. https://doi.org/10.1016/j.jallcom.2012.07.033

[25] Chen. N.S., Yang. X.J., Liu. E.S., Huang. J.L., Sensors and Actuators B: Chemical Reducing gas-sensing properties of ferrite compounds MFe_2O_4 (M=Cu, Zn, Cd and Mg). 2000; 66 178. https://doi.org/10.1016/S0925-4005(00)00368-3

[26] S. B. Kale, S. B. Somvanshi, M. N. Sarnaik, S. D. More, S. J. Shukla and K. M. Jadhav. Enhancement in surface area and magnetization of $CoFe_2O_4$ nanoparticles for targeted drug delivery application. *AIP Conf. Proc.,* 2018; 1953 030193.https://aip.scitation.org/doi/abs/10.1063/1.5032528

[27] M. Kaiser. Electrical conductivity and complex electric modulus of titanium doped nickel–zinc ferrites. *Physica B* 2012; 407 606. https://doi.org/10.1016/j.physb.2011.11.043

[28] A. M. Abdeen, O.M. Hemeda, E.E. Assem, M.M.El-Sehly. Structural, electrical and transport phenomena of Co ferrite substituted by Cd. Journal *of*

Magnetism and Magnetic Materials 2002; 238 75-83. https://doi.org/10.1016/S0304-8853(01)00465-6

[29] S. B. Somvanshi, R. V. Kumar, J. S. Kounsalye, T. S. Saraf and K. M. Jadhav. Investigations of structural, magnetic and induction heating properties of surface functionalized zinc ferrite nanoparticles for hyperthermia applications. AIP Conf. Proc., 2019; 2115 030522. https://doi.org/10.1063/1.5113361

[30] A. R. Chavan, R. R. Chilwar, P. B. Kharat and K. Jadhav. Impact of Trivalent Metal Ion Doping on Structural, Photoluminescence and Electric Properties of $NiFe_2O_4$ Thin Films. *J. Supercond.*, 2018; 1–10. https://doi.org/10.1007/s11664-019-07329-w

[31] D. R. Mane, D. D. Birajdar, Sagar E. Shirsath, R. A. Telugu, R. H. Kadam. Structural and magnetic characterizations of $Ni_{0.7-x}Mn_xZn_{0.3}Fe_2O_4$ ferrite nanoparticles. *Physica Status Solidi (a)* 2010; 207 2355-.2363. https://doi.org/10.1002/pssa.201026079

[32] M. A. Gabal, Y. M. Al Angari and S. S. Al-Juaid. A study on Cu substituted Ni–Cu–Zn ferrites synthesized using egg-white. *Journal of Alloys and Compounds* 2010; 492(1-2) 411-415. https://doi.org/10.1016/j.jallcom.2009.11.124

[33] S. S. Bellad, R. B. Pujar, B. K. Chougule. Structural and magnetic properties of some mixed Li-Cd ferrites. *Materials Physics and Chemistry* 1998; 52 166-169. https://doi.org/10.1016/S0254-0584(98)80019-9

[34] C.B.Kolekar, P.N.Kumble and S.G. Kulkarni. Effect of Gd^{3+} substitution on dielectric behavior of copper-cadmium ferrites, *Journal of Materials Science.* 1995; 30 (22) 5784-5788. https://doi.org/10.1007/BF00356721

[35] A. R. Chavan, M. Babrekar, A. C. Nawle and K. Jadhav. Structural and Optical Properties Of Nanocrystalline Ni–Zn Ferrite Thin Films. *J. Electron. Mater.*, 2019; 48 5184–5194. https://doi.org/10.1016/j.jallcom.2010.07.171

[36] N.F. Mott, E.A. Davis. Electron4. IC Processes in Non-crystalline Material. Oxford, London. 1979;

[37] S. M. Hoque, Md. A. Choudhury, Md. F. Islam. Characterization of Ni-Cu mixed spinel ferrite. *Journal of Magnetism and Magnetic Materials* 2002; 251 292-303. https://doi.org/10.1016/S0304-8853(02)00700-X

[38] V. Vinayak, P. P. Khirade, S. D. Birajdar, R. Alange and K. Jadhav. Electrical and Dielectrical Properties of Low-Temperature-Synthesized Nanocrystalline Mg^{2+}-Substituted Cobalt Spinel Ferrite Novel. Magn., *J. Supercond.*, 2015; 28 3351–3356. https://doi.org/10.1007/s10948-015-3159-6

[39] P. K. Gupta, C.T. Hung. Magnetically controlled targeted micro-carrier systems. *Life Sciences.* 1989; 44 175-186. https://doi.org/10.1016/0024-3205(89)90593-6

[40] Devan, R.S., Kolekar, Y.D., Chougule, B.K. Effect of cobalt substitution on the properties of nickel–copper ferrite. *J. Phys., Condens. Matter.* 2006; 18, 9809–9821 https://iopscience.iop.org/article/10.1088/0953-8984/18/43/004

[41] L. Vergard. "Die Konstitution der Mischkristalle, Die Raumfüllung der Atome" Physik A: Had- rons and Nuclei. 1921; 5(1) 17-26. https://doi.org/10.1007/BF01349680

[42] Compounds, B.I. International Union of Crystallography. Commission on Crystallographic Data List of works containing new data on crystal structures. 1961.. *J Struct Chem*. 1964;

4 739–754. https://doi.org/10.1007/BF00747673

[43] J.B. Li, G.H. Rao, J.K. Liang, Y.H. Liu, J. Luo, J.R. Chen. Magnetic properties of $Bi(Fe_{1-x}Cr_x)O_3Bi$ $(Fe_{1-x}Cr_x)O_3$ synthesized by a combustion method. *Appl. Phys. Lett,.* 2007; 90 162513. https://doi.org/10.1063/1.2720349

[44] S.H. Lee, S.J. Yoon, G.J. Lee, H.S. Kim, C.H. Yo, K. Ahn, D.H. Lee, K.H. Kim. Electrical and magnetic properties of $NiCr_xFe_{2-x}O_4$ spinel $(0 \leq x \leq 0.6)$ *Mater. Chem. Phys.,* 1999; 61 147. https://doi.org/10.1016/S0254-0584(99)00136-4

[45] M. Raghasudha. Characterization of nano-structured magnesium chromium ferrites synthesized by citrate-gel auto combustion method. *Adv. Mat. Lett.,.* 013; 4(12) 910-916. https://DOI:10.5185/amlett.2013.5479

[46] Muhammed Javed Iqbal, Mah Rukh Siddiquah. Electrical and magnetic properties of chromium-substituted cobalt ferrite nanomaterials. *J. Alloys Comp.,* 2008; 453 513. https://doi.org/10.1016/j.jallcom.2007.06.105

[47] M.A. Gabal , W.A. Bayoumy, A. Saeed, Y.M. Al Angari. Structural And Electromagnetic Characterization Of Cr-Substituted Ni–Zn Ferrites Synthesized Via Egg-White Route. *Journal of Molecular Structure.,* 2015; 1097 45–51.https://doi.org/10.1016/j.molstruc.2015.04.032

[48] R. D. Waldron. Infrared Spectra of Ferrites. *Physical Review of Journals Archive.* 1955; 99 1727.https://link.aps.org/doi/10.1103/PhysRev.99.1727

[49] R.M. Mohamed, M.M. Rashad, F.A. Haraz, W. Sigmund. Structure and magnetic properties of nanocrystalline cobalt ferrite powders synthesized using organic acid precursor method. *J. Magn. Magn. Mater.,* 2010; 322 2058.https://doi.org/10.1016/j.jmmm.2010.01.034

[50] Rintu Mary Sebastian, Sheena Xavier. E M Mohammed. Structural and electrical studies of Gd^{3+} substituted zinc ferrite nano particles. *IJESIT.* 2013; 2 4. https://doi.org/10.1080/00150193.2016.1239478

[51] S.A. Mazen And A.M. Abdel-Daiem. IR Spectra And Dielectric Properties Of Cu–Ge Ferrite. *Mater. Chem. Physics,* 2011; 130 847-852. https://doi.org/10.1016/j.matchemphys.2011.09.017

[52] Santosh Bhukal, Tsering Namgyal, S. Mor, S. Bansal, Sonal Singhal. Structural, electrical, optical and magnetic properties of chromium substituted Co–Zn nanoferrites $Co_{0.6}Zn_{0.4}Cr_xFe_{2-x}O_4$ $(0 \leq x \leq 1.0)$ prepared via sol–gel auto-combustion method. *Journal of Molecular Structure.* 2012; 1012 162–167.https://doi.org/10.1016/j.molstruc.2011.12.019

[53] Y.M. Yokovlev, L.B. Rubarikaya, N. Lapovok, Sov. Magnetic and electrical properties of Cr substituted Ni nano ferrites. *Phys.Solid.State.,* 1969; 10 2301. https://doi.org/10.2298/PAC1801001V

[54] S. Chikazumi. Physics of Magnetism. Wiley. New York. 1959;

[55] M. George, A.M. John, S.S. Nair, P. A. Joy, M.R. Anantharaman. Finite size effects on the structural and magnetic properties of sol–gel synthesized $NiFe_2O_4$ powders. *J. Magn. Magn. Mater.,* 2006; 302 190–195. https://doi.org/10.1016/j.jmmm.2005.08.029

[56] F. Li, H. Wang, L. Wang, J. Wang. Magnetic properties of $ZnFe_2O_4$ nanoparticles produced by a low - temperature solid-state reaction method. *J. Magn. Magn. Mater.,* 2007; 309 295. https://doi.org/10.1016/j.jmmm.2006.07.012

[57] Y. Koseoglu, H. Kavas. Size and Surface Effects on Magnetic Properties of Fe_3O_4 Nanoparticles. *J. Nanosci. Nanotechnol.,* 2008; 8 584. https://doi.org/10.1166/jnn.2008.B012

[58] Y. Yafet, C. Kittel. Physical Review. 1952; 87 239.

[59] A.K. Ghatage, S.C. Choudhary, S.A. J. Patil. Mater. X-ray, Structure and magnetic properties of Cd and Ti/Si substituted cobalt ferrites. *Sci. Lett.*, 1996; 15 1548–1550. https://doi.org/10.1016/S0254-0584(98)00202-8

[60] E.J.W. Verwey. The Crystal Structure of γ-Fe_2O_3 and γ-Al_2O_3. *Z. Kristallogr, Kristallgeo.* 1935; 91 65. https://doi.org/10.1524/zkri.1935.91.1.65

[61] M.J. Iqbal, M.R. Siddiquah. Structural, electrical and magnetic properties of Zr–Mg cobalt ferrite. *J. Magn. Magn. Mater.*, 2008; 320 845–850. https://doi.org/10.1016/j.jmmm.2007.09.009

[62] Sagar E. Shirsath, B.G. Toksha, Maheshkumar L. Mane, V.N. Dhage. Frequency, temperature and In^{3+} dependent electrical conduction in $NiFe_2O_4$ powder. *Powder Technology.*, 2011; 212 218–223. https://doi.org/10.1016/j.powtec.2011.05.019

[63] R. Manjula, V.R.K. Murthy, J. Sobhanadri. Electrical conductivity and thermoelectric power measurements of some lithium–titanium ferrites. *J. Appl. Phys.*, 1986; 59 2929. https://doi.org/10.1063/1.336954

[64] S.M. Patange, Sagar E. Shirsath, B. G. Toksha, S.S. Jadhav, K.M. Jadhav. Electrical and magnetic properties of Cr^{3+} substituted nanocrystalline nickel ferrite. *J. Appl. Phys.*, 2009; 106 023914-1. https://doi.org/10.1063/1.3176504

[65] S.M. Ramay, S.A. Siddiqi, S. Atiq, M. S. Awan, S. Riaz, Chin. Structural, Magnetic, and Electrical Properties of Al3+ Substituted CuZn-ferrites. *J. Chem. Phys.*, 2010; 23 591–595. DOI:10.1088/1674-0068/23/05/591-595

[66] M. Azhar Khan, M.U. Islam, M. Ishaque, I.Z. Rahman. Effect of Tb substitution on structural, magnetic and electrical properties of magnesium ferrites. *Ceramics International.*, 2011; 37 2519–2526. https://doi.org/10.1016/j.ceramint.2011.03.063

[67] N. Rezlescu, E. Rezlescu, P.D. Popa, L. Rezlescu. Effects of rare-earth oxides on physical properties of Li–Zn ferrite. *Journal of Alloys and Compounds.*, 1998; 657–659. https://doi.org/10.1016/S0925-8388(98)00413-7

[68] A.B. Gadkari, T.J. Shinde, P.N. Vasmbekar. Structural analysis of Y^{3+}-doped Mg–Cd ferrites prepared by oxalate co-precipitation method, Materials Chemistry and Physics. 2009; 114 505–510.https://doi.org/10.1016/j.matchemphys.2008.11.011

[69] B. Ramesh, D. Ravinder. Electrical properties of Li–Mn ferrites. *Materials Letters,* 2008; 62 2043–2046. https://doi.org/10.1016/j.matlet.2007.11.010

[70] Santosh Bhukal, Tsering Namgyal, S. Mor, S. Bansal. Structural, electrical, optical and magnetic properties of chromium substituted Co–Zn nanoferrites $Co_{0.6}Zn_{0.4}Cr_xFe_{2-x}O_4$ $(0 \leq x \leq 1.0)$ prepared via sol–gel auto-combustion method. *Journal of Molecular Structure,* 2012; 1012 162–167. https://doi.org/10.1016/j.molstruc.2011.12.019

[71] Muhammad Naeem Ashiq, Farah Naz, Muhammad Aslam Malana, R S Gohar.Role of Co–Cr substitution on the structural, electrical and magnetic properties of nickel nano-ferrites synthesized by the chemical co-precipitation method. *Materials Research Bulleti.,* 2012; 47 683–686. https://doi.org/10.1016/j.materresbull.2011.12.017

[72] A.Lakshman, K.H.Rao, R.G. Mendiratta. Magnetic properties of In^{3+} and Cr^{3+} substituted Mg–Mn ferrites. *J. Magn.Magn.Mater.*, 2002; 250 92–97. https://doi.org/10.1016/S0304-8853(02)00359-1

[73] S.J. Haralkar, R.H. Kadam, S.S More, Sagar E. Shirsath, M.L. Mane, Swati Patil. Substitutional effect of Cr^{3+} ions on the properties of Mg–Zn ferrite nanoparticles. Physica B. 2012; 407 4338–4346. https://doi.org/10.1016/j.physb.2012.07.030

[74] Chandra Babu B, Naresh V, Jayaprakash B, Buddhudu S. Structural, thermal and dielectric properties of lithium zinc silicate ceramic powders by sol-gel method. Journal of ferroelectric letters section. 2011; 38 124-130.https://doi.org/10.1080/07315171.2011.623610

[75] J.C. Maxwell. Electricity and Magnetism. Oxford University. Press, London. 1973;.

[76] K.W.Wagner. Ann. Phys. 1993; 40 818.

[77] C.G. Koops. Phys. Rev. 1951; 83 121.

[78] A.M. Shaikh, S.S. Bellad, B.K. Chougule. Temperature and frequency-dependent dielectric properties of Zn substituted Li–Mg ferrites. *J.Magn. Magn. Mater.*, 1999; 195 384-390. https://doi.org/10.1016/S0304-8853(99)00138-9

[79] I.T. Rabinkin, Z.I. Novikova, Ferrites, Izv Acad. Nauk USSR Minsk. 1960; 146.

[80] N. Rezlescu and E. Rezlescu. Dielectric properties of copper containing ferrites. *Phys. Status Solidi A*, 1974; 23 575-582. https://doi.org/10.1002/pssa.2210230229

[81] R.K.Kotnala, M.Abdullah Dar, Vivek Verma, A.P.Singh. Synthesis and characterizations of Ni^{2+} substituted cobalt ferrite nanoparticles. *Journal of Magn. Magn. Matter.* 2010; 322 3714-3719. https://doi.org/10.1016/j.matchemphys.2012.09.019

[82] Mohd Hashim, Alimuddin, Sagar E. Shirsath, Shalendra Kumar, Ravi Kumar. Preparation and characterization chemistry of nano-crystalline Ni–Cu–Zn ferrite. *Journal of Alloys. Comp.*, 2013; 549 348-357. https://doi.org/10.1016/j.jallcom.2012.08.039

[83] R. M. Sebastian. Structural and dielectric studies structural and dielectric studies tural and dielectric studies of Cr^{3+} doped znfe doped $znfe_2o_4$ nanoparticles Nano Studies. *Journal_of_Nano_Research,* 2013; 8 121-130. https://www.researchgate.net/journal/1661-9897

[84] M. Raghasudha, D. Ravinder, P. Veerasomaiah. Influence of Cr^{+3} Ion on the Dielectric Properties of NanoCrystalline Mg-Ferrites Synthesized by Citrate-Gel Method. *Materials Sciences and Applications.* 2013; 4 432-438. https://DOI:10.4236/MSA.2013.47052

[85] D. Elkony, Study of dielectric and impedance properties of Mn ferrites. *Egypt J. solids,* 2004; 27 285-296.

[86] M. Raghasudha, D. Ravinder, P. Veerasomaiah. Electrical and Magnetic Properties of Mg-Cr and Co-Cr Nano Ferrites Synthesized by Citrate-Gel Method. *Solid State Phenomena.,* 2016; 241 69-92. https://doi.org/10.4028/www.scientific.net/SSP.241.69

Chapter 8

The Effect of Ceramic Wastes on Physical and Mechanical Properties of Eco-Friendly Flowable Sand Concrete

Mohamed Guendouz, Djamila Boukhelkhal, Alexandra Bourdot, Oussama Babachikh and Amine Hamadouche

Abstract

This work aims to study the valorization and recycling of ceramic wastes (wall tiles) as a fine aggregate instead of sand in the manufacturing of flowable sand concrete (FSC). For this, the sand is substituted with the ceramic wastes at different dosages (0, 5, 10, 15, 20, and 25% by volume of the sand). The influence of the ceramic wastes addition on the physical (workability, density) and mechanical (compressive, flexural and elastic modulus) properties of FSC was studied. The results show that the use of ceramic waste as partial replacement of sand contributes to reduce the workability, bulk density and improves the mechanical strengths of FSC according to the use of 25% of wall tiles waste.

Keywords: valorization, recycling, ceramic waste, flowable sand concrete, workability, mechanical strengths

1. Introduction

Due to the increase in coarse aggregates consumption and the availability of large quantities of sand in Algeria, as well as the complexity of designs and reinforcement details in modern structural members, producing fluid concretes for easy implementation and no compaction has become one of the main desires and challenges for building industry. Flowable sand concrete is a new type of concrete that make part of the important building materials permitting to valorize some local resources and waste (dune sand, fillers, waste, garbage and under local products).

In the last years, construction and demolition waste provides a substantial source of natural raw materials for building works by using construction site waste. In addition, the amount of ceramic waste, presents the highest fraction of construction and demolition wastes. Therefore, disposal of a variety of ceramic wastes in an ecofriendly way is the thrust area of today's research.

Several studies have been done on the use of ceramic waste as coarse aggregate, powder and filler in the preparation of cement mortar [1–6], concrete [7–13] and self-compacting concretes [14–16], high strength concrete [17, 18] and ultra-high-performance concrete [19, 20]. Many of them, explored that the use of ceramic waste, as a material in concrete composite, leads to a decrease in workability and density of mixture [2, 3]. The mechanical properties of mortar and concrete incorporating ceramic waste have also studied and analyzed by several researchers [3, 10]. The majority of results showed that, for up to optimum percentage replacement of natural sand by ceramic waste, the mechanical strength of concrete is similar or even better than those containing natural aggregates. Tabak et al. [21] studied the effect of ceramic waste as aggregate (CW), dust (FTDA) and combinations of them to produce concrete. They found an increasing of about 13.53, 16.70, 2.91% and 23.21, 0.10, 19.47% for compressive and flexural strength respectively, at 2, 7 and 28 days. The similar results were also stated by Abadou et al. [5]. They studied the effect of partial replacement for natural sand by ceramic waste with different percentage (10, 20, 30, 40 and 50%) on the performance of ordinary concrete. And they observed that the mechanical properties of ceramic waste mortar increase with replacement of natural dune sand by CW.

The durability properties of concrete counting ceramic wastes were also investigated by several researchers. It found in the study of many authors [1, 5, 10, 18] that there is no significant change in the basic trend of permeation characteristics of this recycled aggregate concrete when compared to the conventional concrete. Tabak et al. [21] demonstrated a reduction of about 0.17% in water absorption of concrete made with recycled ceramic aggregate, when compared to conventional concrete. Elçi [10] studied the effect of total replacement of natural sand by ceramic waste on water absorption and shrinkage of cement concrete. It was observed an increasing in the values of drying shrinkage and water absorption compared to the conventional concrete. Abadou et al. [5] found that the addition of CW in dune sand mortar increases its acids resistance property. It was observed that the mortar made with ceramic waste aggregate shows better resistance to sulfuric (H2SO4) and hydrochloric (HCl) acid solution attack than reference mortar. Hence, the mortar with CW performs well in durability aspect, this addition of ceramic waste, improves the behavior of mortars subject to attacks HCl and H2SO4 acids.

The aim of this research is to study the possibility of recycling ceramic waste without any prior treatment except crushing in order to produce low cost flowable sand concrete. For this the influence of partial replacement of natural sand by ceramic waste on the workability and physic-mechanical properties of the new composite material has been studied and compared to the control FSC. And the optimal proportion of ceramic waste substitution which can give the ecofriendly lightweight flowable sand concrete was then determined.

2. Experimental

2.1 Materials

The used sand in this study is a local material, extracted from the south of Algeria. It presents a siliceous nature as demonstrated by its X-ray Diffraction Analysis (**Figure 1**), and a continuous particle size distribution ranging from 0.08 to 4 mm (as given in **Figure 2**). Hence, their physical properties are presented in **Table 1**. The Scanning Electron Micrograph (SEM) view of their grains is given in **Figure 3**.

Portland cement CEM II/A 42.5 from MASCARA Factory in Algeria was used throughout this study, with a density of 3100 kg/m^3.

The use of fillers in flowable sand concrete composition is essential [22]. Its use helps to improve the compactness of concrete by completing the granular distribution of sand in its finest part. As well as to reducing the cement content and produce a low cost concrete. A marble powder (MP) was used in this study as fillers in FSC mixers with a specific density of 2.73 kg/m^3. And a specific

DOI: http://dx.doi.org/10.5772/intechopen.95041

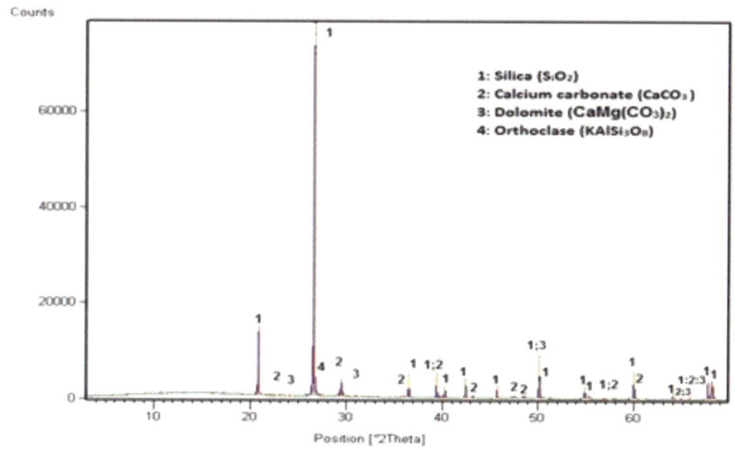

Figure 1.
X-ray diffractogram analysis of sand.

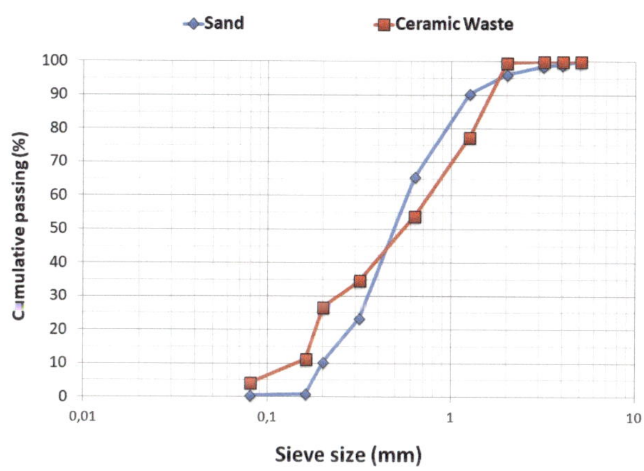

Figure 2.
Particle size distribution of sand and ceramic waste.

Properties	Sand	Ceramic waste
Apparent density (kg/m^3)	1630	1010
Specific density (kg/m^3)	2650	2440
Water absorption (%)	2.00	4.05
Sand equivalent (%)	62.50	—
Fineness modulus	2.25	2.9
Compactness (%)	61	41
Porosity (%)	39	59

Table 1.
Physical properties of sand and ceramic wastes.

surface area measured with the Blaine's permeability meter according to EN 196–6 standard of about 220 m^2/kg.

The ceramic waste used in this study has been obtained from the disposal area of the ceramic factory in Algeria (Ceramic wall tiles). The physical properties of this waste are presented in **Table 1** and their sieve analysis results are shown in **Figure 2**.

After the collection of these wastes, they were crushed and extruded in the form of grains (**Figure 4**), and then used in the manufacturing of FSC by volumetric substitution of natural sand with different percentages (0, 5, 10, 15, 20 and 25%). The Scanning Electron Micrograph (SEM) view of their grains is given in **Figure 5**.

Figure 3.
Scanning electron micrographs of sand, G = 40.

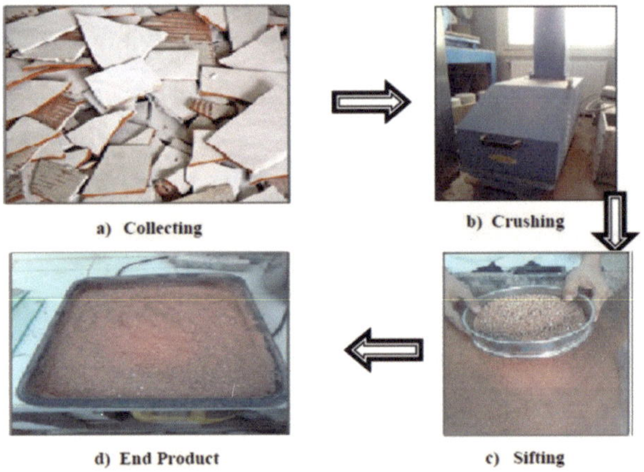

Figure 4.
Stages of obtaining the ceramic wastes.

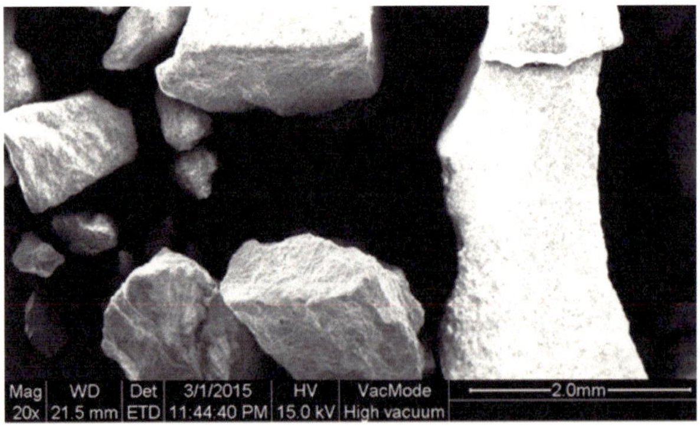

Figure 5.
General aspect SEM micrograph (26X) of ceramic waste.

In this work a polyether–polycarboxylate based superplasticizer 'MEDAFLUID145' in liquid form and chestnut color was used as chemical admixture with a solid content of 30%, specific density of 1.08 g/cm^3, pH equal to 6 and a content of color <1 g/L.

The mixing water used for the different mixes is the distribution drinking water.

2.2 Mix design

In this study, the FSC formulation is based on the theoretical method of Sablocrete project [22]. The CW was incorporated into the mass of flowable sand concrete by partial replacement of sand volume with different percentages from 0 to 25%. The mix proportions of each FSC are given in **Table 2**. As seen from this table and, described below, the mixtures were coded such that, the percentage of CW used were identified in a precise way.

- **CFSC**: Control flowable sand concrete (without waste).
- **FSC CW**: Flowable sand concretes with ceramic waste.

All FSC mixes are manufactured in the laboratory environment by a standard mortar mixer with a capacity of 5 l and all components of FSC mixture were batched by weight. For a better distribution of admixtures within the mass of FSC, superplasticizer was diluted with 40% of mixing water before added to the concrete. It consists to mixes the entire components (aggregate + cement + filler) in the dry state for a half minute. Then, a 60% of mixing water was added and mixed for one minute (1 min) before adding the remaining 40% of water mixed with the superplasticiser and mixed for 1 min. The mixing is stopped after about 3 min before remixing for another one minute (1 min).

All specimens were produced in a laboratory environment at 20°C and 50% relative humidity (RH). After 24 h, they were removed from the molds and placed in water at 20°C and 100% RH until the day of testing.

2.3 Methods

Before casting, the fluidity of FSC was measured by using mini-slump flow diameter test and V-funnel test according to EFNARC (**Figures 6** and 7). For the flow spread test, the truncated cone mold is placed on the plate, filled with the

Constituent	**CFSC**	**FSC CW (%)**				
		5	10	15	20	25
Sand (Kg/m^3)	1330	1260	1190	980	1120	1050
Cement (Kg/m^3)	350	350	350	350	350	350
Fillers (Kg/m^3)	250	250	250	250	250	250
Water (l/m^3)	247	247	247	247	247	247
SP (%)*	1	1.4	1.4	1.60	1.70	2
Ceramic waste (kg/m^3)	0	64.7	129.39	194.09	258.5	528.18

Table 2.
Mix proportion of FSC with ceramic waste.

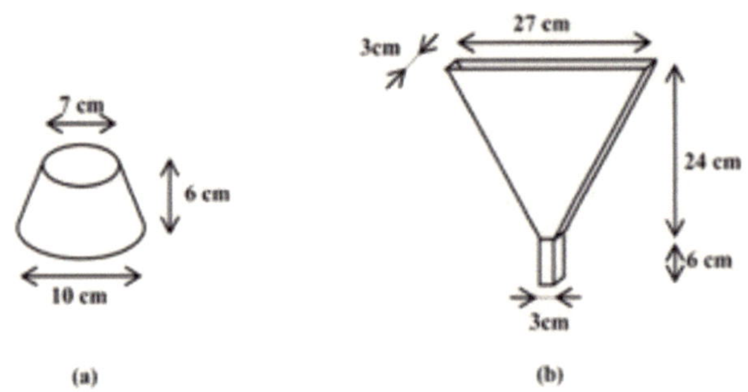

Figure 6.
Workability tests for fresh FSC mixes (a/ mini-slump flow test; b/ V-funnel test).

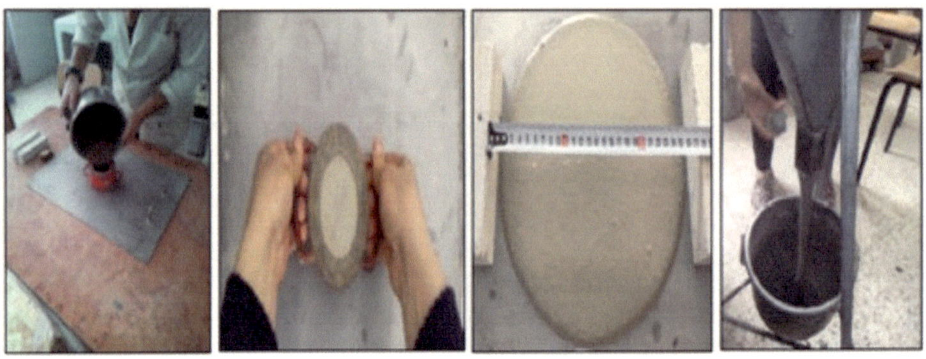

Figure 7.
Fluidity measurement test of FSC.

FSC mixture, and lifted. The subsequent diameter of the mixture is measured in two perpendicular directions, and the mean is taken. For the V-funnel test, the funnel is filled with 1.1 l of FSC mixture, and the V-funnel flow time is that between opening the orifice and the first daylight appearing when looking vertically down through the funnel. The bulk density was evaluated after according to NF EN 12350–6.

The compressive and flexural strength are measured in the hardened state on three 40 × 40 × 160 mm samples at 28 days according to EN 196–1. The flexural strength was measured by a three-point bending test, using a testing machine with a maximum load capacity of 30 kN. The half samples resulting from this test were then submitted to compression test. The modulus of elasticity in compression was measured at the age of 28 days on cylinders of 320 mm in diameter and 160 mm of height by determining the longitudinal deformations during loading using a strain gauge and according to ISO 834.

The microstructure of various FSC mixtures is investigated after 28 days of curing by means of scanning electron microscopy (SEM) for very high magnifications and a video - microscope (Controlab ®) VH-Z25 equipped with a 25x to 175x zoom for low magnifications. The FSC samples were first cut into slices using a diamond saw. From the middle of the mid-slice, a block of 20 × 20 mm was cut. Flat polished epoxy impregnation specimens were used for acquiring backscattered electron images. The SEM observation ware carried out on simple surface after making them conductive by metallization (covering them, under vacuum, with a layer of approximately 10 to 20 nm of gold).

3. Results and discussion

3.1 Characterization of FSC in fresh state

The results of the ceramic waste effect on the slump flow diameter and V-funnel flow time for FSC mixes are given in **Table 3**. The results displayed that the workability of FSC decrease with the increase of ceramic waste. This decrease in workability is maybe due to the high water absorption (**Table 1**), angular shape and rough surface texture of -ceramic waste grain compared to natural sand, which had a rounded shape of the grains (**Figures 3** and **5**). The decrease in workability of concrete, made with CW, was observed also by Abadou et al. [2] Guendouz and Boukhekhal [23] and Daniyal et al. [24].

In order to limit the number of compositions and to be able to compare them in the hardened state on a common basis, the workability was fixed by a constant slump flow diameter, with a value close to 27 mm, and fixed water to binder ratio at 0.4. The workability measure was adjusted by varying the superplasticizer quantity for each mixture contains ceramic waste as reported in **Figure 8**. It is clearly shown from this figure that superplasticizer demand increased with the increase of ceramic waste content in FSC.

Figure 9 shows that the substitution of sand by ceramic waste causes a slight decrease in bulk density of mixes, which is probably due to the lower density of the ceramic waste aggregate grains compared with natural sand (**Table 1**). The decrease in density of concrete made with CW was proved also by many authors [5, 10, 25–29] which reported that the use of ceramic waste as aggregates reduced the concrete density.

Concrete	Slump flow diameter (mm)	V-funnel flow time (s)
CFSC	260	7
FSC 5% CW	250	10
FSC 10% CW	230	13
FSC 15% CW	210	18
FSC 20% CW	190	22
FSC 25% CW	180	26

Table 3.
Properties of FSC mixes at fresh state.

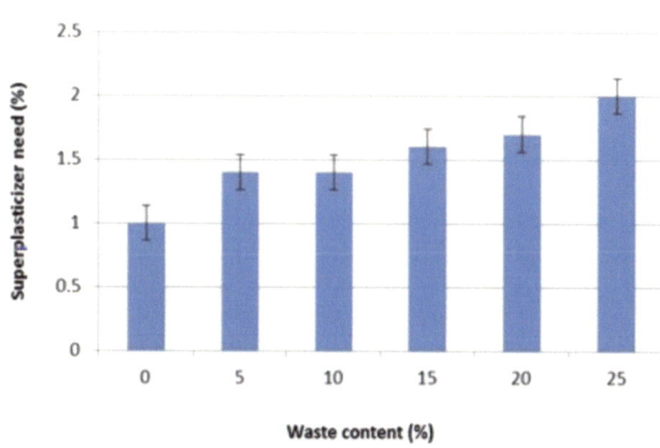

Figure 8.
Superplasticizer need of fresh FSC as function of ceramic wastes content.

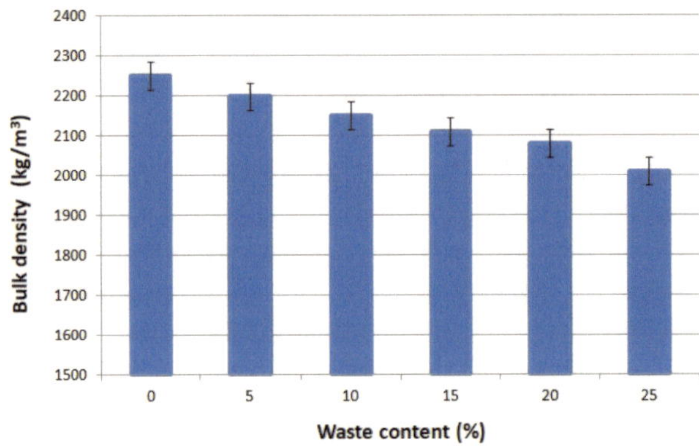

Figure 9.
Bulk density of fresh FSC as function of ceramic wastes content.

3.2 Mechanical properties of FSC in hardened state

The results of compressive and flexural strength for all FSC mixes, at 28 days of age are presented in **Figures 10** and **11** respectively. This results show an improvement in compressive and flexural strength of FSC mixes with ceramic waste for all ages compared to FSC without ceramic waste. An increase of about 30 and 57% was observed at 28 days in compressive and flexural strength respectively, for a replacement ratio of 25% of sand by ceramic waste. This increase in mechanical strength is due to the hardness of ceramic waste grains compared to those of natural sand, and to their rough and irregular shape which fill the void. This later led to higher frictional resistance and improves their good adhesion with the cement paste (**Figure 12**).

Similar results were observed by Abadou et al. [5] for dune sand mortar containing CW. They found an increase in compressive strength of dune sand mortar with 40 and 50% of CW. Elçi [10] has reported similar mechanical properties to those of traditional limestone concrete when using ceramic as recycled aggregates. Anderson et al. [27] studied also the effect of ceramic waste on concrete mechanical strength; they observed an increase in concrete strength with the incorporation of fine ceramic aggregates along with the coarse. Tennich et al. [30] reported that the compressive strength of concrete containing CW is higher than those of concrete made with natural aggregate.

The results of modulus of elasticity tests carried out on the different FSC mixes at 28 days are summarized in **Figure 13**. As shown in this figure, the elastic modulus of all concrete mixes, increases when ceramic waste aggregates content increases.

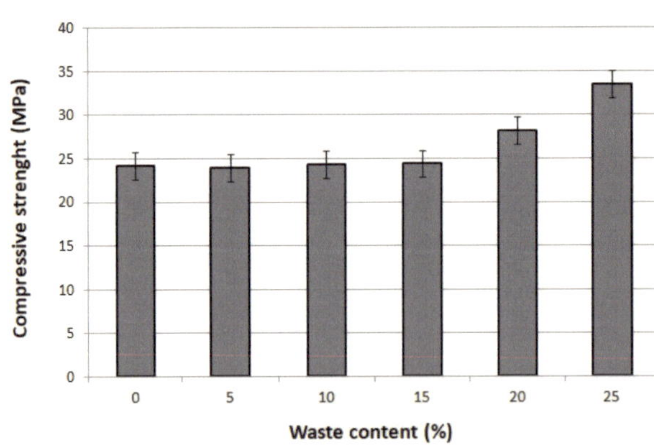

Figure 10.
Effect of ceramic waste on compressive strength of FSC.

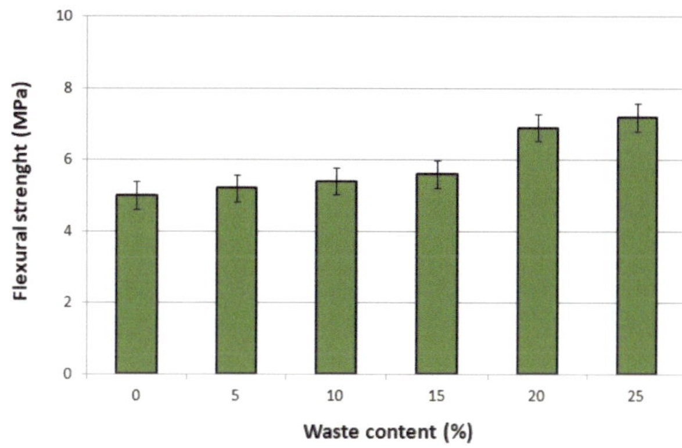

Figure 11.
Effect of ceramic waste on flexural strength of FSC.

Figure 12.
SEM image showing the good adhesion of CW with cement matrix in FSC.

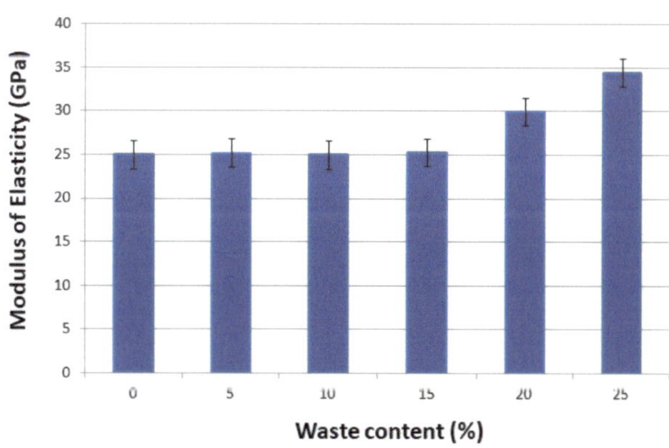

Figure 13.
Elastic modulus of different FSC mixes.

Its value varies between 25 GPa for control flowable sand concrete, and 34.92 GPa for mixtures containing 25% of CW, with a gain of about 28.4%. This increase on modulus of elasticity is due to the high compressive strength of mixtures, the angular particles shape that occupy the voids between sand grains, and to the better adherence between CW and paste. Thus, FSC mixes become more compact and, as a consequence, the modulus of elasticity increased. The modulus of elasticity may be directly related to the compressive strength. The results agree with those in [27] for CW aggregates. Tennich et al. [30] studied the effect of partial replacement of natural sand by ceramic waste with different percentage (0, 20, 35, 50, 65, 80 and 100%) on the elastic modulus of natural concrete. It observed that the elastic modulus of ceramic waste series increased by 27% with 100% replacement. Abadou et al. [5] also reported that the mortar incorporating ceramic waste has a higher modulus of elasticity compared to the ordinary mortar. Contrarily, Elçi [10] found that there was a decrease of in modulus of concrete made with CW compared to reference concrete.

4. Conclusions

From the obtained results in this study, the following conclusions can be drawn:

- The use of ceramic waste as partial replacement of sand in flowable sand concrete lead to decrease its workability.
- The density of flowable sand concrete is decreased when the percentage of ceramic waste increases.
- The use of ceramic waste as sand in FSC improves their mechanical strengths at all ages. An increase of about 30 and 57% in compressive and flexural strength respectively was observed at 28 days when 25% of natural sand were substituted by ceramic waste.

On the basis of the results obtained in this study, it can be concluded that the use of ceramic waste with the local materials in flowable sand concrete manufacturing, is an interesting economic, environmental and technological alternative and is added to the list of materials available in the country.

Author details

Mohamed Guendouz[1*], Djamila Boukhelkhal[1], Alexandra Bourdot[2], Oussama Babachikh[1] and Amine Hamadouche[1]

1 LME Laboratory, University of Medea, Medea, Algeria

2 LMT Laboratory, University of Paris-Saclay, Paris, France

*Address all correspondence to: guen12moh@gmail.com

References

[1] Klimek B, Szulej J. & Ogrodnik P. The effect of replacing sand with aggregate from sanitary ceramic waste on the durability of stucco mortars. Clean Techn Environ Policy.2020; 22: 1929-1941. https://doi.org/10.1007/s10098-020-01932-w

[2] Abadou Y, Kettab R. and Ghreib A. Experimental investigation on the carbonation properties of dune sand ceramic waste mortar. Jour of Eng Des Tech.2018; 16 (3): 501-516. https://doi.org/10.1108/JEDT-05-2017-0047

[3] Farinha CB , de Brito J and Veiga R. Rendering Mortars with Low Sand and Cement Content. Incorporation of Sanitary Ware Waste and Forest Biomass Ashes Appl Sci. 2020; 10(9): 3146. https://doi.org/10.3390/app10093146

[4] Ghrieb A and Abadou Y . The use of Crushed Ceramic Waste as Granular Corrector in the Manufacture of the Mortars Based on Dune Sand. Jour of Eng Sci and Tech Rev. 2018; 11 (4): 126-131. https://doi.org/10.25103/jestr.114.16

[5] Abadou Y, Mitiche-Kettab R, Ghrieb A. Ceramic waste influence on dune sand mortar performance. Constr Build Mater. 2016;125:703-713. https://doi.org/10.1016/j.conbuildmat.2016.08.083

[6] Jimenez JR, Ayuso J, Lopez M et al. Use of fine recycled aggregates from ceramic waste in masonry mortar manufacturing. Constr Build Mater. 2013;40:679-690. https://doi.org/10.1016/j.conbuildmat.2012.11.036

[7] Yuko O, Phuong Trinh B, Kenji K. Ryoichi Sato c Effects of porous ceramic roof tile waste aggregate on strength development and carbonation resistance of steam-cured fly ash concrete Constr Build Mater. https://doi.org/10.1016/j.conbuildmat.2019.117462

[8] Keshavarz Z, Mostofinejad D. Steel chip and porcelain ceramic wastes used as replacements for coarse aggregates in concrete. Jour of Clean Prod. 2019;230: 339-351 https://doi.org/10.1016/j.jclepro.2019.05.010

[9] Awoyera P, Onoja E and Adesina A. Fire resistance and thermal insulation properties of foamed concrete incorporating pulverized ceramics and mineral admixtures. Asian J Civ Eng.2020; 21: 147-156. https://doi.org/10.1007/s42107-019-00203-4

[10] Elci H. Utilisation of crushed floor and wall tile wastes as aggregate in concrete production.J Clean Prod. 2016;112:742-752. https://doi.org/10.1016/j.jclepro.2015.07.003

[11] Miguel CS, Nepomuceno Rui, Isidoro A.S, Catarino JPG. Mechanical performance evaluation of concrete made with recycled ceramic coarse aggregates from industrial brick waste. Constr Build Mater. 2018;165: 284-294 https://doi.org/10.1016/j.conbuildmat.2018.01.052

[12] Keshavarz Z, Mostofinejad D. Porcelain and red ceramic wastes used as replacements for coarse aggregate in concrete. Constr Build Mater. 2019;195:218-230. https://doi.org/10.1016/j.conbuildmat.2018.11.033

[13] Zegardło B , Szela M , Ogrodnik P, Concrete resistant to spalling made with recycled aggregate from sanitary ceramic wastes – The effect of moisture and porosity on destructive processes occurring in fire conditions. Constr Build Mater. 2018;173: 58-68. https://doi.org/10.1016/j.conbuildmat.2018.04.030

[14] Subaşı S, Öztürk H ,Emiroğlu M . Utilizing of waste ceramic powders as filler material in self-consolidating concrete. Constr Build Mater. 2017;149:567-574. https://doi.org/10.1016/j.conbuildmat.2017.05.180

[15] Viramgama PD , Vaniya SR, Parikh KB . Effect of Ceramic Waste Powder in Self Compacting Concrete Properties: A Critical Review. 2016 ;13 (1):08-13

[16] Huseien GF, Mohd Sam AR, Shah KW, Mirza J. Effects of ceramic tile powder waste on properties of self-compacted alkali-activated concrete. Constr Build Mater. 2020;236: 117574. https://doi.org/10.1016/j.conbuildmat.2019.117574

[17] Zareei SA , Ameri F, Shoaei P, Bahrami N. Recycled ceramic waste high strength concrete containing wollastonite particles and micro-silica: A comprehensive experimental study. Const Build Mater. 2019;201 11-32. https://doi.org/10.1016/j.conbuildmat.2018.12.161

[18] Zareei SA, Ameri F, Bahrami N, Shoaei P, Musaeei HR, Nurian F. Green high strength concrete containing recycled waste ceramic aggregates and waste carpet fibers: Mechanical, durability, and microstructural properties. J of Build Eng. 2019;26: 100914. https://doi.org/10.1016/j.jobe.2019.100914

[19] Amin M, Bassam A, Tayeh, Agwa IS. Effect of using mineral admixtures and ceramic wastes as coarse aggregates on properties of ultrahigh-performance concrete.2020;273: 123073. https://doi.org/10.1016/j.jclepro.2020.123073

[20] Soleimani SM, Alaqqad AR, Jumaah A and Majeed A. Examining the Eects of Introducing and Combining Electric-Arc Furnace Slag and Ceramic Waste in a Single Self-Consolidating, High-Strength Concrete Mix. Appl Sci.2020; 10(14):4844. https://doi.org/10.3390/app10144844

[21] Tabak Y, Kara M, Günay E, Yildirim S.T and Yilmaz Ş . Ceramic tile waste as a waste management solution for concrete. in proceeding of the 3 rd International Conference on Industrial and Hazardous Waste Management (CRETE), Turkey. 2012.

[22] Presse de l'Ecole Nationale des Ponts et Chaussées 'Béton de sable, Caractéristiques et pratique d'utilisation', (Projet SABLOCRETE), édition. Association Amicale des Ingénieurs Anciens Elèves de L'Ecole Nationale des Ponts et Chaussées; 1994. p. 15-71.

[23] Guendouz M, Boukhelkhal Dj. Physical and mechanical properties of cement mortar made with brick waste. MATEC Web of Conferences.2018;149: 01077.https://doi.org/10.1051/matecconf/201814901077

[24] Daniyal M and Ahmad S. Application of Waste Ceramic Tile Aggregates in Concrete. Inter Jour of Innov Resea in Sci Eng and Tech. 2015; 4(12): p. 12808-12815.

[25] Pitarch A, Reig L, Tomás AE, López FJ. Effect of Tiles, Bricks and Ceramic Sanitary-Ware Recycled Aggregates on Structural Concrete Properties. Waste and Biomass Valorization. 2017: p. 1-15.

[26] Tavakoli D, Heidari A and Karimian M. Properties of concretes produced with waste ceramic tile aggregate. As Jour of Civ Eng. 2013; 14(3): p. 369-382.

[27] Anderson DJ, Smith ST and Au FT. Mechanical properties of concrete utilising waste ceramic as coarse aggregate. Constr Build Mater. 2016; 117: p. 20-28.

[28] Malik P, Malhotra J, Verma A, Bhardwaj P, Dhoundiyal A and N Yadav. Mix design for concrete with crushed ceramic tiles as coarse aggregate. Inter Jour of Civ Eng Resear. 2014; 5(2): p. 151-154.

[29] Guendouz M, Boukhelkhal Dj : Properties of flowable sand concrete

containing ceramic wastes. J Adhes Sci Technol.2019;33(24): 2661-2683.

[30] Tennich M, Kallel A, and Ouezdou MB. Incorporation of fillers from marble and tile wastes in the composition of self-compacting concretes. Constr build mater. 2015; 91: p. 65-70.

Section 5

Applications of Ceramic Materials

Chapter 9

Ceramics Coated Metallic Materials: Methods, Properties and Applications

Dongmian Zang and Xiaowei Xun

Abstract

Surface coating can allow the bulk materials to remain unchanged, while the surface functionality is engineered to afford a more wanted characteristic. Ceramic coatings are considered as ideal coatings on metal which can significantly improve the surface properties of metal materials including anti-fouling, self-cleaning, corrosion resistance, wear resistance, oil/water separation and biocompatibility. Furthermore, various techniques have been utilized to fabricate a range of different ceramic coatings with more desirable properties on metal materials, which make the materials widely used in service environment. This chapter focus will be on the types, fabrication methods, surface properties and applications of ceramics coated metal materials.

Keywords: ceramic coating, metallic materials, surface physicochemistry

1. Introduction

Metallic materials such as Fe, Cu, Ti, Al, Mg and their alloys have excellent mechanical and physical properties showing tremendous application in architecture, marine, aerospace and biomedicine fields, etc. [1–6]. To a certain extents, the surface properties of the metallic materials are playing irreplaceable roles in operating environments. Surface functionalization can improve corrosion resistance, anti-fouling, self-cleaning, wear resistance, oil/water separation and biocompatibility of metallic materials [7–9]. In this context, surface coating is an efficient and resource saving method to realize the surface functionalization of metallic materials. In addition, ceramic coating is environmentally friendly, and has the advantages of low cost, simple preparation, corrosion and wear resistance, thermal stability, and mechanical durability [10]. As such, constructing a ceramic coating on metallic material surface is a rational strategy to realize the surface multi-function [11, 12].

In this chapter, we briefly introduce the types and the properties of ceramic coatings. Then, we summarize the strategies for preparing ceramic coatings on metallic materials and applications of ceramics coated metallic materials.

2. Ceramics coated metallic materials

Ceramics materials can be divided into oxide ceramics and non-oxide ceramics according to their compositions. Many oxide ceramics are metal oxides forming oxide films on their surfaces, which are used as coating materials for the protection

IntechOpen

and functional layer of metallic materials (for example, aluminum, stainless steel or titanium alloys). Also, diverse non-oxide ceramic materials are used to functionalize the surfaces of metal materials.

2.1 Ceramic coatings types

Ti and its alloy have excellent corrosion resistant to alkali, chloride and some strong acids because of the compact oxide film (Titania, TiO_2) formed spontaneously on surfaces. Therefore, TiO_2 coating is considered to be an ideal corrosion resistant layer to protect the metal substrate from corrosion. Shen *et al.* fabricated a uniform TiO_2 nanoparticle coating on 316 L stainless steel by using sol-gel technology, the electrochemical results showed that the TiO_2 coating on 316 L stainless steel effectively prevent the substrate from corrosion in chloride containing solution at the room temperature [13]. Furthermore, studies exhibited that the TiO_2 coating with nanostructure had excellent photoactive antibacterial property and hemocompatibility [14, 15].

Alumina (Al_2O_3) exhibits exceptional mechanical property and thermostability possessing a broad range of applications in optics, electronic, and biomedical fields. In addition, the corrosion resistance of Al and its alloys is attributed to inherent Al_2O_3 coating, which can effectively improve the corrosion resistance of metallic substrate. Gao *et al.* prepared the Al_2O_3 ceramic coating on AZ31PH Mg alloy by laser remelting plasma-sprayed coating, it was found that the Al_2O_3 ceramic coating exhibited high hardness as well as wear and corrosion resistance properties [16].

Similarly, silica (SiO_2) is also highly desirable coating materials on metallic materials as wear and corrosion resistant coating. The corrosion-resistant SiO_2 ceramic coating on alloys was prepared by metal organic chemical vapor deposition (MOCVD) [17]. In addition, Sadreddini *et al.* revealed that the corrosion rate and porosity of coating decreased with increasing the quantity of the SiO_2 nanoparticles in the bath [18, 19].

As the most stable oxide of manganese, manganese dioxide (MnO_2) has abundant reserves in the earth, and has the advantages of low cost, environmental friendliness and simple preparation, which is widely used in energy, catalysis and sewage treatment. MnO_2 coating with different crystal structure and surface morphology can be prepared by different methods meeting wanted requirements [20, 21]. Inspired by lotus flower, we used an in situ immersion method to fabricate MnO_2 coating on AZ31B Mg alloy, and post-modification with stearic acid to obtain the superhydrophobic MnO_2 coating. The prepared superhydrophobic Mg alloy surface showed excellent self-cleaning property both in air and under oil (shown in **Figure 1**), as well as mechanical durability and chemical stability [22].

As to non-oxide ceramics, Hydroxyapatite (HA) is the main inorganic component of human and animal bones. It is a kind of bioactive ceramic material, which is widely used in bone tissue engineering. The HA ceramic coating was widely used in surface functionalization of metallic biomaterials. Hiromoto *et al.* prepared the HA coatings on AZ31 magnesium alloy, results showed that the HA coatings can remarkably reduce the Mg ion-release and corrosion current density [23]. In addition, it was reported that HA coating on 316 L stainless steel improved the corrosion behavior and biocompatibility of metallic implant and bone Osseointegration simultaneously [24]. Also, Surmeneva *et al.* prepared the HA coatings with different Ti contents on a Ti-6Al-4 V alloy, which was considered to be a possible candidate for biomedical applications [25].

Additionally, non-oxide ceramics materials such as silicon carbide (SiC), monolithic silicon nitride (Si_3N_4), and aluminum nitride (AlN) exhibit superior high-temperature strength and durability indicating their potential in industrial application [26, 27]. Furthermore, Liu *et al.* used non-oxide ceramics coating (bioactive silica-based glasses) on Ti alloys to promote the formation of HA layers in vivo [28].

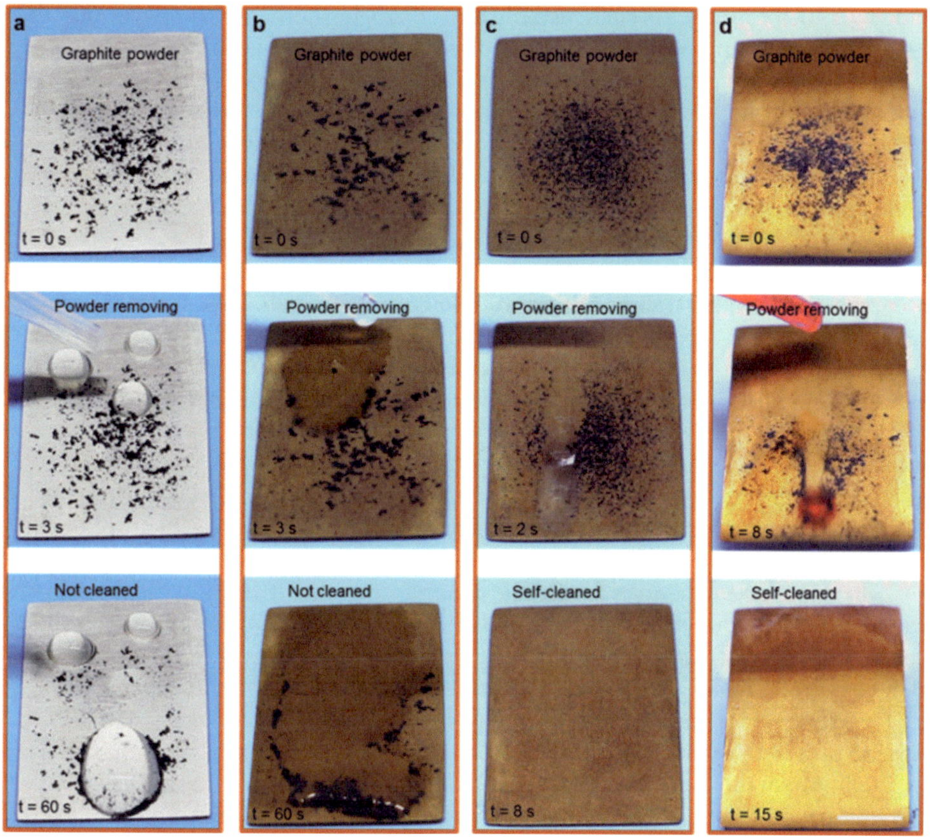

Figure 1.
Self-cleaning tests on AZ31B Mg alloy. (a, b) time-sequence images showing pristine AZ31B Mg alloy and MnO_2 coated AZ31B Mg alloy surface without self-cleaning properties, time-sequence images showing self-cleaning properties on superhydrophobic surface (c) in air and (d) in oil (isooctane). Scale bar, 1 cm. Reproduced with permission [22]. Copyright 2011, Elsevier.

In this context, oxide ceramic coatings and non-oxide ceramic coatings are playing important roles in the field of surface functionalization of metallic materials.

2.2 Properties of ceramic coating

Different metallic materials, in a sense, have different mechanical properties. Hardness and wear resistance are required to expand application prospect when metallic materials are used for industrial engineering. Numerous studies have shown that rare earth silicate barrier coatings can be potentially used for the application in high temperature aero-engines [29]. Bio-inspired by lotus leaf, Wu *et al.* synthesized the wear-resistant MoS_2 coated BN–TiN composite coating [30]. In addition, Xu *et al.* indicated that electrochemical co-deposition of nano SiO_2 and nano-CeO_2 particles with Ni–W–P composite coatings on $15^{\#}$ steel significantly improved the microhardness and abrasion resistance properties of the substrate [31]. Not only that, nano-structured Ni-Al_2O_3 composite coatings on Al plate exhibited the ultrahigh hardness (657 ± 28 Hv) and wear resistance [32]. Impressively, the TiO_2/Al_2O_3 composite coatings were prepared on Ti-6Al-4 V Alloy by micro arc oxidation, and the microhardness up to 11,000 MPa. The wear resistance was increased by 9.5 times than the as-received sample [33].

Metal corrosion is commonly found, hard to prevent, does harm to our environment, and costs several percent of the gross domestic production (GDP) of an industrialized country. As such, establishing corrosion control systems for metallic materials is very important for the sake of environment and economy harmony. The ceramic coating is widely used to protect metallic materials because of its good corrosion resistance. Like other corrosion-resistant coatings, the ceramic coating provides a barrier on the surface of metallic materials effectively isolating the corrosion solution from the substrate [34]. Moreover, the ceramic coating with micro-nano hierarchical structure can be prepared to obtain a superhydrophobic surface after hydrophobic treatment. In this regard, superhydrophobic ceramic coating has favorable corrosion resistance due to its excellent water-repellent property showing great potential application in corrosion protection of metallic materials [35].

To improve the corrosion resistance of mild steel, Tiwari *et al.* fabricated the conversion coating and sol–gel Al_2O_3 coating on mild steel [36]. The electrochemical results indicated that this coating reduced the corrosion current density of the mild steel by 5 orders of magnitude and increased the corrosion potential up to more than 1.0 V_{SCE}. Furthermore, Wang *et al.* used silane coupling agent bonding to the hydrotalcite/hydromagnesite conversion coating on Mg alloy, then the superhydrophobic ceramic coating was obtained, as such, the superhydrophobic ceramic coating had excellent corrosion resistance owing to its anti-water property [37]. In this context, superhydrophobic ceramic coating with hierarchical structure can trapped more air when immersed in the corrosive liquid greatly reducing the corrosive media attacked to the substrate, which provide a new idea for the application of ceramic coating in metallic materials protection.

Owing to their good thermal barrier properties, ceramic coatings are widely used to provide thermal barrier for heat transfer on the surface of metallic material and to improve the thermal stability of the substrate. Ghosh *et al.* evaluated the thermal properties of a thermal barrier coating (TBC) system on nimonic alloy ($BaO–MgO–SiO_2$ based glass-ceramic bond coating, 8% (mass fraction) yttria stabilized zirconia (8YSZ) top coating), the results showed that thermal barrier ceramic coating has extremely low thermal diffusivity and thermal conductivity than the bare substrate [38].

Ceramic materials can be divided into bioinert materials and bioactive materials according to their biological properties. Bioinert materials do not induce any visible tissue reactions; the majority of ceramics belong to this group. Al_2O_3 and ZrO_2 as bioinert materials have inherently low levels of reactivity, which have great potential for medical application owing to nontoxic, non-allergenic, and non-carcinogenic [39].

Some ceramics regarded as bioactive materials favor organ/tissue repairs and the integration of associated devices, which are essentially used in orthopedics, like favor bone repair and the integration of implants in bone tissues. As the most representative bioactive ceramic material, HA is widely used in bone tissue engineering for it is the main component of bones and teeth of human and animal. To improve the biodegradation performance of AZ91D Mg alloy, Song *et al.* prepared the bioactive HA coating electrodeposited on the Mg alloy, which can obviously reduce the biodegradation rate of AZ91D Mg alloy in stimulated body fluid (SBF) [40]. More importantly, HA-coated implants have been used in clinical research [41].

2.3 Fabrication of ceramic coating on metallic materials

The preparation and application of ceramic coatings have been studied for a long time. In order to adapt to different substrates, various technologies have been developed. These technologies of ceramics coated metallic materials enable to expand the application range in many fields.

Sol-gel method can easily prepare the ceramic coatings on metallic materials. Villatte *et al.* prepared TiO_2 antibacterial coating on fixation pins by using sol-gel method. This fabrication involved two steps: to create TiO_2 coating via a sol-gel process, and then to anneal at 500°C for 1 h [15]. In order to improve oxidation resistance, Małecka *et al.* used the sol-gel method to obtain a SiO_2 coating on Ti-46Al-7Nb-0.7Cr-0.1Si-0.2Ni alloy [42]. Moreover, sol-gel nanostructured Al_2O_3 coating can be fabricated on mild steel by hydrolysis and polycondensation of aluminum isopropoxide and catalyzed by HNO_3 [36].

Micro-arc oxidation (MAO) has been used as a critical method for many years to prepare much thicker and harder ceramic coatings on metallic materials. Shen *et al.* used the MAO technology to fabricate the TiO_2/Al_2O_3 composite coatings on Ti-6Al-4 V alloy in the Na_2SiO_3-$(NaPO_3)_6$-$NaAlO_2$ solution. The growth process revealed that O^{2-} reacted rapidly with Al^{3+} and Ti^{4+} (from substrate) to form the Al_2O_3 and Al_2TiO_5 simultaneously, and then Al_2TiO_5 was immediately decomposed into rutile TiO_2 and α-Al_2O_3 [33]. In addition, the porous Cu-TiO_2 coatings can be fabricated on titanium through MAO process under the constant current density of 20 A/dm^2 for 5 min, and the high stability TiO_2 coating formed during MAO process improved the corrosion resistance of titanium [43].

Atomic layer deposition (ALD) is a surface modification method through depositing inorganic species on the surface of different substrates, and the materials with arbitrary shape could be modified through vapor phase ALD. After multiple cycles of deposition, a conformal and uniform ceramic coating with good heat resistance and stiffness would be formed [44]. Huang *et al.* deposited the dense TiO_2 thin coatings on Co-Cr alloy with excellent antifungal activity by using ALD process [45]. Impressively, in order to prevent copper from water corrosion, Abdulagatov *et al.* developed an ultrathin barrier film on Cu. In this context, the barrier film was prepared by utilizing Al_2O_3 ALD and then TiO_2 ALD to protect the substrate [46].

Electrochemical method is usually used to fabricate oxide ceramics coated metallic materials. Notably, the electrochemical method is independent on the shape and the size of substrate. As such, Song *et al.* used electrodeposit technology to obtain the HA coatings on AZ91D Mg alloy [40], and Charlot *et al.* employed anodic electrophoretic deposition (EPD) to fabricate the SiO_2 submicron coatings, and found that the thickness of the film was related to the applied electric field [47]. In addition, the anodizing method is another well-established electrochemistry to form the ceramic coatings. Vengatesh et al. reported an anodic aluminum oxide surface by using anodizing process to prepare the superhydrophobic Al surface [48]. The prepared aluminum anodizing film not only had strong surface adhesion to the substrate, but also enabled fatty acids graft on the substrate ensuring the stability of superhydrophobic surface.

As a surface-deposited technology, plasma treatment is a simple and effective way to obtain ceramics coated metallic materials showing fine adhesion strength of coating-substrate. To improve corrosion resistance and bioactivity, the HA coating was prepared on AZ91HP Mg alloy by using plasma spraying method [49]. In addition, Sun *et al.* fabricated a TiO_2 coating on titanium substrate by using plasma electrolytic oxidation method in a sodium silicate (Na_2SiO_3) aqueous solution. In this regard, the TiO_2 coating was obtained on the titanium substrate with the best quality of density and adhesion by adjusting the duty ratio, frequency, and positive/negative pulse proportion on the microstructure and phase compositions [50].

Magnetron sputtering is also an efficient method to prepare ceramic coatings on the surface of metallic materials. Krishna *et al.* developed a novel process to improve the tribological and corrosion properties of austenitic stainless steels, a titanium coating deposited onto AISI 316 L stainless steel by magnetron sputtering, and then to partially convert the titanium coatings into titanium oxide by thermal oxidation. The resultant coating showed strong adhesion, good corrosion resistance, together with excellent surface hardness and tribological properties [51].

Solution immersion is a conventional method for fabrication of ceramic coatings on the surface of metallic materials. In this context, it is inexpensive and easy to carry out [52, 53]. In order to obtain a HA coating on Mg and its alloy, Hiromoto *et al.* immersed AZ31 Mg alloy and pure Mg in a 250 mmol/L $C_{10}H_{12}N_2O_8Na_2Ca$ aqueous solution of pH 8.9 [23]. Recently, a superhydrophobic MnO_2 coating was fabricated on AZ31B Mg alloy using two-step in situ immersion method, and post-modification with stearic acid. The superhydrophobic surface showed excellent corrosion resistant and anti-bioadhesion [54].

Laser-cladding is considered to be one of the most effective methods to fabricate a ceramic coating on metallic materials because of the powerful energy of laser to accelerate metal oxidation [55]. Boinovich *et al.* fabricated a superhydrophobic surface on Al alloys by nanosecond laser treatment [56]. After laser etching, a thick oxide film with high roughness was formed after several stages of melting and

Method	Ceramic coating	Substrate	Property	Ref.
Sol-gel	TiO_2	Stainless steel	Antibacterial and sufficient Mechanical strength	[15]
	SiO_2	Titanium alloy	Oxidation resistance	[42]
	Al_2O_3	Mild steel	Corrosion resistance	[36]
Micro-arc oxidation	TiO_2/Al_2O_3	Ti-6Al-4 V alloy	Wear resistance	[33]
	TiO_2	Titanium	Corrosion resistance	[43]
Atomic layer deposition	TiO_2	Co-Cr	Antifungal	[45]
	Al_2O_3/TiO_2	Copper	Corrosion resistance	[46]
Electrochemical	HA	Mg alloy	Biodegradation performance	[40]
	SiO_2	Platinum		[47]
	Al_2O_3	Aluminum	Corrosion resistance	[48]
Plasma treatment	HA	Mg alloy	Corrosion resistance and bioactivity	[49]
	TiO_2	Titanium	Corrosion resistance	[50]
Magnetron sputtering	HA	Titanium	Corrosion resistance	[25]
	TiO_2	Stainless steel	Tribological properties and corrosion resistance	[51]
Solution immersion	HA	Mg alloy	Corrosion resistance	[23]
	MnO_2	Mg alloy	Self-cleaning	[54]
Laser-cladding	Al_2O_3	Aluminum	Corrosion resistance	[56]
	$Al_2O_3/TiB_2/TiC$	Carbon steel	Microhardness and wear resistance	[57]
Metal organic chemical vapor deposition	SiO_2	Alloys	/	[17]
Dip-coating	Na_2SiO_3/Al_2O_3	Stainless steel	High temperature oxidation inhibition and corrosion resistance	[60]

Table 1.
Summary of fabrication methods of ceramic coated metallic materials.

solidifying. Similarly, through laser cladding, Al_2O_3-TiB_2-TiC ceramic coatings can be fabricated on carbon steel surface providing high microhardness and good wear resistance due to the results that the cladding thin film was uniformly and densely organized on the substrate [57].

Chemical vapor deposition can produce the ceramic coatings with controlled surface topography. Hofman *et al.* deposited the SiO_2 coatings on alloys by metal organic chemical vapor deposition (MOCVD) in sulphidizing high-temperature environments. The results indicated that the presence of silanol groups in SiO_2 coatings reduced the viscosity of the coating and enhanced the stress relaxation, thereby improving the coating performance [17].

Dip-coating is a time-saving and low-cost method for preparation of ceramic coatings [58, 59]. In 2017, Yu *et al.* produced a chemically robust and corrosion resistant Na_2SiO_3/Al_2O_3 composite coating on the surface of the 304 stainless steel, on which Na_2SiO_3 was incorporated into the nanopore of porous alumina layer by dip-coating heat treatment [60].

The fabrication methods of ceramic coated metallic materials are summarized in **Table 1**.

3. The applications of ceramics coated metallic materials

Up to now, the ceramics coated metallic materials have great potential in a wide variety of applications due to its unusual properties, such as good mechanical properties, corrosion resistance, thermal stability, and biological properties. It is worth noted that hydrophobic treatment of ceramic coatings on metallic materials ensuring superhydrophobic surfaces with special surface physicochemistry has recently received much attention in many fields.

It is well known that metallic material is irreplaceable in industrial application. The ceramic coatings bestow numerous unusual properties to metallic materials. Early in 1987, Ceramic coating as thermal barrier coating was tested on turbine blades in a research engine. Today, thermal barrier ceramic coatings are used in a low risk location within the turbine section of certain gas turbine engines [11]. In addition, Qin *et al.* reported that multiphase ceramic coatings significantly improved the hardness and wear resistance properties of 5052 Al alloy, which is conducive to industrial application [61]. In 2018, an alumina-titania ceramic coating was fabricated on carbon steel for corrosion protection [62].

Recently, superhydrophobic surface has been extensively developed due to its unique property including corrosion protection, self-cleaning, oil water separation, anti-fouling, anti-icing, and drag reduction [63]. Superhydrophobic ceramic coating was obtained by hydrophobic treatment of ceramic coating with hierarchical rough structure, which greatly expanded the application range of metal materials [64, 65]. In 2020, Emarati *et al.* fabricated a superhydrophobic nano-TiO_2/TMPSi ceramic composite coating on 316 L steel by using a one-step electrophoretic deposition method, the results indicated that the superhydrophobic ceramic nanocomposite coating had excellent corrosion resistance [66]. Also, the water shear stress and drag can be reduced on superhydrophobic ceramic coated metallic materials surfaces resulting from the air pockets present between the liquid and solid substrate. In this context, the rolling-off droplets can remove contamination particles displaying self-cleaning feature [22]. Furthermore, a superhydrophobic ceramic coating is also reported as an emerging material exhibiting their promising diverse applications for anti-fogging, anti-fouling, and oil water separation [67–69]. **Figure 2** shows the oil/water separation of *1H, 1H, 2H, 2H*-perfluorodecyltriethoxysilane-modified CuO-grown copper foam (PCCF).

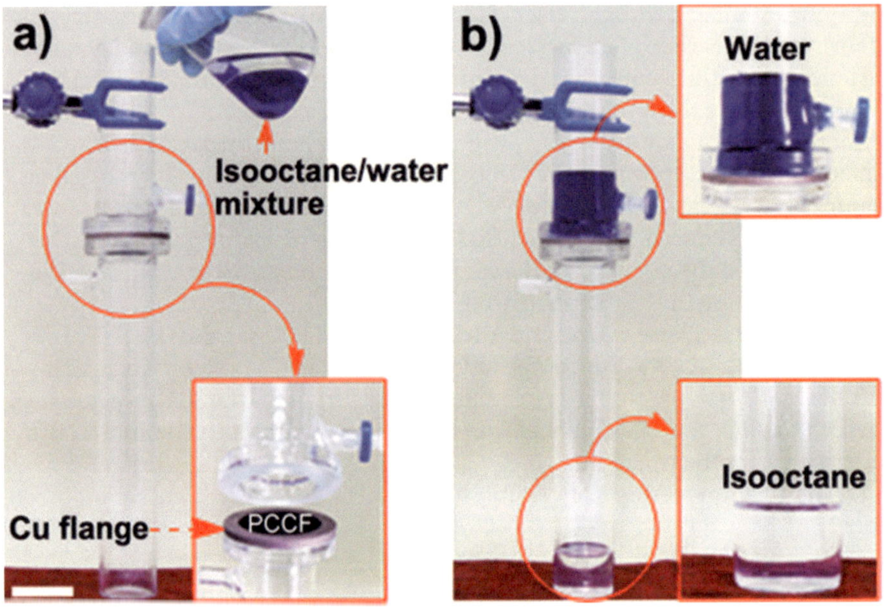

Figure 2.
Separation apparatus with an 18:25 v:v isooctane/water mixture above PCCF. Inset, PCCF was fixed in Cu flange and then sandwiched between two glass tubes (a). Isooctane passed through PCCF whereas water was retained (b). Water is dyed blue. Scale bar, 3 cm. Reproduced with permission [69]. Copyright 2013, Royal Society of Chemistry.

In addition, ceramic coatings have numerous applications in the field of biomedical engineering, mainly because of their biological properties. The bioinert properties of ceramic coatings help them with biocompatibility, and good hardness and wear-resistance properties make them suitable for substitution of hard tissues (bones and teeth). On the contrary, bioactive ceramic coatings such as HA coating have been clinically used onto the metallic implant surfaces combining the mechanical strength of metals and their alloys with the excellent biological properties of ceramics for the enhancement of new bone osteogenesis [70, 71].

Importantly, researching work shows that superhydrophobic surfaces can dramatically reduce the contact between fouling organisms and substrate surfaces

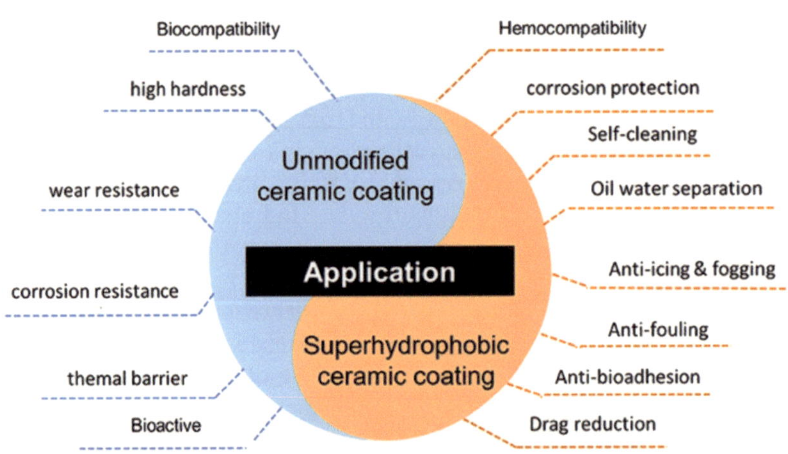

Figure 3.
The comparison of properties of unmodified ceramic coating and superhydrophobic ceramic coating on metallic materials.

exhibiting excellent anti-fouling and hemocompatibility properties [72, 73]. Hu *et al.* designed a superhydrophobic SiO_2 biodegradable coating with exceptional anti-bioadhesion through one-step co-electrospraying poly(L-lactide) (PLLA) modified with silica nanoparticles [74]. It was revealed that the superhemophobic TiO_2 surface with a robust Cassie–Baxter state displayed more hemocompatible compared to hemophobic or hemophilic TiO_2 surface [75]. The comparison of properties of unmodified ceramic coating and superhydrophobic ceramic coating on metallic materials is shown in **Figure 3**.

4. Conclusion

In this chapter, we introduce and discuss various techniques utilized to fabricate a range of different ceramic coatings on metal materials with desirable properties such as good mechanical property, corrosion resistance, thermal stability, and biological property. It is not surprising that superhydrophobic ceramic coatings on metallic materials can make the materials be attractive for applications in anti-fouling, self-cleaning, corrosion protection, wear resistance, oil/water separation and biotechnology.

Acknowledgements

This work was supported by the Natural Science Foundation of Jiangxi Province (20192BAB203008).

Conflict of interest

The authors declare no conflict of interest.

Author details

Dongmian Zang* and Xiaowei Xun
School of Materials Science and Engineering, East China Jiaotong University, Nanchang, P.R. China

*Address all correspondence to: zdmian@iccas.ac.cn

References

[1] Gurrappa I. Characterization of titanium alloy Ti-6Al-4V for chemical, marine and industrial applications. Materials Characterization. 2003; 51:131-139. DOI: 10.1016/j.matchar.2003.10.006

[2] Hosseinalipour S M, Ershadlangroudi A, Hayati A N, et al. Characterization of sol–gel coated 316L stainless steel for biomedical applications. Progress in Organic Coatings. 2010;67:371-374. DOI: 10.1016/j.porgcoat.2010.01.002

[3] Collins W, Sherman R J, Leon R T, et al. Fracture toughness characterization of high-performance steel for bridge girder applications. Journal of Materials in Civil Engineering. 2019;31:04019027. DOI: 10.1061/(asce)mt.1943-5533.0002636

[4] Singh J, Chauhan A. Characterization of hybrid aluminum matrix composites for advanced applications–A review. Journal of materials research technology. 2016;5:159-169. DOI: 10.1016/j.jmrt.2015.05.004

[5] Gupta M, Tay A A O, Vaidyanathan K, et al. An investigation of the synthesis and characterization of copper samples for use in interconnect applications. Materials Science and Engineering: A. 2007;454-455:690-694. DOI: 10.1016/j.msea.2006.11.099

[6] Cipriano A F, Lin J, Miller C, et al. Anodization of magnesium for biomedical applications–Processing, characterization, degradation and cytocompatibility. Acta Biomaterialia. 2017;62:397-417. DOI: 10.1016/j.actbio.2017.08.017

[7] Fu Y. Review on the corrosion behavior of metallic materials influenced by biofilm(ii). Development and Application of Materials. 2006;21:38-43. DOI: 10.3969/j.issn.1003-1545.2006.02.009

[8] Cai X, Ma K, Zhou Y, et al. Surface functionalization of titanium with tetracycline loaded chitosan–gelatin nanosphere coatings via EPD: fabrication, characterization and mechanism. RSC Advances. 2016;6:7674-7682. DOI: 10.1039/C5RA17109A

[9] Su Y, Luo C, Zhang Z, et al. Bioinspired surface functionalization of metallic biomaterials. Journal of The Mechanical Behavior of Biomedical Materials. 2018;77:90-105. DOI: 10.1016/j.jmbbm.2017.08.035

[10] Spear K E. Diamond–Ceramic coating of the future. Journal of the American Ceramic Society. 1989;72:171-191. DOI: 10.1111/j.1151-2916.1989.tb06099.x

[11] Miller R A. Current status of thermal barrier coatings–An overview. Surface and Coatings Technology. 1987;30 DOI: 10.1016/0257-8972(87)90003-X

[12] Best S M, Porter A E, Thian E S, et al. Bioceramics: Past, present and for the future. Journal of the European Ceramic Society. 2008;28:1319-1327. DOI: 10.1016/j.jeurceramsoc.2007.12.001

[13] Shen G X, Chen Y C, Lin L, et al. Study on a hydrophobic nano-TiO2 coating and its properties for corrosion protection of metals. Electrochimica Acta. 2005;50:5083-5089. DOI: 10.1016/j.electacta.2005.04.048

[14] Jiang J Y, Xu J L, Liu Z H, et al. Preparation, corrosion resistance and hemocompatibility of the superhydrophobic TiO2 coatings on biomedical Ti-6Al-4V alloys. Applied Surface Science. 2015;347:591-595. DOI: 10.1016/j.apsusc.2015.04.075

[15] Villatte G, Massard C, Descamps S, et al. Photoactive TiO2 antibacterial

coating on surgical external fixation pins for clinical application. International Journal of Nanomedicine. 2015;10:3367-3375. DOI: 10.2147/IJN.S81518

[16] Gao Y, Wang C, Yao M, et al. The resistance to wear and corrosion of laser-cladding Al2O3 ceramic coating on Mg alloy. Applied Surface Science. 2007;253:5306-5311. DOI: 10.1016/j.apsusc.2006.12.001

[17] Hofman R, Westheim J, Pouwel I, et al. FTIR and XPS studies on corrosion-resistant SiO2 coatings as a function of the humidity during deposition. Surface and Interface Analysis. 1996;24:1-6. DOI: 10.1002/(SICI)1096-9918(199601)24:13.0.CO;2-I

[18] Sadreddini S, Afshar A. Corrosion resistance enhancement of Ni-P-nano SiO2 composite coatings on aluminum. Applied Surface Science. 2014;303:125-130. DOI: 10.1016/j.apsusc.2014.02.109

[19] Wang Y, Zhou Q, Li K, et al. Preparation of Ni–W–SiO2 nanocomposite coating and evaluation of its hardness and corrosion resistance. Ceramics International. 2015;41:79-84. DOI: 10.1016/j.ceramint.2014.08.034

[20] Hunter J C. Preparation of a new crystal form of manganese dioxide: λ-MnO2. Journal of Solid State Chemistry. 1981;39:142-147. DOI: 10.1016/0022-4596(81)90323-6

[21] Carpino L A. Simple preparation of active manganese dioxide from activated carbon. Journal of Organic Chemistry. 1970;35:3971-3972. DOI: 10.1021/jo00836a091

[22] Zang D, Xun X, Gu Z, et al. Fabrication of superhydrophobic self-cleaning manganese dioxide coatings on Mg alloys inspired by lotus flower. Ceramics International. 2020;46:20328-20334. DOI: 10.1016/j.ceramint.2020.05.121

[23] Hiromoto S, Tomozawa M. Hydroxyapatite coating of AZ31 magnesium alloy by a solution treatment and its corrosion behavior in NaCl solution. Surface and Coatings Technology. 2011;205:4711-4719. DOI: 10.1016/j.surfcoat.2011.04.036

[24] Parsapour A, Khorasani S N, Fathi M H. Corrosion behavior and biocompatibility of hydroxyapatite coating on H2SO4 passivated 316L SS for human body implant. Acta Metallurgica Sinica. 2013;26:409-415. DOI: 10.1007/s40195-012-0212-3

[25] Surmeneva M A, Vladescu A, Surmenev R A, et al. Study on a hydrophobic Ti-doped hydroxyapatite coating for corrosion protection of a titanium based alloy. RSC Advances. 2016;6:87665-87674. DOI: 10.1039/c6ra03397k

[26] Smialek J L, Robinson R C, Opila E J, et al. SiC and Si3N4 recession due to SiO2 scale volatility under combustor conditions. Advanced Composite Materials. 1999;8:33-45. DOI: 10.1163/156855199x00056

[27] Lee K N, Fox D S, Bansal N P. Rare earth silicate environmental barrier coatings for SiC/SiC composites and Si3N4 ceramics. Journal of the European Ceramic Society. 2005;25:1705-1715. DOI: 10.1016/j.jeurceramsoc.2004.12.013

[28] Pazo A, Saiz E, Tomsia A P. Silicate glass coatings on Ti-based implants. Acta Materialia. 1998;46:2551-2558. DOI: 10.1016/S1359-6454(98)80039-6

[29] Xu Y, Hu X, Xu F, et al. Rare earth silicate environmental barrier coatings: Present status and prospective. Ceramics International. 2017;43:5847-5855 DOI: 10.1016/j.ceramint.2017.01.153

[30] Wu J H, Phillips B S, Jiang W, et al. Bio-inspired surface engineering and tribology of MoS2 overcoated cBN–TiN

composite coating. Wear. 2006;261:592-599. DOI: 10.1016/j.wear.2006.01.027

[31] Xu R, Wang J, He L, et al. Study on the characteristics of Ni–W–P composite coatings containing nano-SiO2 and nano-CeO2 particles. Surface and Coatings Technology. 2008;202:1574-1579. DOI: 10.1016/j.surfcoat.2007.07.012

[32] Yazdani A, Isfahani T. Hardness, wear resistance and bonding strength of nano structured functionally graded Ni-Al2O3 composite coatings fabricated by ball milling method. Advanced Powder Technology. 2018;29:1306-1316. DOI: 10.1016/j.apt.2018.02.025

[33] Yizhou S, Haijun T, Yuebin L, et al. Fabrication and wear resistance of TiO2/Al2O3 coatings by micro-arc oxidation. Rare Metal Materials and Engineering. 2017;46:23-27. DOI: 10.1016/s1875-5372(17)30071-1

[34] Kim W B, Kwon S C, Cho, S H, et al. Effect of the grain size of YSZ ceramic materials on corrosion resistance in a hot molten salt CaCl2-CaF2-CaO system. Corrosion Science. 2020;170:108664. DOI: 10.1016/j.corsci.2020.108664

[35] Xun X, Zhu R, Dong J, et al. Superhydrophobic light alloy materials with corrosion-resistant surfaces. Research and Application of Materials Science. 2020;2. DOI: 10.33142/MSRA.V2I1.1972

[36] Tiwari S K, Sahu R K, Pramanick A K, et al. Development of conversion coating on mild steel prior to sol gel nanostructured Al2O3 coating for enhancement of corrosion resistance. Surface and Coatings Technology. 2011;205:4960-4967. DOI: 10.1016/j.surfcoat.2011.04.087

[37] Wang J, Li D, Liu Q, et al. Fabrication of hydrophobic surface with hierarchical structure on Mg alloy and its corrosion resistance. Electrochimica Acta. 2010;55:6897-6906. DOI: 10.1016/j.electacta.2010.05.070

[38] Ghosh S. Thermal properties of glass-ceramic bonded thermal barrier coating system. Transactions of Nonferrous Metals Society of China. 2015;25:457-464. DOI: 10.1016/s1003-6326(15)63624-x

[39] Marti A. Inert bioceramics (Al2O3, ZrO2) for medical application. Injury-International Journal of the Care of the Injured. 2000;31:S33-S36.

[40] Song Y W, Shan D Y, Han E H. Electrodeposition of hydroxyapatite coating on AZ91D magnesium alloy for biomaterial application. Materials Letters. 2008;62:3276-3279. DOI. 10.1016/j.matlet.2008.02.048

[41] Sun L, Berndt C, Gross K A, et al. Material fundamentals and clinical performance of plasma sprayed hydroxyapatite coatings: A review. Journal of Biomedical Materials Research. 2001;58:570-592. DOI: 10.1002/jbm.1056.abs

[42] Małecka J, Krzak-Roś J. Preparation of SiO2 coating by sol-gel method, to improve high-temperature corrosion resistance of a γ-TiAl phase based alloy. Advances in Materials Sciences. 2013;12:1-12. DOI: 10.2478/v10077-012-0011-6

[43] Wu H, Zhang X, Geng Z, et al. Preparation, antibacterial effects and corrosion resistant of porous Cu–TiO2 coatings. Applied Surface Science. 2014;308:43-49. DOI: 10.1016/j.apsusc.2014.04.081

[44] Lu J, Li Y, Song W, et al. Atomic layer deposition onto thermoplastic polymeric nanofibrous aerogel templates for tailored surface properties. ACS Nano. 2020. DOI: 10.1021/acsnano.9b09497

[45] Huang L, Jing S, Zhuo O, et al. Surface hydrophilicity and antifungal properties of TiO2 films coated on a Co-Cr substrate. BioMed Research International. 2017;2017:1-7. DOI: 10.1155/2017/2054723

[46] Abdulagatov A I, Yan Y, Cooper J R, et al. Al2O3 and TiO2 atomic layer deposition on copper for water corrosion resistance. ACS Appllied Materials Interfaces. 2011;3:4593-4601. DOI: 10.1021/am2009579

[47] Charlot A, Deschanels X, Toquer G. Submicron coating of SiO2 nanoparticles from electrophoretic deposition. Thin Solid Films. 2014;553:148-152. DOI: 10.1016/j.tsf.2013.11.064

[48] Vengatesh P, Kulandainathan M A. Hierarchically ordered self-lubricating superhydrophobic anodized aluminum surfaces with enhanced corrosion resistance. ACS Applied Materials & Interfaces. 2015;7:1516-1526. DOI: 10.1021/am506568v

[49] Gao Y L, Liu Y, Song X Y. Plasma-sprayed hydroxyapatite coating for improved corrosion resistance and bioactivity of magnesium alloy. Journal of Thermal Spray Technology. 2018;27:1381-1387. DOI: 10.1007/s11666-018-0760-9

[50] Sun C, Hui R, Qu W, et al. Effects of processing parameters on microstructures of TiO2 coatings formed on titanium by plasma electrolytic oxidation. Journal of Materials Science. 2010;45:6235-6241. DOI: 10.1007/s10853-010-4718-7

[51] Krishna D S R, Sun Y. Thermally oxidised rutile-TiO2 coating on stainless steel for tribological properties and corrosion resistance enhancement. Applied Surface Science. 2005;252: 1107-1116. DOI: 10.1016/j.apsusc.2005.02.046

[52] Song J, Lu Y, Huang S, et al. A simple immersion approach for fabricating superhydrophobic Mg alloy surfaces. Applied Surface Science. 2013;266:445-450. DOI: 10.1016/j.apsusc.2012.12.063

[53] Qu M, Zhang B, Song S, et al. Fabrication of superhydrophobic surfaces on engineering materials by a solution-immersion process. Advanced Functional Materials. 2007;17:593-596. DOI: 10.1002/adfm.200600472

[54] Xun X, Wan Y, Zhang Q, et al. Low adhesion superhydrophobic AZ31B magnesium alloy surface with corrosion resistant and anti-bioadhesion properties. Applied Surface Science. 2020;505:144566. DOI: 10.1016/j.apsusc.2019.144566

[55] Boinovich L B, Modin E B, Sayfutdinova A R, et al. Combination of Functional Nanoengineering and Nanosecond Laser Texturing for Design of Superhydrophobic Aluminum Alloy with Exceptional Mechanical and Chemical Properties. ACS Nano. 2017;11:10113-10123. DOI: 10.1021/acsnano.7b04634

[56] Boinovich L B, Emelyanenko A M, Modestov A D, et al. Synergistic effect of superhydrophobicity and oxidized layers on corrosion resistance of aluminum alloy surface textured by nanosecond laser treatment. ACS Applied Materials Interfaces. 2015;7:19500-19508. DOI: 10.1021/acsami.5b06217

[57] Li Z, Wei M, Xiao K, et al. Microhardness and wear resistance of Al2O3-TiB2-TiC ceramic coatings on carbon steel fabricated by laser cladding. Ceramics International. 2019;45:115-121. DOI: 10.1016/j.msec.2020.110847

[58] Procopio A M S, Carvalho J D L, Silveira T H R, et al. CeO2 thin film

supported on TiO2 porous ceramics Materials Letters. 2020;276:128224. DOI: 10.1016/j.matlet.2020.128224

[59] Naveas, N, Manso-Silvan M, Pulido R, et al. Fabrication and characterization of nanostructured porous silicon-silver composite layers by cyclic deposition: dip-coating vs spin-coating. Nanotechnology. 2020;31:365704. DOI: 10.1088/1361-6528/ab96e5

[60] Yu J, Liu S, Li F, et al. Na2SiO3/Al2O3 composite coatings on 304 stainless steels for enhanced high temperature oxidation inhibition and chloride-induced corrosion resistance. Surface and Coatings Technology. 2017;309:1089-1098. DOI: 10.1016/j.surfcoat.2016.10.003

[61] Qin D, Xu G, Yang Y, et al. Multiphase ceramic coatings with high hardness and wear resistance on 5052 aluminum alloy by a microarc oxidation method. ACS Sustainable Chemistry & Engineering. 2018;6:2431-2437. DOI: 10.1021/acssuschemeng.7b03883

[62] Pinzón A V, Urrego K J, González-Hernández A, et al. Corrosion protection of carbon steel by alumina-titania ceramic coatings used for industrial applications. Ceramics International. 2018;44:21765-21773. DOI: 10.1016/j.ceramint.2018.08.273

[63] Zhang X, Shi F, Niu J, et al. Superhydrophobic surfaces: from structural control to functional application. Journal of Materials Chemistry. 2008;18:621-633. DOI: 10.1039/B711226B

[64] Luo S, Zheng L, Luo H, et al. A ceramic coating on carbon steel and its superhydrophobicity. Applied Surface Science. 2019;486:371-375. DOI: 10.1016/j.apsusc.2019.04.235

[65] Xu P, Coyle T W, Pershin L, et al. Superhydrophobic ceramic coating: Fabrication by solution precursor plasma spray and investigation of wetting behavior. Journal of Colloid Interface Science. 2018;523:35-44. DOI: 10.1016/j.jcis.2018.03.018

[66] Emarati S M, Mozammel M. Efficient one-step fabrication of superhydrophobic nano-TiO2/TMPSi ceramic composite coating with enhanced corrosion resistance on 316L. Ceramics International. 2020;46:1652-1661. DOI: 10.1016/j.ceramint.2019.09.137

[67] Zhang F, Robinson B, de Villiers-Lovelock H, et al. Wettability of hierarchically-textured ceramic coatings produced by suspension HVOF spraying. Journal of Materials Chemistry A. 2015;3:13864-13873. DOI: 10.1039/c5ta02130h

[68] Lai Y, Tang Y, Gong J, et al. Transparent superhydrophobic/superhydrophilic TiO2-based coatings for self-cleaning and anti-fogging. Journal of Materials Chemistry. 2012;22:7420-7426. DOI: 10.1039/c2jm16298a

[69] Zang D, Wu C, Zhu R, et al. Porous copper surfaces with improved superhydrophobicity under oil and their application in oil separation and capture from water. Chemical Communications. 2013;49: 8410-8412. DOI: 10.1039/c3cc43536a

[70] Heimann R B. Osseoconductive and corrosion-inhibiting plasma-sprayed calcium phosphate coatings for metallic medical implants. Metals. 2017;7:468. DOI: 10.3390/met7110468

[71] Surmenev R A, Surmeneva M A, Ivanova A. Significance of calcium phosphate coatings for the enhancement of new bone osteogenesis–A review. Acta Biomaterialia. 2014;10:557-579. DOI: 10.1016/j.actbio.2013.10.036

[72] Jokinen V, Kankuri E, Hoshian S, et al. Superhydrophobic blood-repellent

surfaces. Advanced Materials. 2018;30:e1705104. DOI: 10.1002/adma.201705104

[73] Zhang P, Lin L, Zang D, et al. Designing bioinspired anti-biofouling surfaces based on a superwettability strategy. Small. 2017;13:1503334. DOI: 10.1002/smll.201503334

[74] Hu C, Liu S, Li B, et al. Micro-/nanometer rough structure of a superhydrophobic biodegradable coating by electrospraying for initial anti-bioadhesion. Advanced Healthcare Materials. 2013;2:1314-1321. DOI: 10.1002/adhm.201300021

[75] Movafaghi S, Leszczak V, Wang W, et al. Hemocompatibility of superhemophobic titania surfaces. Advanced Healthcare Materials. 2017;6: 1600717. DOI: 10.1002/adhm.201600717

Chapter 10

Nanostructured Multilayer Composite Coatings for Cutting Tools

Sergey Grigoriev, Alexey Vereschaka, Marina Volosova, Caterina Sotova, Nikolay Sitnikov, Filipp Milovich and Nikolay Andreev

Abstract

The chapter deals with the specific features concerning the application of wear-resistant coatings to improve the performance properties of ceramic cutting tools. The paper discusses the theoretical background associated with the specific operation conditions and wear of ceramic cutting tools and influencing the choice of the compositions and structures of wear-resistant coatings. The studies were focused on the application of the Ti-(Ti,Al)N-(Zr,Nb,Ti,Al)N multilayer composite coating with a nanostructured wear-resistant layer, as well as the (Cr,Al,Si)N–(DLC–Si)–DLC–(DLC–Si) and (Cr,Al,Si)N–DLC composite coatings in order to improve the cutting properties of ceramic tools. The chapter presents the results of the comparative cutting tests for the tools with the coatings under study, uncoated tools, and tools with the Ti-(Ti,Al)N commercial coating. The wear mechanisms typical for ceramic cutting tools with coatings of various compositions have been investigated.

Keywords: nanocomposite functional coating, diamond-like carbon (DLC), ceramic cutting tool, tool wear

1. Introduction

Ceramic cutting tools are more and more widely used due to their high hardness, wear resistance, and relatively low cost [1–6]. The main specific feature of cutting ceramics is the absence of a binder phase, which significantly reduces the degree of softening in ceramic cutting tools during heating and increases their plastic strength. Due to the above, the cutting process can imply high cutting speeds, which significantly exceeds the cutting speeds typical for the machining with carbide cutting tools [1, 2, 5, 6]. While for a carbide cutting tool, the limiting level of cutting speeds is 500,600 m/min, then for a tool equipped with cutting ceramics, this level increases up to 9,001,000 m/min and higher [1]. However, the absence of the binder phase also has negative influence on the performance properties of ceramic cutting tools. In particular, their brittle strength, impact toughness, and resistance to crack formation decrease [1–4]. This fact significantly influences the wear patterns on ceramic cutting tools. For example, low crack resistance provokes the formation of a crack front, which, due to the absence of a plastic binder phase, encounters no barriers to slow down or stop the crack development. The above is the main reason for micro- and macrochipping on contact pads of a ceramic cutting tool already at the stages of running-in or initial steady-state wear, causing failures because of brittle fracture. The noted wear mechanism prevails on ceramic cutting tools, and it actually does not depend on the cutting speed, because the temperature factor does not have a noticeable influence on the transformation of the wear mechanism. To a large extent, it is this mechanism of wear which determines the scope of application of the ceramic cutting tools [6–12].

At present, ceramic cutting tools are usually recommended for the finishing of gray, malleable, high-strength, and chilled cast irons, low- and high-alloy steels, including improved and heat-treated (up to HRC 55–60) non-ferrous alloys and structural polymer materials (K01 K05, P01 P05) [1–6]. Under the above conditions, tools equipped with ceramic cutting inserts are noticeably superior to carbide cutting tools in terms of working efficiency.

During the intermittent cutting, the use of ceramic tools in machining with advanced values of cross-section of cut ($a_p \times f$) sharply reduces their efficiency due to the high probability of sudden failure because of the brittle fracture of the cutting parts of the tools [1, 2]. This fact largely explains the relatively low volume of ceramic tools used in production sector [1, 2].

In this regard, the main direction of improving the performance activities of cutting ceramics is an enhancement of its strength characteristics to expand the area of the technological application in cutting. Recently, a new class of tool materials has appeared attributed to the group of cutting ceramics with increased strength, toughness, and crack resistance (silicon nitride, reinforced ceramics), and this fact indicates that the scope of application of ceramic cutting tools expands noticeably [1–6].

2. Theoretical background

The operation conditions for ceramic cutting tools significantly differ from those under which tools of high-speed steel and carbides are used. A substantial increase in cutting speed changes the mechanism of chip formation and contact processes during the cutting, as well as the nature and level of power and temperature loads, thermomechanical stress, and mechanisms of tool wear [13–15]. The results of the studies focused on the stress state in the cutting part of a ceramic tool [16] indicate the presence of tensile and compressive stresses (**Figure 1**). At the

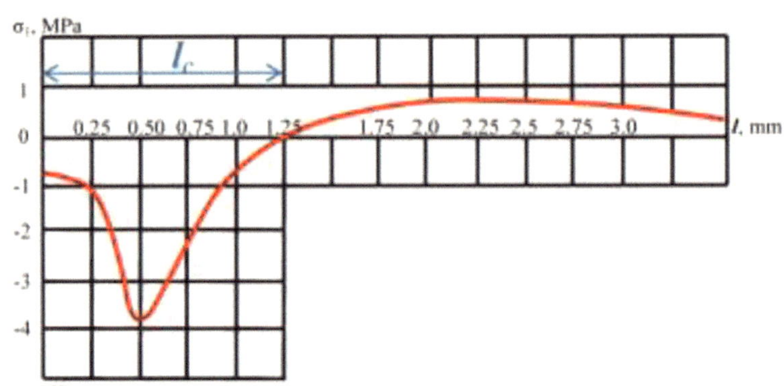

Figure 1.
Distribution of principal stresses σ_1 and σ_2 along the rake face of a cutting tool [1, 2].

same time, in most cases, the area of tensile stresses begins at the end of the contact area of the chips with the rake face. When section thickness is small, compressive stresses prevail, while at larger thicknesses, tensile stresses begin to play a significant role [17, 18].

The most stressed section in the tensile area is located on the rake face of the tool at a distance equal to (2–2.5)*lc*, where *lc* is the length of contact between the chips and the rake face of the tool.

With a decrease in the rake angle γ, the compression area extends, and the tensile area decreases or disappears at all. As a result, the negative angle γ is typical for a ceramic cutting tool and makes it possible to achieve a change in the stress state in the direction of the predominantly compressive stresses [17, 18]. Deposition of a coating on a cutting tool significantly changes the nature of the interaction between the material being machined and the tool. In [19, 20], the studies revealed that the coating parameters had a significant influence on the characteristics of the contact processes and the chip formation. To prevent a sudden failure of a ceramic cutting tool as a result of brittle fracture, it is necessary to control the processes of the contact interaction between the tool material and the material being machined by depositing coatings on the working surfaces of the tool. The composition and structure of such coatings will increase the length of the full contact between the chips and the rake face of the tool through enhancement of the adhesion to the material being machined and the improvement of heat removal from the cutting area due to increased thermal conductivity of the tool material. Thus, the specific thermomechanical loads on the cutting edge of the ceramic tool can be reduced (see **Figure 2**).

There are a number of studies considering the use of coatings to improve the performance properties of ceramic cutting tools, with both oxide and nitride ceramics as the ground. In [21], the investigation is focused on PVD of the (Ti,Zr)N-(TiN/ZrN) and TiN-(TiAl)N-(TiN/(TiAl)N) nanostructured multilayer coatings, deposited on ceramic cutting inserts of Al_2O_3 + ZrO_2 + Ti(C,N) and Al_2O_3 + TiC, with the external layer formed by the alternating nanolayers of TiN and ZrN or TiN and (Ti,Al)N, respectively. With the total coating thickness of 3–5 μm and the microhardness of about HV 29 GPa, the above coatings prolonged the tool life by 20–80% during the dry cutting of NC6 steel (HRC 48–52), at v_c = 150 m/min, f = 0.10 mm/rev, a_p = 0.5 mm.

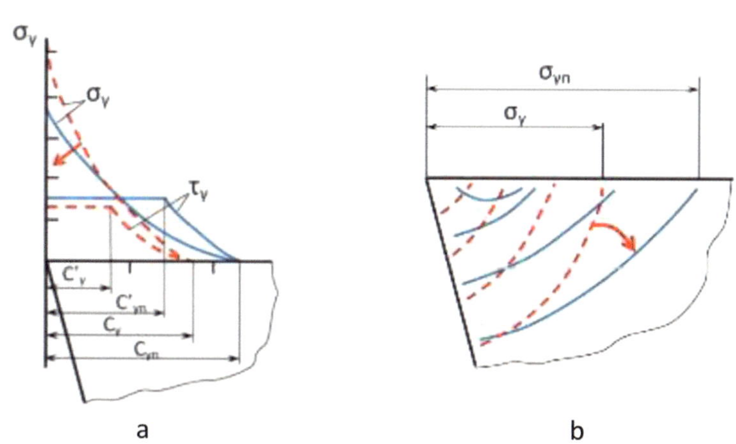

Figure 2.
Differences in (a) stresses and (b) isotherms for a coated and uncoated tool: C_γ and $C_{\gamma n}$ - the total lengths of the contact between the chips and the rake face of the uncoated (the dashed line) and coated (solid line) tools, respectively; changes in stresses and isotherms occur in the direction of the arrows [1, 2].

In [22], the studies considered the properties of a ceramic cutting tool with the PVD coating of TiN-(Ti,Al,Si)N-TiN with the thickness of 2–4 μm with a nano-structured layer of (Ti,Al,Si)N and the CVD coating of TiN-Al_2O_3 with the thickness of 2.6–10 μm, when inserts of nitride ceramics of Si_3N_4 and oxide ceramics of Al_2O_3 + ZrO_2 were used as substrates. The cutting properties were used during the turning of EN-GJL-250 gray cast iron and C45E steel. The cutting process was carried out under the following conditions: f = 0.10, 0.15, and 0.20 mm/rev; a_p = 1 and 2 mm; v_c = 200 and 400 m/min. The cutting tools with coatings of all types demonstrated the longer tool life compared to that of the uncoated tools, while the longest tool life was detected for the tools with the PVD coating of TiN-(Ti,Al,Si)N-TiN with the nanostructured middle layer. In [23], the studies considered the cutting properties of tools made of mixed ceramics of Al_2O_3 + TiCN with the TiN commercial coating during the turning of hardened American Iron and Steel Institute (AISI) 52,100 (HRC 63) steel under the following cutting conditions: f = 0.07, 0.11, and 0.14 mm/rev, a_p = 0.5 mm, v_c = 100, 150, 200, 250, and 300 m/min. The studies found that for the uncoated cutting inserts, wear in the form of cracking and chipping was more typical, while for the coated tools, the formation of a wear crater on the rake face was typical. The cutting path for the coated tool was about 8 times longer, and the temperature in the cutting area was substantially lower than for the uncoated tool.

According to the results of [24], which studied the cutting properties of tools made of silicon nitride (Si_3N_4) with the CVD coating of TiN-Al_2O_3 during the continuous turning of gray cast iron with various depths of cut, the prevailing failure mechanism for the above cutting tools was abrasive wear during the continuous turning and a combination of abrasive wear and brittle fracture during the machining with variable depths of cut. The coated cutting tools demonstrated much longer tool life compared to the uncoated cutting tools: the length and time of cutting were about 3.5 times longer at the cutting speed of 300 m/min and 2 longer at the cutting speed of 380 m/min.

The cutting properties of ceramic tools of Si_3N_4 with the PVD coating of (Ti,Al)N-(Al,Cr)O during the turning of HT250 gray cast iron and AISI 4340 steel were studied in [25]. The thicknesses of the (Ti,Al)N and (Al,Cr)O layers were about 2.0 and 0.6 μm, respectively. The tool life of a tool made of silicon nitride with the (Ti,Al)N-(Al,Cr)O coating was longer compared to uncoated inserts during the turning of gray cast iron and steel. In [26], authors investigated the cutting properties of tools based on silicon nitride with the PVD coatings of $(Ti_{0.5},Al_{0.5})N$ and $(Cr_{0.3},Al_{0.7})N$ during the dry turning of gray cast iron. For the tools with the $(Ti_{0.5},Al_{0.5})N$ and $(Cr_{0.3},Al_{0.7})N$ coatings, the tool life was at least 2 times longer compared to the uncoated tools. The tools with the $(Ti_{0.5},Al_{0.5})N$ coating demonstrated longer tool life compared to the tools with the $(Cr_{0.3},Al_{0.7})N$ coating. In [27], the studies are focused on the cutting properties of tools made of the Al_2O_3 + TiC mixed ceramics with the TiN-(Al,Cr)N multilayer coating during the dry turning of AISI 4340 (HRC 46) hardened steel at v_c = 125–175 m/min, a_p = 0.25–0.63 mm, f = 0.10–0.25 mm/rev. After 9 minutes of cutting, the wear VB was on average 45% higher for the uncoated tool.

The application of diamond-like carbon (DLC) coatings for ceramic cutting tools should be considered separately. While several studies consider the properties of DLC, deposited on a ceramic substrate, there are hardly any investigations focused on ceramic cutting tools with DLC coatings. For example, in [28, 29], the studies consider the challenges of improving the performance properties of ceramic tribological pairs (sliding bearings). A significant decrease in the coefficient of friction (COF) was noted with the use of samples with DLC coatings. Ceramic products made of SiC with DLC coatings demonstrate excellent chemical stability, low COF, and very good wear resistance [30]. The properties of DLC coatings deposited on the Si_3N_4 substrate were also considered. Gomes et al. [31] studied the tribological properties of uncoated samples and samples with the DLC and DLC-Si coatings under friction, paired with counterbodies made of stainless steel. Both coatings demonstrated good tribological properties, but samples with the DLC-Si coating separated from the substrate, and the wear coefficient for samples with the DLC coating was much lower compared to samples with the DLC-Si coating. As a result of the studies considering the properties of mechanical face seals of nitride ceramics with the DLC and DLC-Si coatings, it was found that the use of these coatings significantly reduced the COF and improved wear resistance of products. At the same time, the DLC coatings look more preferable compared to DLC-Si coatings [32]. Following the results of the investigation focused on the properties of the DLC coating, deposited on the substrate of Si_3N_4 and M50 steel, it was found that the normal stresses on the boundary of the "coating–substrate" interface were higher (by about 10%) for the ceramic substrate, which could be explained by the higher value of the elastic modulus of Si_3N_4 [33]. It has also been found that as the coating thickness grows from 200 up to 400 nm, the stresses decrease at the boundary of the "coating–substrate" interface in accordance with the quadratic expression, and such a decrease slows down with the growth of the coating thickness [34]. A sample with the DLC coating demonstrates a lower COF compared to a sample with the MoS_2 coating [35]. Following the investigation focused on the tribological properties of a sample with the Cr-DLC coating under friction, paired with uncoated counterbodies of Al_2O_3, ZrO_2, Si_3N_4, and WC in air and in the helium atmosphere, it was found that the tribological properties of the samples with the DLC coating were significantly higher in air than in the helium atmosphere [36]. Two-dimensional finite element modeling of the properties of the DLC coating, deposited on the substrate of Al_2O_3 exhibited that a growth of the DLC coating thickness led to an increase in its hardness and crack resistance [37]. The studies revealed the ability of the DLC coating to minimize surface defects on the substrate and significantly reduce the intensity of oxidation processes [38]. The DLC coatings deposited on the substrate of β-SiAlON increase the surface hardness and improve the surface quality [39]. The comparison of the tribological properties of the Cr_2O_3-based samples with the DLC, TiN, and TiAlN coatings and of the uncoated samples in contact with cast iron counterbodies found that the samples with the DLC coatings demonstrated the highest scuffing resistance and the lowest coefficient of friction (COF) [40]. The studies carried out in air and in water environment, in nitrogen atmosphere, and in vacuum revealed a significant decrease in the COF, an increase in wear and oxidation resistance after deposition of the DLC coatings on the substrates of SiC, Si_3N_4, and ZrO_2 [41, 42]. Close values of the elastic modulus of the coating and the substrate is an important factor able to reduce internal stresses and thus improve the service life and reliability of products with the DLC coatings. At the same time, the deposition of coatings, in particular, the DLC coatings, on ceramic products increases their wear resistance, significantly reduces the COF, and enhances the oxidation resistance. Another important factor is also a leveling effect of a coating, which minimizes the influence of microconcentrators of stresses (pores, microcracks, etc.) on the reliability and service life of a product, while reducing the surface roughness value is important. Meanwhile, many studies note such a problem as the low strength of adhesive bonds between a DLC coating and a ceramic substrate, which leads to failure of the coating due to its separation from the substrate. Another important challenge is to study the influence of Si on the properties of the DLC coatings and the cutting properties of tools with such coatings.

3. Materials and methods

Multicomponent coatings of Ti-(Ti,Al)N-(Zr,Nb,Ti,Al)N with three-layer architecture, including adhesion, transition, and wear-resistant layers [43–50] were studied. The Ti-(Ti,Al)N coating widely used as a coating for metal-cutting tools was assumed as an object of comparison. The coating was chosen with the multi-layer architecture, including an (Cr,Al,Si)N adhesive-smoothing layer, a DLC-Si transition layer, and a DLC wear-resistant layer, in order to secure high adhesion to the ceramic substrate and release the smoothing effect of the coating. Following some studies [51–53], it was found that the adhesion of the coating to the substrate was enhanced due to a DLC-Si layer included in the coating. Furthermore, in [51], the studies found that the DLC-Si demonstrated lower hardness and wear resistance compared to those of the DLC coating, and that fact contributed to the selection of DLC-Si as a transition layer ensuring high adhesion and a smooth transition of properties.

The adhesion layer of (Cr,Al,Si)N can demonstrate the extremely high hardness (up to 55 GPa [54]) in a combination with significant toughness [55]. High thermal stability is another important feature of the above compound [55, 56]. Therefore, there is a possibility of a transition from a ceramic substrate (with the hardness of 15 to 20 GPa) through the (Cr,Al,Si)N layer to DLC-Si layers and a DLC coating (with the hardness from 30 to 80 GPa). Furthermore, due to its greater toughness compared to the DLC coating, the (Cr,Al,Si)N layer is able to "heal" microcracks and micropores by penetrating them on the surface of a ceramic substrate. In [57–59], the studies found that due to its nanocomposite structure, the (Cr,Al,Si)N layer improved the crack resistance while retaining high hardness.

The filtered cathodic vacuum arc deposition (FCVAD) was used to deposit the coatings of Ti-(Ti,Al)N-(Zr,Nb,Ti,Al)N and Ti-(Ti,Al)N in the VIT-2 unit [43, 60–66].

The lateral rotating cathode technology (LARC; developed by PLATIT – BCI Group, Switzerland) was applied to deposit DLC coatings on a PLATIT π-311 unit.

A nitride coating was deposited using cathodes containing Cr, Al, or Al–Si (88:12 at%). Argon (Ar) ions were subjected to purification using a beam of Ar ions at an anode voltage of 800/200 V and a current of 0.5 A for 20 min. A coating of Si-DLC was deposited using a mixture of gases, including 90% acetylene (C_2H_2), 8% Ar, and 2% tetramethylsilane ($Si(CH_3)_4$). A similar mixture of gases, except for $Si(CH_3)_4$, was used to deposit a pure DLC coating.

The microstructural investigation of samples involved a scanning electron microscope (SEM; Field Electron and Ion Company) Quanta 600 FEG.

The micro- and nanostructures of the samples were analyzed with a JEM 2100 high-resolution transmission electron microscope (HR-TEM), at the accelerating voltage of 200 kV. The energy-dispersive X-ray spectroscopy (INCA Energy) was applied to study the chemical composition of the samples.

The nanoindentation technique on an Instron Wilson Hardness Group Tukon device at the load of 0.01 N was applied to find the microhardness of the coatings. A CU 500 MRD lathe (Sliven) with a ZMM CU500MRD variable speed drive was used during the turning of workpieces made of AISI 52100 (HRC 56–58) hardened steel to study the cutting properties of the coated tool and the dynamics of its wear, at $f = 0.1$ mm/rev, $a_p = 0.5$ mm, and $v_c = 320$ m/min (for the tool with the DLC coatings) and ASTM T31507 hardened steel (DIN 1.2419, HRC 58–60) and ASTM X153CrMoV12 (DIN 1.2379, HRC 60–61) hardened steel $v_c = 80$–350 m/min; $f = 0.1$–0.25 mm/rev; $a_p = 0.5$–1.0 mm for the tool with the Ti-(Ti,Al)N-(Zr,Nb,Ti,Al)N and Ti-(Ti,Al)N coatings.

4. Results and discussion

4.1 Turning of X153CrMoV12 hardened steel with tool made of Al_2O_3-TiC and Al_2O-SiCw mixed ceramics with Ti-(Ti,Al)N-(Zr,Nb,Ti,Al)N coating

The structure of the Ti-(Ti,Al)N-(Zr,Nb,Ti,Al)N coating is depicted in **Figure 3** [67]. The total thickness of the coating is about 4 μm, and the nanolayer binary period λ is about 120 nm.

The chemical compositions of the coatings under study are presented in **Table 1**.

The coatings contain a similar amount of aluminum; however, the presence of Zr and Nb in the second coating causes some decrease in hardness with an increase in ductility, which makes it possible to provide a good level of resistance to brittle fracture [44, 46].

Figure 4a and **b** exhibit the results of the studies focused on the cutting properties of a tool equipped with (1) uncoated ceramic inserts made of Al_2O_3-TiC and Al_2O_3-SiCw ceramics, (2) Ti-(Ti,Al)N coating, and (3) Ti-(Ti,Al)N-(Zr,Nb,Ti,Al)N coating [67]. During the longitudinal turning of X153CrMoV12 hardened steel, the wear rate of ceramic tools based on mixed ceramics of Al_2O_3-TiC and Al_2O_3-SiCw depends rather largely on the coating composition. In particular, the maximum increase in the wear resistance of ceramic tools was provided by coatings based on the complex composite nanostructured system of Ti-(Ti,Al)N-(Zr,Nb,Ti,Al)N. At VB = 0.4 mm, the above system increases the tool life of a ceramic tool up to 1.5 times compared to an uncoated tool and by 1.3 times compared to a tool with the Ti-(Ti,Al)N commercial coating.

According to the studies of [1–4, 14–18], for a ceramic tool, the brittle fracture of its cutting edge is the most probable mechanism of failure. This fact can be explained by the lower brittle strength of the ceramic tool material in comparison with the same parameters of the carbide tool material. Ceramic tools also tend to stochastic brittle fracture because of the higher contact stresses, in particular, normal stresses. In general, the above contact stresses exceed the same values for carbide tools because of the considerable decrease in the total length of the contact between the chips and the rake face of the ceramic tool, at a small decrease in the normal load.

There are is a balance in the nature of wear centre development on the rake and flank faces of the Al_2O_3-TiC ceramic cutting insert with the Ti-(Ti,Al)N and Ti-(Ti,Al)N-(Zr,Nb,Ti,Al)N coatings during the longitudinal turning of X153CrMoV12 hardened steel, with no visible chips and microchipping.

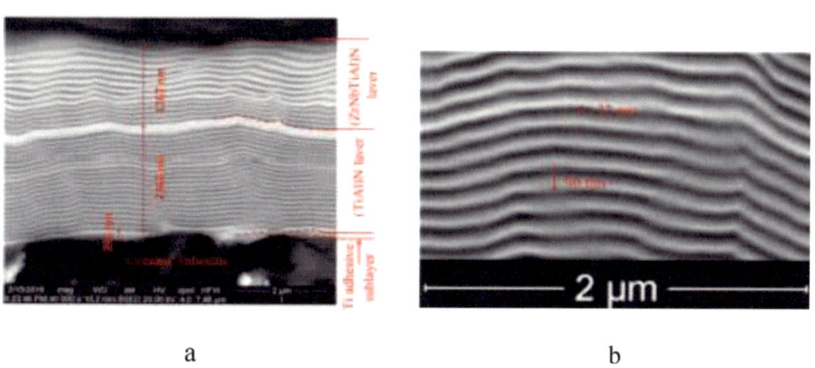

Figure 3.
Micro (a) and nano (b) structure of the cross-section for cutting Al_2O_3–TiC with Ti-(Ti,Al)N-(Zr,Nb,Ti,Al)N coatings ceramic inserts [67] (SEM).

	Ti, at%	Al, at%	Zr, at%	Nb, at%
Ti-(Ti,Al)N	70	30	—	—
Ti-(Ti,Al)N-(Zr,Nb,Ti,Al)N	42	26	29	3

Table 1.
Chemical compositions of the coatings under study.

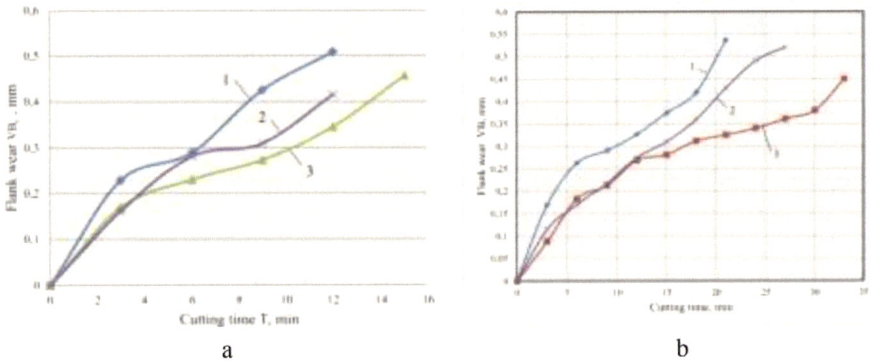

Figure 4.
Relationship between wear VBmax and cutting time for (1) uncoated inserts, (2) tools with Ti-(Ti,Al)N coating, and (3) Ti-(Ti,Al)N-(Zr,Nb,Ti,Al)N coating, during the longitudinal turning of X153CrMoV12 hardened steel at v_c = 250 m/min, f = 0.05 mm/rev, a_p = 0.5 mm, inserts made of (a) Al_2O_3-TiC and (b) Al_2O_3-SiCw at v_c = 300 m/min, f = 0.1 mm/rev, a_p = 0.5 mm, inserts made of (b) Al_2O-SiCw [67].

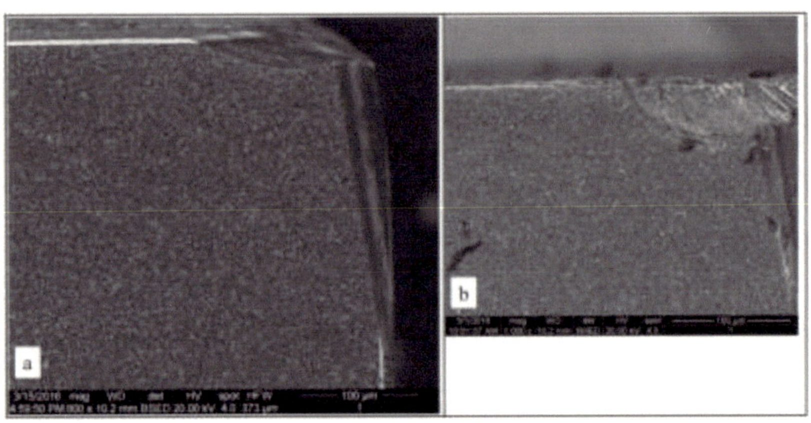

Figure 5.
Wear pattern after 25 minutes in the longitudinal turning of X153CrMoV12 hardened steel at v_c = 250 m/min, f = 0.05 mm/rev, a_p = 0.5 mm for cutting inserts made of Al_2O_3-TiC with Ti-(Ti,Al)N [67] (SEM).

Undisturbed residues of coatings at the edges of wear centres both on the rake and flank faces of the ceramic cutting insert are also typical for the process described above (**Figure 5a** and **b**) [67].

Both on the rake and flank faces of the tool, the wear mechanism typical for the Ti-(Ti,Al)N-(Zr,Nb,Ti,Al)N coating is primarily an abrasive interaction with the material being machined (**Figures 6** and 7) [67]. It should be noted that the coating and the substrate work as a unified system, where cracks and chipping hardly occur. No adherents of the material being machined are detected on the coating surface, which may relate to the low adhesion between the external (wear-resistant) layer and the material being machined.

During the cutting, good adhesion retains between the coating and the ceramic substrate (**Figure 7**) [67].

For the Ti-(Ti,Al)N-(Zr,Nb,Ti,Al)N coating, the typical mechanism of the coating failure is the formation of longitudinal cracks in the areas immediately adjacent to the cutting area (**Figure 8**) [67]. However, such cracks related to the delamination of the coating under the influence of the compressive residual stresses are much less dangerous compared to transverse cracks, often formed in monolithic coatings.

During the process of cutting with the ceramic tool with the Ti-(Ti,Al)N coating, massive adherents of the material being machined are formed, both on the rake and flank faces of the tool (**Figure 9**) [67]. The mechanism of tool wear relates to the adhesive-fatigue processes, which is confirmed by the nature of the coating failure with clear tear-outs of the coating elements (see **Figure 9**).

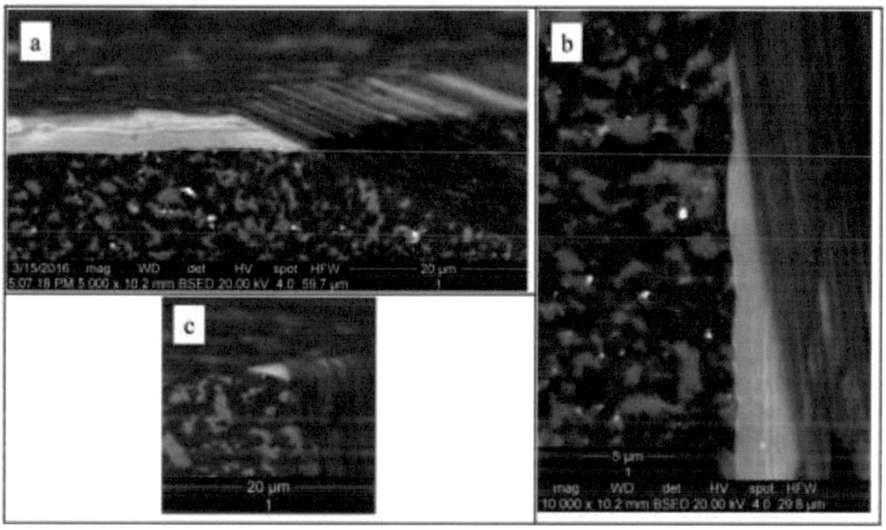

Figure 6.
Wear pattern on (a) rake face, (b) flank face, and (c) corner of the Al_2O_3-TiC ceramic insert with the Ti-(Ti,Al)N-(Zr,Nb,Ti,Al)N coating [67] (SEM).

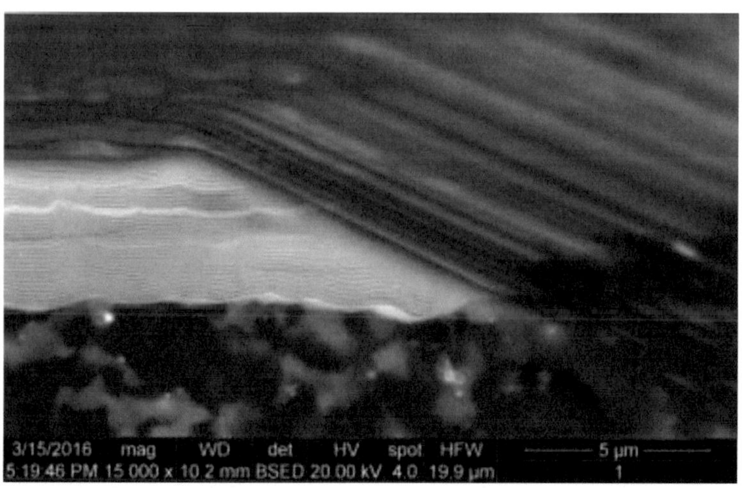

Figure 7.
Wear pattern on the rake face of the Al_2O_3-TiC ceramic insert with Ti-(Ti,Al)N-(Zr,Nb,Ti,Al)N coating [67] (SEM).

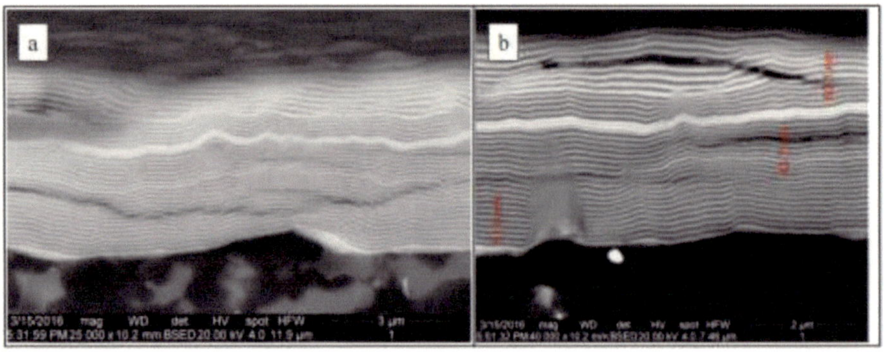

Figure 8.
Ceramic insert of Al_2O_3-TiC with the Ti-(Ti,Al)N-(Zr,Nb,Ti,Al)N coating [67] (SEM).

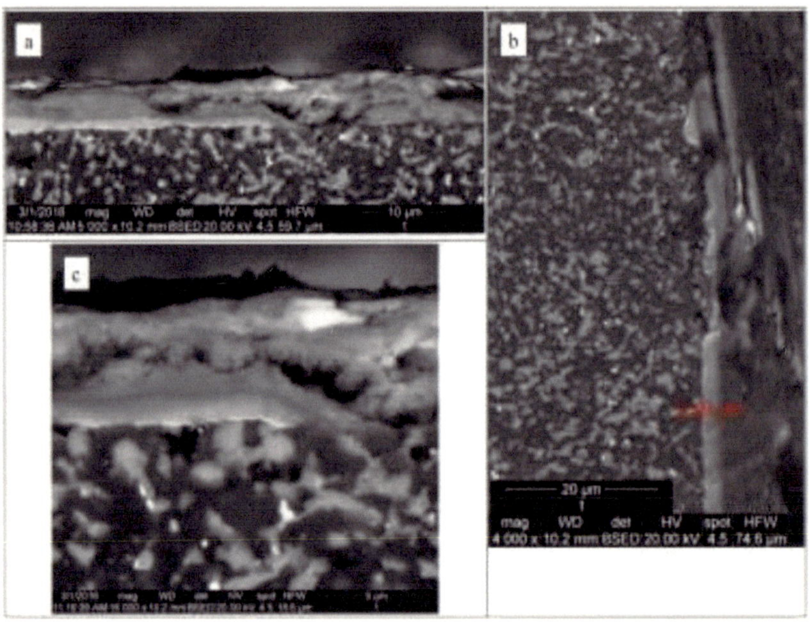

Figure 9.
Wear pattern (a) on the rake face of the Al_2O_3-TiC ceramic insert with the Ti-(Ti,Al)N coating and wear pattern (b) on the flank face of insert with particles of the material being machined after 20 minutes in longitudinal turning of X153CrMoV12 hardened steel at v_c = 250 m/min, f = 0.05 mm/rev, a_p = 0.5 mm [67] (SEM).

4.2 Turning of AISI 52100 (HRC 56–58) hardened steel with tools with (Cr,Al,Si)N-(DLC-Si)-DLC-(DLC-Si), and (Cr,Al,Si)N-DLC coatings

The investigation of the DLC-1 coating structure using TEM reveals the presence of a wear-resistant layer in the amorphous DLC and a (Cr,Al,Si)N transition layer with the columnar structure (**Figure 10a** and **b**) [68]. In its turn, the DLC layer structure includes sublayers of DLC-Si at the border of the (Cr,Al,Si)N layer and the coating surface. The (Cr,Al,Si)N layer is about 0.4 μm thick, and the thickness of the DLC layer is about 1.4 μm.

The DLC-2 coating also has a two-layer structure with a (Cr,Al,Si)N transition layer and a DLC wear-resistant layer (**Figure 10c** and **d**). The thickness of the transition layer is about 0.3 μm, while the thickness of the DLC layer is about 1.1 μm.

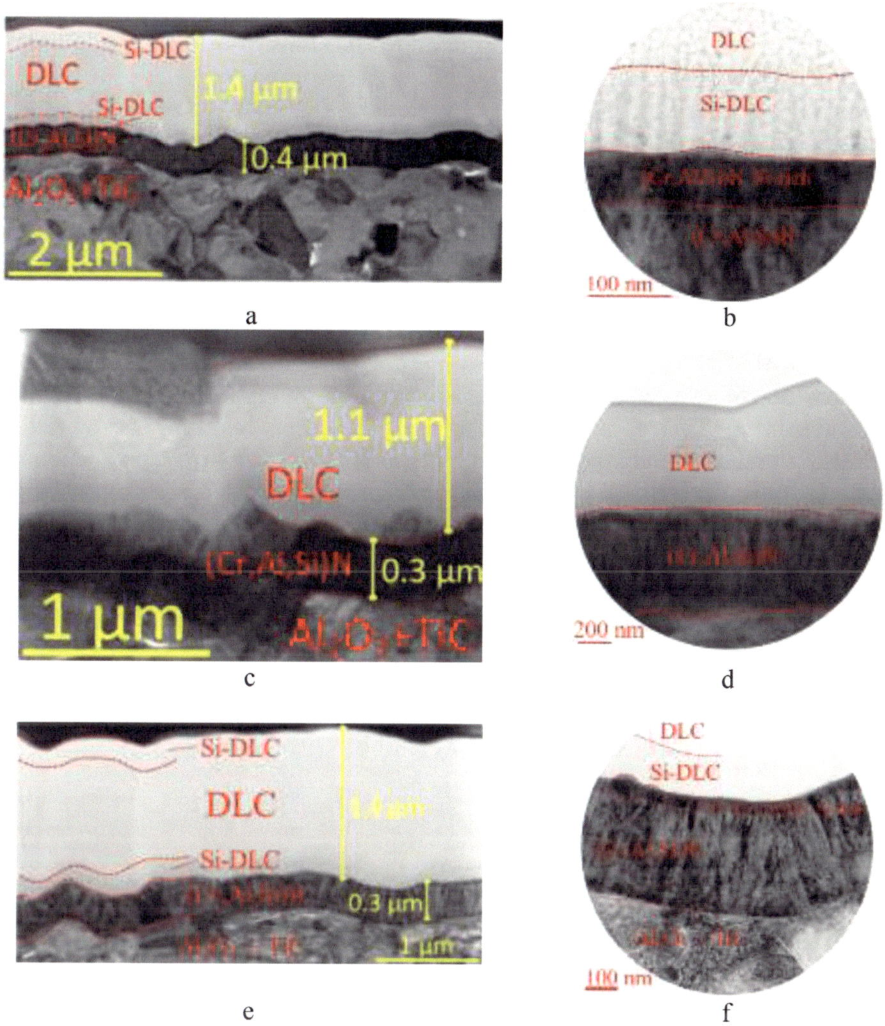

Figure 10.
Internal structure of the coatings under study: (a and b) DLC-1; (c and d) DLC-2; and (e and f) DLC-3 obtained using TEM [68].

The TEM analysis of the cross-section image for the DLC-1 coating (**Figure 10a**) reveals the absence of any clear structure in the DLC layer. The electron diffraction patterns depict a broadened halo, typical for an amorphous structure, while a structure close to columnar can be noticed in the DLC-Si sublayer (**Figure 10b**). The chemical composition of the DLC-Si sublayer includes about 56 at% Si + 43 at% C + 1 at% O. The DLC layer includes about 1 at% Si + 97 at% C + 2 at% O (**Figure 10b**). The (Cr,Al,Si)N layer includes 70 at% Cr + 23at% Al + 7 at% Si. The surface layer of the ceramic substrate Al_2O_3 exhibits signs of diffusion of Cr in the volume of 0.3–0.4 at%. The area around the (Cr,Al,Si)N layer, adjacent to the border of the DLC-Si layer is characterized by an increase content of Si (about 10 at%) to ensure better adhesion with the DLC-Si layer. The high (above 50 at%) content of Si is detected in the transition layer and the surface layer of DLC.

The earlier studies have found that the structure of (Cr,Al,Si)N is characterized by a face-centered NaCl-type lattice with various crystal orientations: (111), (200), and (220) [26, 27]. In the presence of the CrN phase, the phases of Cr, Cr_2N, $CrSi_2$, and Si_3N_4 were also detected [27]. The investigation has also revealed that Si is present either in the form of a substitutional solid solution in the Cr-Al-N lattice, or in the form of an amorphous Si-N compound that accumulates at the grain boundaries of Cr-Al-N [27, 31, 32]. **Figure 11** exhibits the nanostructure of the (Cr,Al,Si) N layer, characterized by the presence of various phases differing with interplanar spacings and orientations of crystal planes [68].

The formation of a fan-shaped network of microdroplets in the DLC layer, caused by microdroplets of (Cr,Al,Si), embedded in the structure of the DLC layer, is depicted in **Figure 12** [68]. During the process of cutting with a coated tool, active fracture can occur in those areas of the DLC layer. So, during the deposition of the coatings with the structures under study, the minimized number of microdroplets is a key condition to ensure the required working efficiency of the coating.

The (Cr,Al,Si)N layer with its smoothing function plays an essential role during deposition of a DLC-Si layer with regard to the elimination of possible stress concentrators on the surface. In particular, a microdefect which appeared on the surface of the ceramic substrate as a result of conjunction of two grains is depicted in **Figure 13** [68]. When the DLC-Si layer is deposited directly onto a ceramic substrate, such a microdefect can provoke cracking in the DLC-Si structure under the above-described mechanism. The above microdefect is being smoothed by the (Cr,Al,Si)N layer, which forms a smooth surface without serious defects that would be able to act as stress concentration in the DLC layer. With the prime function of the (Cr,Al,Si)

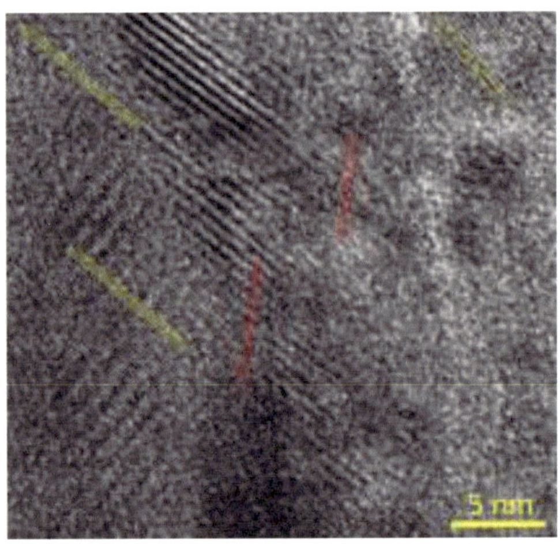

Figure 11.
High-resolution TEM image showing the crystalline structure of the (Cr,Al,Si)N transition layer for DLC-1 [68].

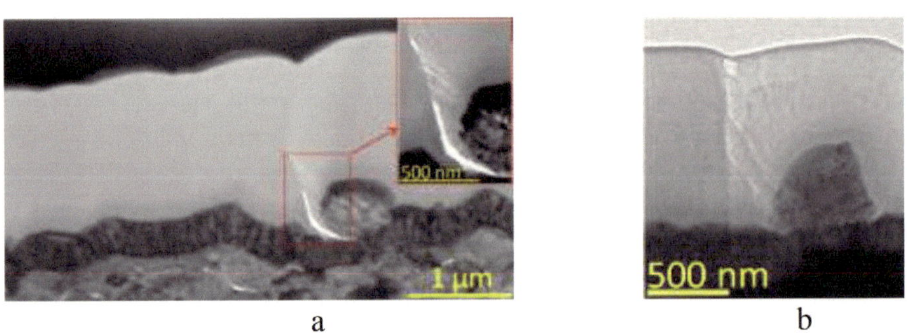

Figure 12.
Influence of embedded microdroplets on forming cracks in the DLC layer for (a) DLC-3 and (b) DLC-2 coatings [68] (TEM).

N layer consisting in the provision of good adhesion between the ceramic substrate and the DLC-Si coating and the formation of a composite structure with a combination of high hardness and brittle fracture resistance, the smoothing functions of the (Cr,Al,Si)N layer are also crucial to secure good coating performance.

Cutting tests are crucial in the assessment of the working efficiency of a coated tool and the performance properties of the very coating. The tests were also conducted to study the dynamics of wear on the rake faces of the uncoated cutting tools made of ceramics and the tools with the DLC-1 and DLC-2 coatings. The tool with the DLC-2 coating proved to have the best resistance to the wear crater formation on the rake face (see **Figure 14**) [68].

The most active formation of a wear crater was detected on the uncoated tool. The active formation of adherents during the cutting is a typical feature of the tool with the DLC-1 coating. This proves high adhesion between the DLC-1 coating and the material being machined. As a result, strong adhesive bond bridges are formed and broken, which leads to a high adhesive-fatigue wear of the tool. At the same time, an abrasive wear with significantly lower adherent formation is typical for the tool with the DLC-2 coating.

The wear pattern on the rake face of the tool was considered in detail (**Figure 10**). The first typical feature was a much more active formation of an adherent of the material being machined for the tool with the DLC-1 coating. At the same time, the uncoated tool and the tool with the DLC-2 coating demonstrated an insignificant formation of such adherents. Based on the obtained results, it can be concluded that there is an

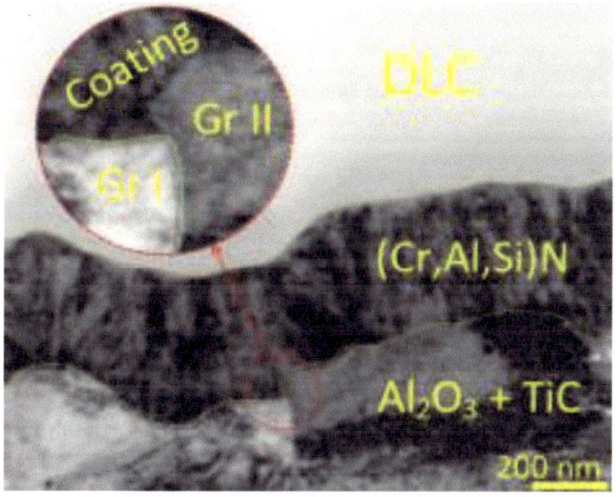

Figure 13.
Example of the smoothing effect of the (Cr,Al,Si)N layer on the DLC-3 coating [68] (TEM).

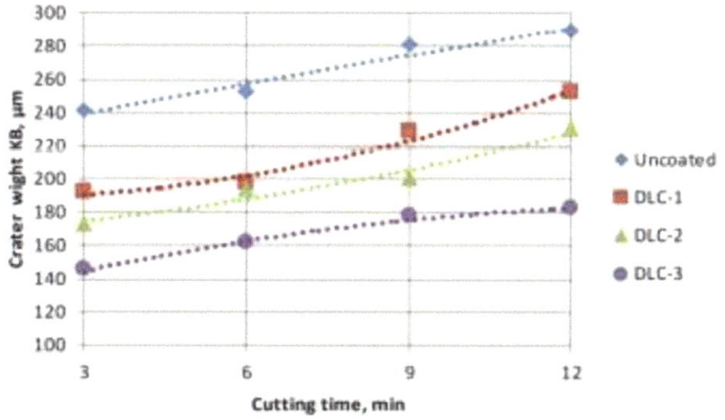

Figure 14.
Dynamics of changes in the crater on the rake face of the tool [68].

increased adhesion between the material of the DLC-1 coating (with a surface sublayer with the high Si content) and the material being machined. The uncoated tool and the tool with the DLC-2 coating are characterized by primarily abrasive wear, accompanied by the formation of typical grooves in the direction of chip flow. The tool with the DLC-1 coating also has similar grooves, but there are also signs of the cutting edge chipping.

Since the flank wear is usually assumed as a limiting factor, its dynamics was considered (**Figure 15**) [68]. The most active flank wear was detected on the tool with the DLC-1 coating. As already stated, this fact can be explained by the high adhesion between the coating and the material being machined. As a result, the tool demonstrated the increased adhesive-fatigue wear. Meanwhile, less active adhesion to the material being machined and lower tendency to chipping were detected on the uncoated tool. At the same time, there is minor chipping in the area of the cutting edge. Finally, the tool with the DLC-2 coating demonstrated a low tendency to the formation of adherents and no visible signs of brittle fracture. Thus, the purely abrasive wear mechanism was typical for the tool with the DLC-2. The tool with the coating DLC-2 exhibited a 17% higher resistance to the flank wear. In combination with its noticeably higher resistance to the formation of a wear crater on the rake face and the more favorable wear pattern (implying the balanced abrasive wear instead of chipping and brittle fracture), the above proves the good prospects of the DLC-2 coating.

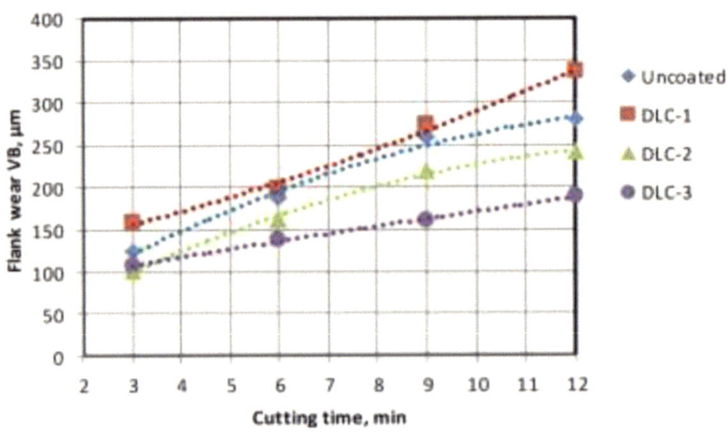

Figure 15.
Tool wear on the flank face depending on the cutting time [68].

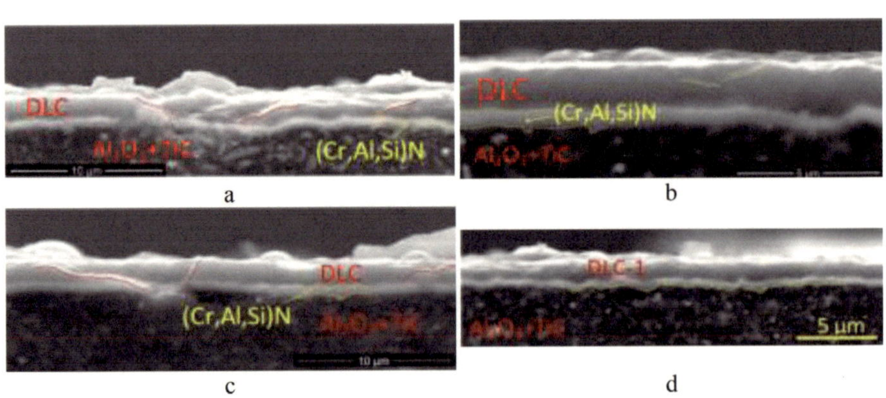

Figure 16.
Investigation of the wear pattern on a cross-section of the tool with the DLC-1 coating [68] (SEM). Through cracks and tears (a, c), minor cracks and intact coating (b, d).

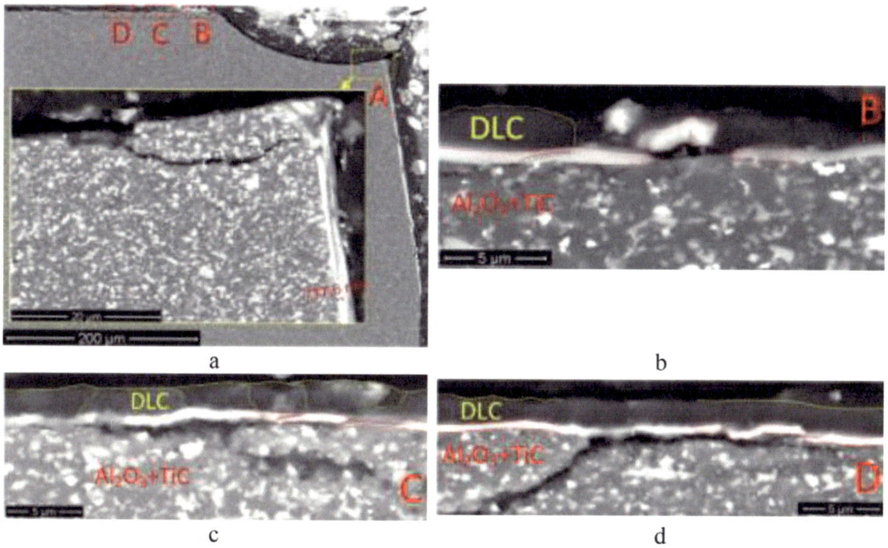

Figure 17.
Investigation of the fracture pattern on a cross-section of the tool with the DLC-2 coating [68] (SEM). General view and a crack passing into a ceramic substrate (a), tearing out of a coating fragment (b), cracks in a coating turning into a substrate (c, d).

For a better understanding of the wear patterns on the samples, cross-sections were made passing through the centre of the wear craters.

Figure 16 demonstrates that the sample with the DLC-1 coating bears clear signs of cutting edge chipping. It is also clear that the coating is retained on the rake face of the tool in the area adjacent to the boundary of the wear crater. On the flank face, in the area adjacent to the flank wear land, the DLC layer failed while the (Cr,Al,Si) N layer was preserved. In general, the coating retains good adhesion to the substrate, and it does not separate from the substrate on the sample with the DLC-1 coating.

The wear process on the DLC-1 coating, especially its upper layer, is characterized by the formation of inclined cracks (**Figure 16a–c**), which in some cases also penetrate into the (Cr,Al,Si)N transition layer (**Figure 16a** and **c**). It is also clear that the surface of the ceramic substrate has a rather complex relief, and the (Cr,Al,Si)N transition layer fills in the microroughness of the surface and thus forms a basis for the DLC layer (**Figure 16d**) [68].

During the consideration of the wear process on the tool with the DLC-2 coating, it can be noticed that the transition layer of this coating demonstrates a higher tendency to brittle fracture compared to the transition layer of the DLC-1 coating (**Figure 17**) [68]. The DLC layer of the DLC-2 coating also bears signs of active brittle fracture. In the area, adjacent to the wear crater (Area B, **Figure 17**), almost complete failure of the DLC layer and partial failure of the transition layer are detected (**Figure 17b**). The brittle fracture of the ceramic substrate, accompanied by microchipping, also takes place. The signs of such brittle fracture are detected throughout the whole contact area on the rake face of the tool.

5. Conclusions

The chapter considered the specific features of the use of wear-resistant coatings deposited to improve the performance properties of ceramic cutting tools. The application of the Ti-(Ti,Al)N-(Zr,Nb,Ti,Al)N multilayer composite coating with a nanostructured wear-resistant layer and the (Cr,Al,Si)N-(DLC-Si)-DLC-(DLC-Si) and (Cr,Al,Si)N-DLC composite coatings increase the tool life of a ceramic cutting tool by 50–80%.

Following the studies, it has been found that the intensity of the thermal effect on structures of the ceramic substrate may be reduced through the deposition of the developed coatings under study on contact areas of ceramic cutting tools, since such coatings provide better heat removal from the cutting area due to an increased length of the plastic contact.

The use of the DLC-based coatings can both prolong (due to an enhanced resistance to abrasive wear) and shorten (due to higher adhesion) the tool life. It is not advisable to deposit a coating with an increased Si content in the surface layers of the coating, at least for the considered cutting conditions, despite a bit higher hardness of such coating. At the same time, a coating with the high Si content demonstrated a lower tendency to brittle fracture compared to the DLC-2 coating. The use of the (Cr,Al,Si)N transition layer is advisable, since it improves the adhesion between the ceramic substrate and the DLC coating.

The results of the conducted studies prove the prospects for the application of multilayer composite coatings, including those with DLC layers introduced to improve the cutting properties of ceramic cutting tools (both assembled and one-piece tools).

Acknowledgements

This study was supported by a grant of the Russian Science Foundation [Agreement No. 18-19-00312 dated 20 April 2018].

Conflict of interest

The authors declare no conflict of interest.

Author details

Sergey Grigoriev[1], Alexey Vereschaka[2*], Marina Volosova[1], Caterina Sotova[1], Nikolay Sitnikov[3], Filipp Milovich[4] and Nikolay Andreev[4]

1 Moscow State Technological University STANKIN, Moscow, Russia

2 IDTI RAS, Moscow, Russia

3 National Research Nuclear University MEPhI, Moscow, Russia

4 National University of Science and Technology MISiS, Moscow, Russia

*Address all correspondence to: dr.a.veres@yandex.ru

References

[1] Riedel R. Handbook of Ceramic Hard Materials. WILEY-VICH: Weinheim–New York; 2000

[2] López de Lacalle LN, Lamikiz A. Machine Tools for High Performance Machining. Springer, 2009

[3] Sun S, Brandt M, Dargusch MS. Thermally enhanced machining of hard-to-machine materials. A review. International Journal of Machine Tools and Manufacture. 2010;**50**(8):663-680

[4] Rahman M, Wang Z-G, Wong Y-S. A review on high-speed machining of titanium alloys. JSME International Journal, Series C: Mechanical Systems, Machine Elements and Manufacturing. 2006;**49**(1):11-20

[5] Ezugwu EO. High speed machining of aero-engine alloys. Journal of the Brazilian Society of Mechanical Sciences and Engineering. 2004;**26**(1):1-11

[6] Senthil Kumar A, Raja Durai A, Sornakumar T. Machinability of hardened steel using alumina based ceramic cutting tools. International Journal of Refractory Metals and Hard Materials. 2003;**21**(3-4):109-117

[7] Kumar AS, Durai AR, Sornakumar T. Wear behaviour of alumina based ceramic cutting tools on machining steels. Tribology International. 2006;**39**(3):191-197

[8] Grigoriev SN, Vereschaka AA, Vereschaka AS, Kutin AA. Cutting tools made of layered composite ceramics with nano-scale multilayered coatings. Procedia CIRP. 2012;**1**(1):301-306

[9] Vereschaka AS, Grigoriev SN, Sotova ES, Vereschaka AA. Improving the efficiency of the cutting tools made of mixed ceramics by applying modifying nano-scale multilayered coatings. Advanced Materials Research 2013;712-715: 391-394

[10] Vereschaka AS, Grigoriev SN, Tabakov VP, Sotova ES, Vereschaka AA, Kulikov MY. Improving the efficiency of the cutting tool made of ceramic when machining hardened steel by applying nano-dispersed multi-layered coatings. Key Engineering Materials. 2014;**581**:68-73

[11] Vereshchaka AA, Sotova ES, Batako AD, Sedykh MI, Vereshchaka AS. A study of the cutting properties and wear mechanism of ceramic edge tools with nanostructure multilayer composite coatings. Journal of Friction and Wear. 2014;**35**(6):483-488

[12] Vereschaka A, Prilukova J, Vereschaka A, Bublikov J, Aksenenko A. Control of temperature in cutting zone in machining of alloyed casehardened steels by applying a ceramic tool with wear-resistant coatings. Materials Science Forum. 2016;**857**:199-205

[13] Ezugwu EO, Bonney J, Yamane Y. An overview of the machinability of aeroengine alloys. Journal of Materials Processing Technology. 2003;**134**(2):233-253

[14] Kitagawa T, Kubo A, Maekawa K. Temperature and wear of cutting tools in high-speed machining of Inconel 718 and Ti-6Al-6V-2Sn. Wear. 1997;**202**(2):142-148

[15] Devillez A, Schneider F, Dominiak S, Dudzinski D, Larrouquere D. Cutting forces and wear in dry machining of Inconel 718 with coated carbide tools. Wear. 2007;**262**(7-8):931-942

[16] Vereschaka AA, Volosova MA, Krapostin AA, Batako A, Seleznev AE. Increased operating properties of

cutting ceramics by application of nanostructured multilayer wear-resistant coating. Journal of Nano Research. 2017;**50**:90-104

[17] Merchant ME. Mechanics of the metal cutting process. II. Plasticity conditions in orthogonal cutting. Journal of Applied Physics. 1945;**16**(6):318-324

[18] Tounsi N, Vincenti J, Otho A, Elbestawi MA. From the basic mechanics of orthogonal metal cutting toward the identification of the constitutive equation. International Journal of Machine Tools and Manufacture. 2002;**42**(12):1373-1383

[19] Volosova MA, Grigor'ev SN, Kuzin VV. Effect of Titanium Nitride Coating on Stress Structural Inhomogeneity in Oxide-Carbide Ceramic. Part 4.Action of Heat Flow. Refract Ind Ceram 2015;56:91-96

[20] Kuzin VV, Grigor'ev SN, Volosova MA. Effect of a TiC coating on the stress-strain state of a plate of a high-density nitride ceramic under nonsteady thermoelastic conditions. Refractories and Industrial Ceramics. 2014;**54**:376-380

[21] Czechowski K, Pofelska-Filip I, Królicka B, Szlosek P, Smuk B, Wszołek J, et al. Effect of nitride nano-scale multilayer coatings on functional properties of composite ceramic cutting inserts. Bulletin of the Polish Academy of Sciences Technical Sciences. 2005;**53**(4):425-431

[22] Sokovic M, Kopac J, Dobrzanski LA, Mikula J, Golombek K, Pakula D. Cutting characteristics of PVD and CVD - coated ceramic tool inserts. Tribology in Industry. 2006;**28**:3-8

[23] Aslantas K, Ucun I, Cicek A. Tool life and wear mechanism of coated and uncoated Al2O3/TiCN mixed ceramic tools in turning hardened alloy steel. Wear. 2012;**274-275**:442-451

[24] Qin J, Long Y, Zeng J, Wu S. Continuous and varied depth-of-cut turning of gray cast iron by using uncoated and TiN/Al2O3 coated silicon nitride-based ceramic tools. Ceramics International. 2014;**40**(8):12245-12251

[25] Long Y, Zeng J, Yu D, Wu S. Cutting performance and wear mechanism of Ti-Al-N/Al-Cr-O coated silicon nitride ceramic cutting inserts. Ceramics International. 2014;**40**(7):9889-9894

[26] Long Y, Zeng J, Yu D, Wu S. Microstructure of TiAlN and CrAlN coatings and cutting performance of coated silicon nitride inserts in cast iron turning. Ceramics International. 2014;**40**(7):9889-9894

[27] Mali NM, Mahender T. Wear analysis of single point cutting tool with and without coating. IJREAT International Journal of Research in Engineering & Advanced Technology. 2015;**3**(3):49-57

[28] Enke K. Upgrading of technical ceramics by DLC-layers. Keramische Zeitschrift. 1998;**50**(12):1022-1025

[29] Enke K. Dry machining and increase of endurance of machine parts with improved doped DLC coatings on steel, ceramics and aluminium. Surface and Coatings Technology. 1999;**116-119**:488-491

[30] Fukuhara T, Kato M, Nishimura S, Ikegami N. Application of ceramics to seal materials. Toraibarojisuto/Journal of Japanese Society of Tribologists. 2018;**63**(12):824-829

[31] Gomes JR, Camargo SS Jr, Simao RA, Carrapichano JM, Achete CA, Silva RF. Tribological properties of silicon nitride ceramics coated with DLC and DLC-Si

against 316L stainless steel. Vacuum. 2007;**81**:1448-1452

[32] Vila M, Carrapichano JM, Gomesc JR, Camargo SS Jr, Achete CA, Silva RF. Ultra-high performance of DLC-coated Si3N4 rings for mechanical seals. Wear. 2008;**265**:940-944

[33] Nie C, Zheng D, Gu L, Zhao X, Wang L. Comparison of interface mechanics characteristics of DLC coatingdeposited on bearing steel and ceramics. Applied Surface Science. 2014;**317**:188-197

[34] Zhang C, Gu L, Zheng D. Tribological Property Analyses of DLC Films on Ceramic Ball Surfaces With 3-D FEA Method and Experiments. ASME/STLE 2012 International Joint Tribology Conference Denver, Colorado, USA, October 7-10, 2012

[35] Gu L, Tang G, Zhang C, Jing C, Wang L. Self-Lubricated Modification for Silicon Nitride Ceramic Ball Surface. STLE/ASME 2010 International Joint Tribology Conference, 2010 Paper No. IJTC2010-41161, 59-60

[36] Wua D, Ren S, Pu J, Lu Z, Zhang G, Wang L. A comparative study of tribological characteristics of hydrogenated DLC film sliding against ceramic mating materials for helium applications. Applied Surface Science. 2018;**441**:884-894

[37] Jian-guang, Zhai Yi-qi, Wang Tae-gyu, Kim Jung-il Song. Finite element and experimental analysis of Vickers indentation testing on Al_2O_3 with diamond-like carbon coating. Journal of Central South University, 2012;19(5):1175-1181

[38] Hamdy AS. Electrochemical behavior of diamond-like-carbon coatings deposited on AlTiC (Al_2O_3 + TiC) ceramic composite substrate in HCl solution. Electrochimica Acta. 2011;**56**:1554-1562

[39] Bisht A, Chockalingam S, Tripathi RK, Dwivedi N, Dayal S, Kumar S, et al. Improved surface properties of β-SiAlON by diamond-like carbon coatings. Diamond & Related Materials. 2013;**36**:44-50

[40] Cho DH, Lee YZ. Evaluation of ring surfaces with several coatings for friction, wear and scuffing life. Transactions of Nonferrous Metals Society of China (English Edition). 2009;**19**(4):992-996

[41] Kato K. Tribology of advanced ceramics and hard coatings. Key Engineering Materials 2004;264-268(I):449-452

[42] Kato K. Tribology of ceramics and hard coatings. Materialwissenschaft und Werkstofftechnik. 2003;**34**(10-11):1003-1007

[43] Vereshchaka AA, Vereshchaka AS, Mgaloblishvili O, Morgan MN, Batako AD. Nano-scale multilayered-composite coatings for the cutting tools. Int. J. Adv. Manuf. Tech. 2014;**72**(1-4):303-317

[44] Vereschaka A, Tabakov V, Grigoriev S, Sitnikov N, Oganyan G, Andreev N, et al. Investigation of wear dynamics for cutting tools with multilayer composite nanostructured coatings in turning constructional steel. Wear. 2019;**420-421**:17-37. DOI: 10.1016/j.wear.2018.12.033

[45] Vereschaka A, Tabakov V, Grigoriev S, Aksenenko A, Sitnikov N, Oganyan G, et al. Effect of adhesion and the wear-resistant layer thickness ratio on mechanical and performance properties of ZrN - (Zr,Al,Si) N coatings. Surface and Coating Technology. 2019;**357**:218-234

[46] Vereschaka AA, Grigoriev SN, Sitnikov NN, Batako AD. Delamination and longitudinal cracking in multi-layered composite nano-structured

coatings and their influence on cutting tool life. Wear 2017;390-391:209-219

[47] Vereschaka A, Tabakov V, Grigoriev S, Sitnikov N, Andreev N, Milovich F. Investigation of wear and diffusion processes on rake faces of carbide inserts with Ti-TiN-(Ti,Al,Si)N composite nanostructured coating. Wear 2018;416-417:72-80

[48] Vereschaka A, Tabakov V, Grigoriev S, Sitnikov N, Milovich F, Andreev N, et al. Investigation of the influence of the thickness of nanolayers in wear-resistant layers of Ti-TiN-(Ti,Cr,Al)N coating on destruction in the cutting and wear of carbide cutting tools. Surface and Coating Technology. 2020;**385**:125402

[49] Vereschaka A, Aksenenko A, Sitnikov N, Migranov M, Shevchenko S, Sotova C, et al. Effect of adhesion and tribological properties of modified composite nano-structured multi-layer nitride coatings on WC-Co tools life. Tribology International. 2018;**128**:313-327

[50] Grigoriev S, Vereschaka A, Milovich F, Tabakov V, Sitnikov N, Andreev N, et al. Investigation of multicomponent nanolayer coatings based on nitrides of Cr, Mo, Zr, Nb, and Al. Surface and Coatings Technology. 2020;**401**:126258

[51] Lubwama M, Corcoran B, Sayers K, Kirabira JB, Sebbit A, McDonnell KA, et al. Adhesion and composite micro-hardness of DLC and Si-DLC films deposited on nitrile rubber. Surface and Coating Technology. 2012;**206**:4881-4886

[52] Qiang L, Bai C, Gong Z, Liang A, Zhang J. Microstructure, adhesion and tribologicalbehaviors of Si interlayer/Si T doping diamond-like carbon film developed on nitrile butadiene rubber. Diamond and Related Materials. 2019;**92**:208-218

[53] Hofmann D, Kunkel S, Bewilogua K, Wittorf R. From DLC to Si-DLC based layer systems with optimized properties for tribological applications. Surface and Coating Technology. 2013;**215**:357-363

[54] Park I-W, Kang DS, Moore JJ, Kwon SC, Rha JJ, Kim KH. Microstructures, mechanical properties, and tribological behaviors of Cr–Al–N, Cr–Si–N, and Cr–Al–Si–N coatings by a hybrid coating system. Surface and Coating Technology. 2007;**201**:5223-5227

[55] Miyake T, Kishimoto A, Hasegawa H, Tribological properties and oxidation resistance of (Cr,Al,Y)N and (Cr,Al,Si)N films synthesized by radio-frequency magnetron sputtering method, Surface and Coating Technology 2010;205:290-294.

[56] Ding J, Zhang T, Yun JM, Kang MC, Wang Q, Kim KH. Microstructure, mechanical, oxidation and corrosion properties of the Cr-Al-Si-N coatings deposited by a hybrid sputtering system. Coat. 2017;7:119

[57] Vereschaka A, Tabakov V, Grigoriev S, Sitnikov N, Oganyan G, Andreev N, et al. Investigation of wear dynamics for cutting tools with multilayer composite nanostructured coatings in turning constructional steel. Wear. 2019;**420-421**:17-37

[58] Vereschaka A, Aksenenko A, Sitnikov N, Migranov M, Shevchenko S, Sotova C, et al. Effect of adhesion and tribological properties of modified composite nano-structured multi-layer nitride coatings on WC-Co tools life. Tribology International. 2018;**128**:313-327

[59] Vereschaka AA, Grigoriev SN, Sitnikov NN, Batako A. Delamination and longitudinal cracking in multi-layered composite nano-structured coatings and their influence on cutting tool life. Wear. 2017;**390-391**:209-219

[60] Kuzin VV, Grigoriev SN, Fedorov MY. Role of the thermal factor in the wear mechanism of ceramic tools. Part 2: Microlevel. Journal of Friction and Wear. 2015;**36**(1):40-44

[61] Metel A, Bolbukov V, Volosova M, Grigoriev S, Melnik Y. Equipment for deposition of thin metallic films bombarded by fast argon atoms. Instruments and Experimental Techniques. 2014;**57**(3):345-351

[62] Metel AS, Grigoriev SN, Melnik YA, Bolbukov VP. Characteristics of a fast neutral atom source with electrons injected into the source through its emissive grid from the vacuum chamber. Instruments and Experimental Techniques. 2012;**55**(2):288-293

[63] Metel A, Bolbukov V, Volosova M, Grigoriev S, Melnik Y. Source of metal atoms and fast gas molecules for coating deposition on complex shaped dielectric products. Surface and Coating Technology. 2013;**225**:34-39

[64] Grigoriev SN, Sobol OV, Beresnev VM, Serdyuk IV, Pogrebnyak AD, Kolesnikov DA, et al. Tribological characteristics of (TiZrHfVNbTa)N coatings applied using the vacuum arc deposition method. Journal of Friction and Wear. 2014;**35**(5):359-364

[65] Vereschaka AA, Grigoriev SN, Vereschaka AS, Popov AY, Batako AD. Nano-scale multilayered composite coatings for cutting tools operating under heavy cutting conditions. Procedia CIRP. 2014;**14**:239-244

[66] Vereschaka AA, Volosova MA, Grigoriev SN, Vereschaka AS. Development of wear-resistant complex for high-speed steel tool when using process of combined cathodic vacuum arc deposition. Procedia CIRP. 2013;**9**:8-12

[67] Vereschaka AA, Volosova MA, Batako A, Vereschaka AS, Sitnikov NN, Seleznev AE. Nano-scale multi-layered coatings for improved efficiency of ceramic cutting tools. International Journal of Advanced Manufacturing Technology. 2017;**90**:27-43

[68] Grigoriev SN, Volosova MA, Vereschaka AA, Sitnikov NN, Milovich F, Bublikov JI, et al. Properties of (Cr,Al,Si)N-(DLC-Si) composite coatings deposited on a cutting ceramic substrate. Ceramics International. 2020;**46**(11):18241-18255

Chapter 11

Three-Dimensionally Ordered Macroporous-Mesoporous Bioactive Glass Ceramics for Drug Delivery Capacity and Evaluation of Drug Release

Reedwan Bin Zafar Auniq, Namon Hirun
and Upsorn Boonyang

Abstract

Bioactive glass ceramics (BGCs) have been used in orthopedic and dentistry due to having better osteoconductive and osteostimulative properties. This study aimed to evaluate and compare the drug release properties of two different BGCs; 45S5 and S53P4. The BGCs were composed with four phases of $SiO_2 – CaO – Na_2O – P_2O_5$ system, synthesized by sol–gel method using dual templates; a block-copolymer as mesoporous templates and polymer colloidal crystals as macroporous templates, called three-dimensionally ordered macroporous-mesoporous bioactive glass ceramics (3DOM-MBGCs). *In vitro* bioactivity test performed by soaking the 3DOM-MBGCs in simulated body fluid (SBF) at 37°C. The results indicated that, the 45S5 have the ability to grow hydroxyapatite-like layer on the surfaces faster than S53P4. Gentamicin drug was used to examine *in vitro* drug release properties in phosphate buffer solution (PBS). The amount of drug release was quantified through UV/Vis spectroscopy by using *o*-phthaldialdehyde reagent. S53P4 showed high drug loading content. The outcome of drug release in PBS showed that both S53P4 and 45S5 exhibited a slowly continuous gentamicin release. The resultant drug release profiles were fitted to the Peppas-Korsmeyer model to establish the predominant drug release mechanisms, which revealed that the kinetics of drug release from the glasses mostly dominated by Fickian diffusion mechanism.

Keywords: macroporous, mesoporous, ceramics, bioactive glasses, drug release, sol–gel process

1. Introduction

Bone is the second most widely transplanted tissue after blood. More than 2.2 million bone graft operations are performed annually worldwide in order to repair bone defects in orthopedics and dentistry [1]. Bioactive glass ceramics (BGCs) are one of the most promising synthetic bone replacements come regeneration material which has the ability to chemically bond with living bone tissue and stimulate bone growth without promoting inflammation or toxicity, developed by Larry Hench at the University of Florida in 1969 [1–4]. In early the heated glass powder, the micro-composite between apatite and β-wollastonite ($CaO{\cdot}SiO_2$) within a homogenous glassy phase showed not only a bioactivity but also high mechanical strength [4]. This BGCs was called A/W derived from the names of crystalline phase. The bioactivity of glass ceramics is believed to be due to the dissolution of calcium from wollastonite and/or the glassy phase. In case of treating bone defect, the bone regeneration rate depends on the material's composition [5]. Hydroxyapatite (HA) is the inorganic part of human bone [6]. The bonding with bone process is associated with the formation of HA layer on the implant's surface [7]. BGCs can either be synthesized by the melt quenching or sol–gel method. Early BGCs were prepared by the melt quenching method. Sol–gel processing was started practicing in early 1990s for bioactive glass synthesis. Sol–gel derived bioactive glasses are made of a colloidal silica solution synthesized by the hydrolysis of alkoxide precursor to form a sol. Tetraethyl orthosilicate (TEOS) is commonly used as silica precursor, Triethyl phosphate (TEP) is used to add phosphate, salt calcium nitrate used to introduce calcium and Na_2O included to decrease the melting temperature [8, 9]. Sol–gel derived bioactive glasses can provide higher surface Si-OH groups, which promote active places for more functionalization. The greater specific surface area that enhance the rate of hydroxyapatite formation is considerably higher degree of bioactivity compare to the melt quenching process [10–12]. Mesoporous bioactive glass ceramics (MBGCs) are considered the third-generation bioactive glasses were developed in 2004 by the combination of sol–gel method. MBGCs can possess more optimal surface area, ordered mesoporous structure, variable pore size and volume, improved in *in vitro* apatite mineralization in simulated body fluid (SBF) comparing with non-mesoporous bioactive glasses (NBG) [13]. However, BGCs having higher specific surface area and pore volume accelerates the hydroxyapatite formation and increase prolong the bioactive behavior [9]. MBGCs also get focused because of having more potential applications, such as catalysis, adsorption/separation, nanomaterial synthesis and also in biomaterial science as bone scaffolds for drug delivery and bone regeneration [9].

Mesoporous bioactive glass ceramic (MBGC) has brought a significant revolution in material science in terms of drug delivery. MBGC has some important properties which make itself more potential for drug delivery, such as well-ordered pores, large pore volumes and high specific surface area. As a result, MBGCs can easily entrap the drug molecules with its highly ordered mesoporous channel with a pore range of 2 to 50 nm [12, 14–17]. These characteristics greatly enhance MBGC for bone forming bioactivity, higher drug loading efficiency and lower drug release kinetics comparing with conventional BGCs [18–21]. Moreover, the mesosized pore are too small to promote cell growth. To overcome this limitation, the macroporous networks was studied and it suitable for tissue scaffolds that mimic the structure of porous bone structure [1].

Sol–gel technology is a wonderful progression in science with various applications since 1800s [12]. It is the process of making ceramic and glass materials using relatively low temperature hydrolysis and condensation reaction followed aging, drying and thermal stabilization [1]. Use of different surfactants (eg: P123, F127) during MBGCs preparation amplify the pore volume and surface area, which enhance the drug loading efficiency [16]. 45S5 and S53P4 bioactive glasses with a system of SiO_2-Na_2O-CaO-P_2O_5 considered more attention for bone tissue regeneration and regeneration properties due to their excellent bioactivity, biocompatibility, osteogenic and angiogenic effects [4, 18, 22–24]. Perioglas® was the first commercial product of 45S5 glasses, later reestablished by NovaBone® and BoneAlive® commercialized with composition of S53P4 [4].

Conventional treatment of bone infections like osteomyelitis involves surgery to remove necrotic bone tissue and repeated irrigations combined with the use of systemic antibiotics administration, wound drainage and implant removal. Systemic therapy of antibiotics has various adverse effects and risk of developing bacterial resistance to drugs. Local drug delivery system solves the problems by providing more advantages including high drug delivery efficiency, continuous action, reduced toxicity and convenience to the patients. Administration of single dose of localized drug with desired therapeutic range can reduce the need for follow-up care, reduce the risk of side effects, toxicity and increase patient compliance [19, 20, 25].

In our study, the hierarchically macroporous structured and mesosized pores of 45S5 and S53P4 BGCs were synthesized by sol–gel method and evaluated their *in vitro* bioactivity. The bioactivity effects of both bioactive glasses were investigated in SBF solution. The *in vitro* drug release properties in PBS were evaluated. Gentamicin sulfate (GS) was chosen as a model drug to encapsulation in the MBGCs to obtain a drug delivery system. GS is a broad-spectrum bactericidal antibiotic belonging aminoglycoside class; antibacterial activity is due to its ability to irreversibly bind ribosomes and half bacterial protein synthesis. GS vastly used in orthopedic treatments [26–28].

2. Experimental

2.1 Materials

Calcium nitrate tetrahydrate (CNT; $Ca(NO_3)_2.4H_2O$), tetraethyl orthosilicate (TEOS; $Si(OC_4H_9)_4$, 98%), triethyl phosphate (TEP; $P(OEt)_3$, 99%) were purchased from Acros Organic, sodium nitrate ($NaNO_3$), nitric acid (HNO_3) were purchased from Merck, Germany, the surfactant Pluronic P123 (EO_{20}-PO_{70}-EO_{20}, average Mn ~ 5800) were produced from Sigma-Aldrich, Germany. Gentamicin was obtained from Fuan Pharmaceutical group Yantai Co., LTD. Polymethylmethacrylate (PMMA) colloidal crystals with 300 nm size were prepared by our previous worked [29].

2.2 Preparation of bioactive glass ceramics

Two different 3DOM-MBGCs; 45S5 and S53P4 were synthesized *via* sol–gel method follow by published work [30]. The synthesis process went through hydrolysis and polycondensation of TEOS, TEP, CNT and $NaNO_3$ (**Figure 1**) with the appropriate mol ratio mentioned in **Table 1**. HNO_3 was used to catalyze the hydrolysis process and non-ionic block copolymer P123 was chosen as structure-directing agent. The solution was vigorously stirred (700 rpm) at room temperature to obtain a clear sol (hydrolysis reaction) and further stirred to reach the gel point (condensation reaction). To obtain 3DOM structure, a monolithic piece of the PMMA colloidal crystal templates were completely immersed in the sol and excess solution was removed. The products were aged in sealed vials at 45°C for 24 h to allow the polycondensation reaction followed by a drying process at 45°C for 24 h to eliminate excess solvents and by products. Finally, the samples were stabilized at 600°C in air for 4 h with a fixed heating rate of 2°C/min. The morphology of the bioactive glasses was analyzed in detail by scanning electron microscopy (SEM). Their element compositions were characterized by energy dispersive spectroscopy (EDS) attached to the SEM with Silicon crystal detectors. Fourier Transform Infrared Spectroscopy (FTIR) was carried out in the transmission mode with mid-infrared range 400–4000 cm^{-1} at a resolution of 4 cm^{-1} by using the KBr pellet method. The N_2 adsorption–desorption measurements carried out by Brunauer–Emmett–Teller (BET) for surface area analysis and Barrett–Joyner–Halenda (BJH) for pore size and volume analysis.

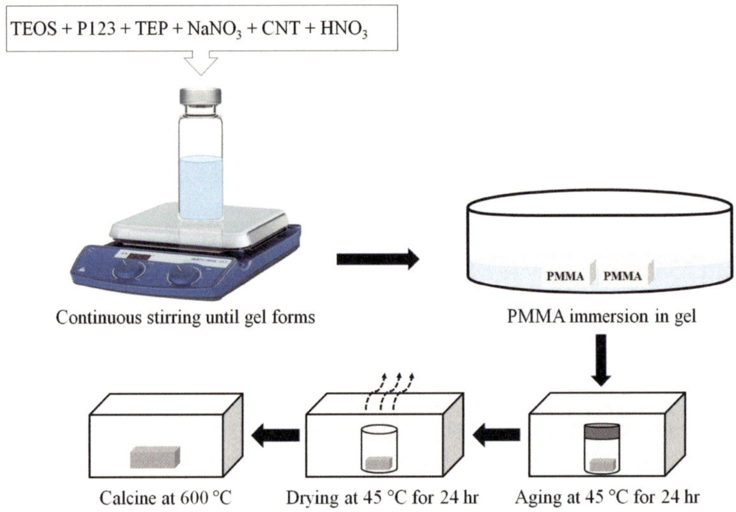

Figure 1.
Synthesis of 3DOM-MBGCs via sol–gel method.

Reagents	Compositions	
	45S5	S53P4
TEOS ($SiC_8H_{20}O_4$)	45% SiO_2	53% SiO_2
TEP ($C_6H_{15}O_4P$)	6% P_2O_5	4% P_2O_5
$Ca(NO_3)_2.4H_2O$	24.5% CaO	20% CaO
$NaNO_3$	24.5% Na_2O	23% Na_2O

Table 1.
Chemicals used in synthesis of 45S5 and S53P4 3DOM-MBGCs.

2.3 Assessment of *in vitro* bioactivity test

The bioactivity of the obtained samples was examined using *in vitro* test by immerging the samples in SBF solution at body temperature, 37°C at pH 7.40 following Kokubo method [31]. The *in vitro* bioactivity was performed by soaking grainy 3DOM-MBGCs in the SBF solution at a temperature of 37.0°C for 1, 2, 3 and 7 days with daily refreshing of the SBF solution. The ratio of the glass powders weight to SBF volume was 1.5 mg/mL. After soaking, samples were removed from the SBF solution and washed with deionized water several times and air-dried at room temperature. The changes of the bioactive glass surfaces were examined by SEM and FTIR techniques.

2.4 *In vitro* study of drug release

2.4.1 Determination of drug concentration

Gentamicin concentration was analyzed by measuring the UV/Vis absorbance of gentamicin-*o*-phthaldialdehyde complex at 333 nm [14]. The *o*-phthaldialdehyde reagent was prepared according to Huang et al., 2017 [17]. 1 mL of gentamicin solution, 1 mL of isopropanol and 1 mL of *o*-phthaldialdehyde reagent were reacted for 45 min at room temperature to prepare the sample to examine in UV/visible spectroscopy [32]. Before determination, a calibration curve (R^2 = 0.99) was made for each set of measurements and determined by taking absorbance vs. drug concentration between 1 to 150 ppm as parameters.

2.4.2 Encapsulation of gentamicin in the bioactive glass ceramics

Encapsulation of gentamicin into the MBGCs (45S5 and S53P4) were carried out in PBS pH 7.4 at room temperature for 24 h. 200 mg of each MBGCs was immersed in 10 mL of PBS containing gentamicin with a concentration of 10 mg/mL and stirred for 24 h. After that, the drug loaded MBGCs were filtered and then the drug loading efficiency and drug loading content were determined. Drug loading efficiency was measured by depletion method, by determining the difference in gentamicin concentration in the loading medium before and after loading [14]. The drug-loading experiments were carried out in triplicate, and the statistical computations were performed with the IBM SPSS Statistics version 25. Drug loading efficiency and drug loading content are two important parameters for drug delivery study with MBGCs. Drug loading content represents the mass ratio of drugs in drug loaded MBGCs and drug loading efficiency reflects the utilization of drugs in feed during drug loading [33]. The drug loading efficiency and drug loading content are expressed according to the following Equations [33].

$$\text{Drug loading efficiency}(wt.\%) = \frac{\text{Mass of the drug in MBGCs}}{\text{Mass of the drug in feed}} x100 \quad (1)$$

$$\text{Drug loading content}(wt.\%) = \frac{\text{Mass of the drug in MBGCs}}{\text{Initial mass of MBGCs}} x100 \quad (2)$$

2.4.3 In vitro drug release

50 mg of drug loaded MBGCs were placed into 10 mL of PBS and subsequently agitated in a horizontal shaking incubator at 37°C. 2 mL of release medium was withdrawn at predetermined time intervals and replaced with fresh release medium (2 mL) at each measurement. The triplicate samples of each drug loaded MBGCs (45S5 and S53P4) were used to determine the drug release profile.

2.4.4 Kinetic analysis drug release profile

To study the drug release kinetic from 45S5 and S53P4 MBGCs, the *in vitro* drug release data of gentamicin was fitted in Peppas-Korsmeyer kinetic model mentioned below;

$$\frac{M_t}{M_\infty} = Kt^n \quad (3)$$

Where K is the Peppas-Korsmeyer constant, $\frac{M_t}{M_\infty}$ is the fractional solute release at time t, and n is the exponent indicative of the release mechanism. An exponential value in the range of 0.45 or less and 0.89 or above indicate respectively Fickian diffusion and case II transport (typical zero-order release). Values between 0.45 and 0.89 indicate non-Fickian or anomalous release by both diffusion and erosion release [34].

3. Results and discussion

3.1 Morphology and microstructure of bioactive glass ceramics

Figure 2 shows the SEM image and EDS spectra of (a) 45S5 and (b) S53P4 3DOM-MBGCs. Bioactive glass with hierarchical porosity was formed through the PMMA and Pluronic P123 dual templating system. The SEM image of 53S4P (**Figure 2(b)**) shows more well ordered macroporous structure with spherical pores are around 300 nm. While, 45S5 shows distorted 3DOM structure (**Figure 2(a)**). The EDS spectrum of 45S5 and S53P4 in **Figure 2** shows the peaks corresponding to Si, Ca, Na and O that represent the preservation of the elements in the precursors without impurity elements. The FTIR spectra of 45S5 and S53P4 3DOM-MBGCs in **Figure 3** exhibits the characteristic peaks of Si-O-Si bending, symmetric and asymmetric stretching vibration at 467, 802 and 1086 cm^{-1}, respectively. In the peak at 564 and 950 cm^{-1} corresponding to the P-O bending and stretching vibration, respectively. In addition, the peak at 1635 and 3450 cm^{-1} correlates to O-H bonds indicates the water trapped inside the sample. The narrow band near 1384 cm^{-1} indicates the characteristic of the carbonate group (CO_3^{2-}) [35]. **Figure 4** indicates that all the BET curves of the 45S5 and S53P4 MBGCs presented a type IV isotherms

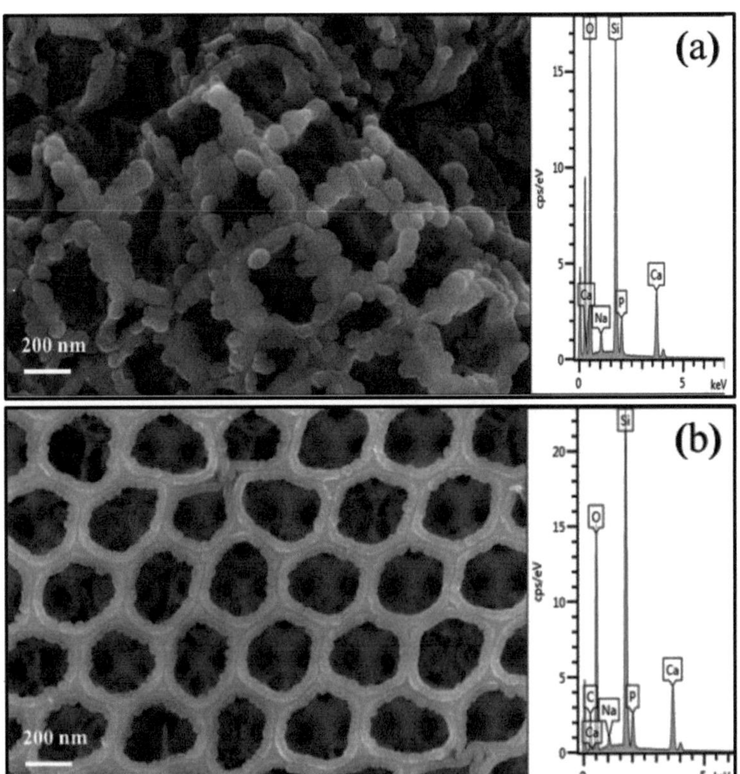

Figure 2.
SEM and EDS spectra of (a) 45S5 and (b) S53P4 3DOM-MBGCs.

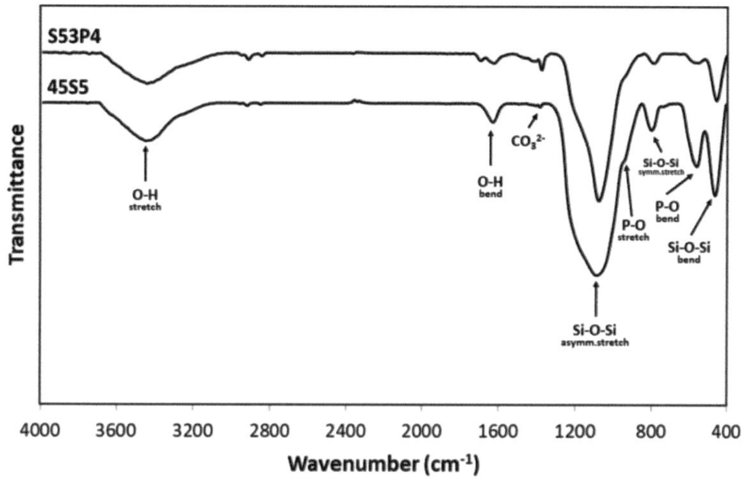

Figure 3.
FTIR spectra of (a) 45S5 and (b) S53P4 3DOM-MBGCs. Copyright [30].

pattern with type H4 hysteresis loops, characterizing mesoporous materials with narrow slit-like pores, with internal voids of irregular shape and broad size distribution [36]. This is confirmed by the average data of specific surface area, pore volume and pore diameter of 45S5 and S53P4 3DOM-MBGCs listed in **Table 2**. The bioactive glasses show specific surface area in range of 96.54 to 116.76 m^2/g. The 45S5 glasses shows a relatively wide pore size distribution calculated from the adsorption branch using the BJH model, and the average pore size is around 15.158 nm, while the S53P4 glasses gained average pore size around 11.230 nm.

3.2 Assessment of *in vitro* bioactivity test

The *in vitro* bioactivity of 3DOM-MBGCs was tested at body temperature of 37°C by using the SBF solution whose composition and ionic concentration similar to human blood plasma. **Figure 5** shows the SEM images of 45S5 bioactive glasses having different soaking time in SBF solutions. Compared with the morphology of the prepared bioactive glasses in **Figure 5a**, the nucleation of hydroxyapatite occurred on the glass surfaces after soaking in the SBF solution for 2 days (**Figure 5c**). The surface of 45S5 glasses were covered by precipitation of apatite-like layer more than 3 days soaking in SBF solution (**Figure 5d-f**). **Figure 6**, the formation of hydroxyapatite-like on the surface of S53P4 glasses started after 3 days of immersion in SBF solution. Within 7 days, most of the glass surfaces were covered by the apatite-like layer (**Figure 6d**). 45S5 showed fast hydroxyapatite-like precipitation than S53P4. However, the hydroxyapatite formation depends on the incorporation of Ca^{2+} and PO_4^{3-} on the MBG's glass surfaces during bioactivity test. Lower SiO_2 and higher CaO, P_2O_5 content in 45S5 could amplify the rate of hydroxyapatite formation on MBG's glass surfaces. The chemical composition and the microstructural morphology of 3DOM-MBGCs directly related to their bioactivity. The SBF can easier penetrate the larger macropores in the 3DOM bioactive glass compared to the mesopores in bioactive glasses [37]. Therefore, the minor difference average surface area and pore size of 3DOM-MBGCs (**Table 3**) has no effect in determining apatite growth. Due to both 3DOM-MBGCs using the same size of PMMA spheres for macroporous and same surfactant for mesoporous.

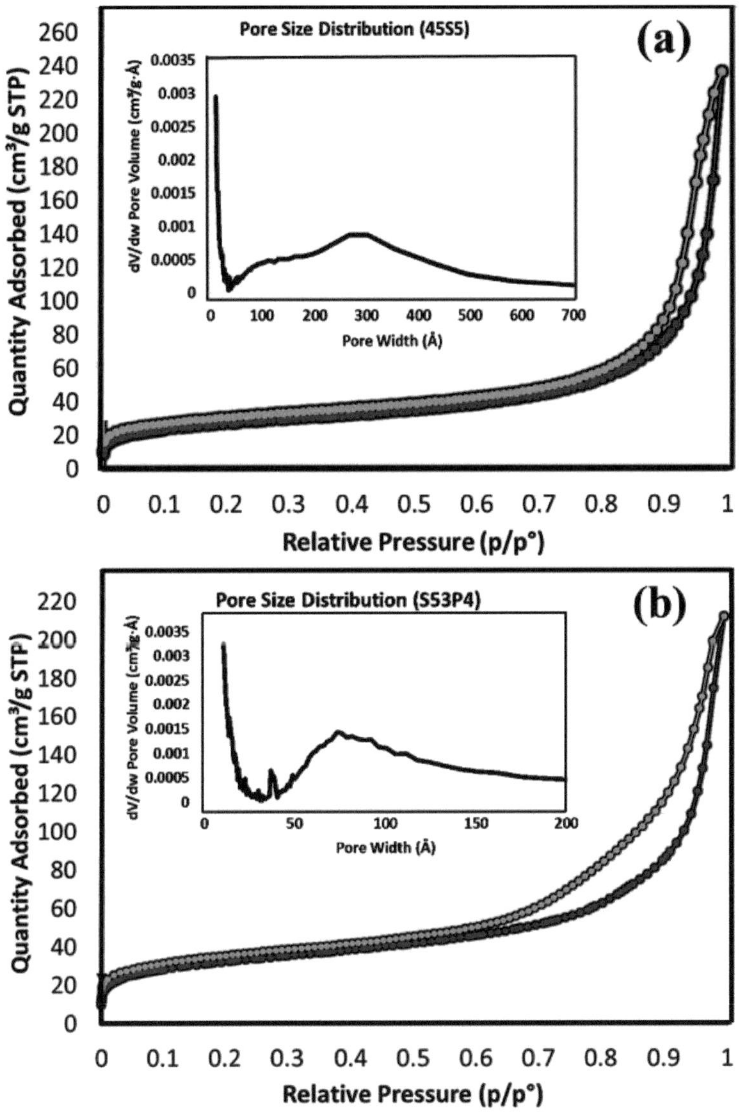

Figure 4.
N_2 adsorption–desorption isotherms and pore size distribution of (a) 45S5 and (b) S53P4 3DOM-MBGCs. Copyright [30].

Samples	Surface area (m^2/g)	Pore Volume (cm^3/g)	Pore Diameter (Å)
45S5	96.54	0.365	151.58
S53P4	116.76	0.327	112.30

Table 2.
The average data of specific surface area, pore volume and pore diameter of 45S5 and S53P4 3DOM-MBGCs. Copyright [30].

The FTIR spectra of 45S5 in **Figure 7**, at below spectrum, the sample before soaking in SBF solution exhibits the peaks at 467, 802 and 1086 cm^{-1} corresponding to the vibration of Si-O-Si bond, bending, symmetric and asymmetric stretching vibration, respectively. In vibrational peak at 564 and 950 cm^{-1} correlates to the P-O vibrational peak. In addition, the O-H bonds of the water trapped inside

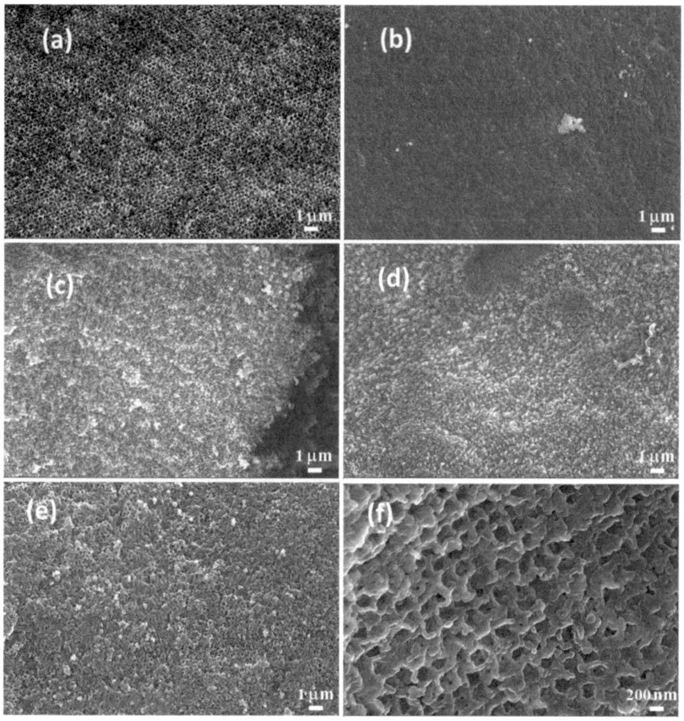

Figure 5.
SEM images of 45S5 bioactive glass ceramics (a) before soaking in SBF solution and after soaking in SBF solution for (b) 1 day (c) 2 days (d) 3 days (e) 7 days and (f) 7 days with higher magnification. Copyright [30].

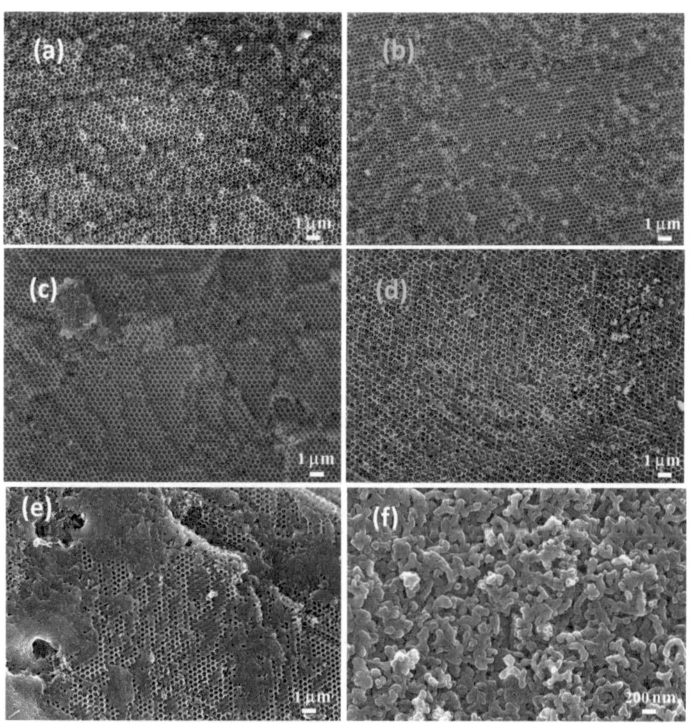

Figure 6.
SEM images of S53P4 bioactive glass ceramics (a) before soaking in SBF solution and after soaking in SBF for (b) 1 day (c) 2 days (d) 3 days (e) 7 days and (f) 7 days with higher magnification. Copyright [30].

MBGC Samples	Drug Loading efficiency (%)	Drug Loading Content (wt. %)
45S5	18.00 ± 3.16	8.74 ± 1.09
S53P4	22.14 ± 2.53	11.91 ± 2.09

Table 3.
The drug loading efficiency and content of 45S5 and S53P4.

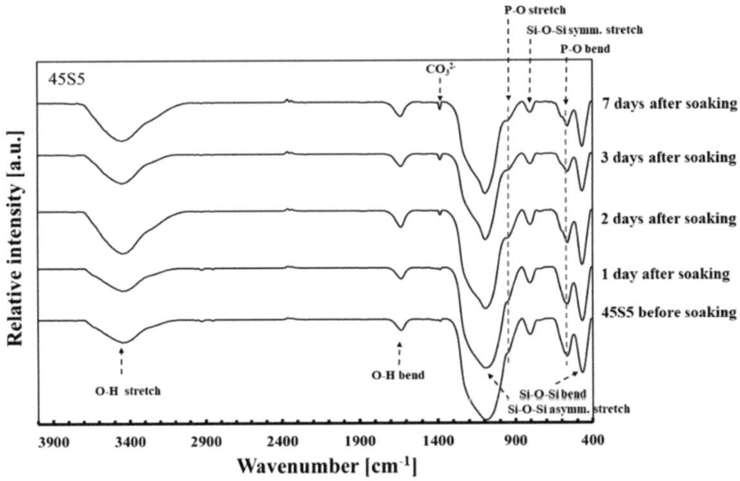

Figure 7.
FTIR spectra of 45S5 bioactive glass ceramics with different soaking time in SBF solution. Copyright [30].

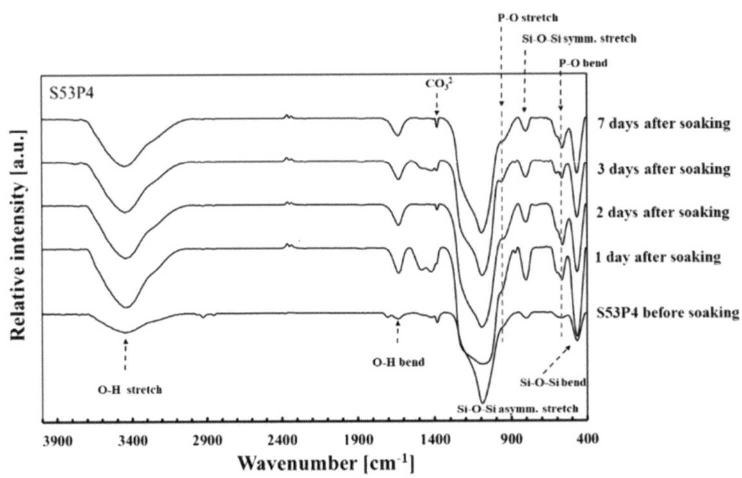

Figure 8.
FTIR spectra of S53P4 bioactive glass ceramics with different soaking time in SBF solution. Copyright [30].

the sample was shown at 1635 and 3450 cm^{-1}. The narrow band near 1384 cm^{-1} indicates the characteristic of the carbonate group (CO_3^{2-}) [35]. After the soaking in SBF, all the characteristic peaks are still observed. The P-O peak at 564 splits into doublet peak at 586 and 564 cm^{-1} which normally appears after immersion of the bioactive glass in SBF solution15. All the bands corresponding to the P-O represent the formation of hydroxyapatite on the surface of MBGCs. **Figure 8** represents the FTIR spectra of S53P4 bioactive glasses. The sample before soaking in SBF solution shows peaks at 467, 1087 and the shoulder at 1087–1250 cm^{-1}

correlates to the vibration of Si-O-Si bond. The peaks at 1385 and 1401 cm^{-1} indicates the characteristics of carbonate group (CO_3^{2-}) [35]. In the peak around 576 and 966 cm^{-1} corresponding to the P-O bending and stretching vibration, respectively. In addition, the peak at 1631 and 3445 cm^{-1} correlated to O-H bonds. After S53P4 MBGCs were soaked in SBF for 1 day, the aforementioned vibrational peaks are still observed. The P-O peak at at 607 and 567 cm^{-1}, which confirmed that the formation of amorphous phosphate phase on the glass surface [29]. Although, the splitting P-O peak of S53P4 appears in 1 day after soaking in SBF, while in the case of 45S5 after soaking for 2 days. However, the formation of hydroxyapatite-like on the surface of S53P4 glasses started after 3 days of soaking in SBF solution, slowly growth comparing with 45S5, indicated the better bioactivity of 45S5 than S53P4 bioactive glasses.

3.3 *In vitro* study of drug release

3.3.1 Drug loading

The drug loading efficiency and drug loading content of both MBGCs are summarized in **Table 3**. The drug loading efficiency in 45S5 was 18.00 ± 3.16%, while S53P4 showed quite higher drug loading efficiency of about 22.14 ± 2.53%. However, the loading efficiency of S53P4 was not statistically different from that of 45S5 (independent t-test, $p > 0.05$). The drug loading content of the MBGCs was found to be 8.74 ± 1.09 wt% for 45S5 and 11.91 ± 2.09 wt% for S53P4 glasses. The significant difference was not observed for the drug loading content of 45S5 from S53P4 (independent t-test, $p > 0.05$). S53P4 provides high average drug-loading content compared with other inert carrier materials that generally have low drug-loading content (less than 10 wt%) [33]. The porous materials can be developed to fabricate high drug-loading carriers due to their promising intrinsic properties, such as large hollow interior, porous surface, high surface area and large pore volume [33]. The good drug-loading capacity obtained in this study could be related to high surface area of the carrier with porous structure as supported by the results obtained from N_2 adsorption desorption analysis.

3.3.2 In vitro drug release

The release profiles for gentamicin from the MBGCs to the PBS are represented in **Figure 9**. For both MBGCs, the release of gentamicin showed an initial fast release followed by a relatively slow subsequent release. An initial fast release of the antibiotic was observed during the first 24 hours of soaking, reaching the mean gentamicin release values of 34.53% (45S5) and 41.21% (S53P4). The subsequent release rate was quite low in comparison with the first period. However, the S53P4 bioactive glasses showed a higher initial drug release behavior than 45S5 glasses. But later, both bioactive glasses reached the same point after 96 hours of release values of 64.27% (45S5) and 64.53% (S53P4). Both S53P4 and 45S5 bioactive glasses showed a slowly continuous gentamicin release.

To study the mechanism of drug release from the MBGCs, the first 60% of gentamicin release profile was fitted in Peppas-Korsmeyer model. In this model, the value of n characterizes the release mechanism of drug as described in **Table 4**. As observed in **Table 4**, the n values of the release data of 45S5 and S53P4 glasses are 0.3992 and 0.3004, respectively. This indicated that the drug release from both systems can be described by Fickian diffusion [34]. Both MBGCs possessed porous structures with hollow interiors. The diffusion through channel might dominate the drug release from these mesoporous materials [13, 38].

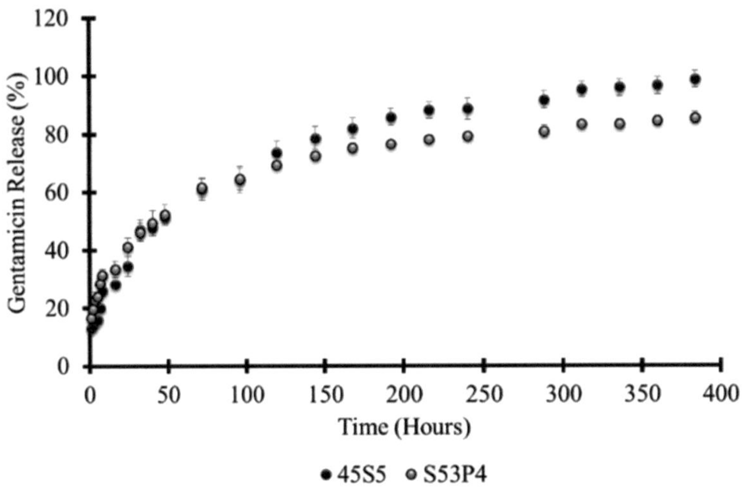

Figure 9.
In vitro gentamicin release from 45S5 and S53P4 bioactive glass ceramics.

MBG Samples	Peppas-Korsmeyer factors		
	K	n	R^2
45S5	0.0204	0.3992	0.95
S53P4	0.0466	0.3004	0.99

Table 4.
Kinetic assessment of gentamicin release data of 45S5 and S53P4 in PBS (Peppas-Korsmeyer model).

4. Conclusions

3DOM-MBGCs were synthesized successfully by the sol–gel method using spherical PMMA colloidal crystals of 300 nm and non-ionic block copolymer P123 as cotamplates. The morphology of S53P4 bioactive glass revealed well-ordered macroporous structure with larger surface area. While, 45S5 bioactive glass had shown distorted 3DOM structure with a bit higher pore diameter. Base on the SEM and FTIR results indicated the better bioactivity of 45S5 than S53P4 bioactive glass ceramics which was able to initiate the formation of hydroxyapatite-like layer on glass surface after soaking in SBF solution within 2 days (45S5) and 3 days (S53P4). The drug delivery system based on 45S5 and S53P4 3DOM-MBGCs have been synthesized, and the release behavior of both porous bioactive glasses were studied. The results indicated that the S53P4 glasses showed higher drug loading efficiency and gave relatively initial fast release compared to the 45S5 due to its high surface area. Even though, the drug loading content was not significant different from that of both bioactive glass ceramics. The resultant drug release mechanism was occupied from the first 60% of gentamicin release profile fitted to the Peppas-Korsmeyer model, which clarified that the kinetics of drug release from the bioactive glass ceramics mostly occurred by Fickian diffusion mechanism.

Therefore, the results indicated the bioactivity and drug release profile of mesoporous bioactive glass ceramics which can accelerate the bone growth or new bone formation and could be a use as a promising drug release system for bone implant materials preparation.

Acknowledgements

This work has been supported by Walailak University Postgraduate Scholarship for Outstanding Students (Contract no. 02/2560), financially supported by Walailak University Graduate Studies Research Fund (Contract no. 14/2561).

Conflict of interest

The authors declare no conflict of interest.

Author details

Reedwan Bin Zafar Auniq[1], Namon Hirun[2] and Upsorn Boonyang[1*]

1 Functional Materials and Nanotechnology Center of Excellence, School of Science, Walailak University, Nakhon Si Thammarat, Thailand

2 Division of Pharmaceutical Sciences, Faculty of Pharmacy, Thammasat University, Pathumthani, Thailand

*Address all correspondence to: upsorn.bo@mail.wu.ac.th

References

[1] Martin, R. A.; Yue, S.; Hanna, J. V.; Lee, P. D.; Newport, R. J.; Smith, M. E.; Jones, J. R., Characterizing the hierarchical structures of bioactive sol-gel silicate glass and hybrid scaffolds for bone regeneration. Philosophical Transactions of the Royal Society A: Mathematical, Physical and Engineering Sciences 2012, 370, 1422-1443. DOI: 10.1098/rsta.2011.0308

[2] Arcos, D.; Vallet-Regí, M., Bioceramics for drug delivery. Acta Materialia 2013, 61 (3), 890-911. DOI: 10.1016/j.actamat.2012.10.039

[3] Yan, X.; Huang, X.; Yu, C.; Deng, H.; Wang, Y.; Zhang, Z.; Qiao, S.; Lu, G.; Zhao, D., The *in-vitro* bioactivity of mesoporous bioactive glasses. Biomaterials 2006, 27 (18), 3396-3403. DOI: 10.1016/j.biomaterials.2006.01.043

[4] Jones, J. R., Reprint of: review of bioactive glass: From Hench to hybrids. Acta Biomaterialia 2015, 23, S53-S82. DOI: 10.1016/j.actbio.2015.07.019

[5] Hench, L. L., Chronology of bioactive glass development and clinical applications. New Journal of Glass and Ceramics 2013, 3 (2), 67-73. DOI: 10.4236/njgc.2013.32011

[6] Li, X.; Wang, X.; Chen, H.; Jiang, P.; Dong, X.; Shi, J., Hierarchically porous bioactive glass scaffolds synthesized with a PUF and P123 cotemplated approach. Chemistry of Materials 2007, 19 (17), 4322-4326. DOI:10.1021/cm0708564

[7] Li, R.; Clark, A. E.; Hench, L. L., An investigation of bioactive glass powders by sol-gel processing. Journal of Applied Biomaterials 1991, 2 (4), 231-239. DOI: 10.1002/jab.770020403

[8] Jones, J. R.; Lin, S.; Yue, S.; Lee, P. D.; Hanna, J. V.; Smith, M. E.; Newport, R. J., Bioactive glass scaffolds for bone regeneration and their hierarchical characterisation. Proceedings of the Institution of Mechanical Engineers, Part H 2010, 224 (12), 1373-87. DOI: 10.1243/09544119JEIM836

[9] Vallet-Regí, M.; Ragel, C. V.; Salinas, Antonio J., Glasses with medical applications. European Journal of Inorganic Chemistry 2003, 2003 (6), 1029-1042. DOI: 10.1002/ejic.200390134

[10] Sepulveda, P.; Jones, J. R.; Hench, L. L., Characterization of melt-derived 45S5 and sol-gel–derived 58S bioactive glasses. Journal of Biomedical Materials Research 2001, 58 (6), 734-740. DOI: 10.1002/jbm.10026

[11] Sharifianjazi, F.; Parvin, N.; Tahriri, M., Formation of apatite nano-needles on novel gel derived SiO_2-P_2O_5-CaO-SrO-Ag_2O bioactive glasses. Ceramics International 2017, 43 (17), 15214-15220. DOI: 10.1016/j.ceramint.2017.08.056

[12] Gupta, R.; Kumar, A., Bioactive materials for biomedical applications using sol–gel technology. Biomedical Materials 2008, 3 (3), 034005. DOI: 10.1088/1748-6041/3/3/034005

[13] Wu, C.; Chang, J., Mesoporous bioactive glasses: structure characteristics, drug/growth factor delivery and bone regeneration application. Interface Focus 2012, 2 (3), 292-306. DOI: 10.1098/rsfs.2011.0121

[14] Xia, W.; Chang, J., Well-ordered mesoporous bioactive glasses (MBG): a promising bioactive drug delivery system. Journal of Controlled Release 2006, 110 (3), 522-530. DOI: 10.1016/j.jconrel.2005.11.002

[15] Liu, Y.-Z.; Li, Y.; Yu, X.-B.; Liu, L.-N.; Zhu, Z.-A.; Guo, Y.-P., Drug delivery property, bactericidal property and cytocompatibility of magnetic mesoporous bioactive glass.

Materials Science and Engineering: C 2014, 41, 196-205. DOI: 10.1016/j.msec.2014.04.037

[16] Zhang, X.; Zhang, J.; Shi, B., Mesoporous bioglass/silk fibroin scaffolds as a drug delivery system: Fabrication, drug loading and release *in vitro* and repair calvarial defects *in vivo*. Journal of Wuhan University of Technology-Mater. Sci. Ed. 2014, 29 (2), 401-406. DOI: 10.1007/s11595-014-0929-0

[17] Huang, C.-Y.; Huang, T.-H.; Kao, C.-T.; Wu, Y.-H.; Chen, W.-C.; Shie, M.-Y., Mesoporous calcium silicate nanoparticles with drug delivery and odontogenesis properties. Journal of Endodontics 2017, 43 (1), 69-76. DOI: 10.1016/j.joen.2016.09.012

[18] Wu, C.; Chang, J., Multifunctional mesoporous bioactive glasses for effective delivery of therapeutic ions and drug/growth factors. Journal of Controlled Release 2014, 193, 282-295. DOI: 10.1016/j.jconrel.2014.04.026

[19] Yan, X.; Yu, C.; Zhou, X.; Tang, J.; Zhao, D., Highly ordered mesoporous bioactive glasses with superior *in vitro* bone-forming bioactivities. Angewandte Chemie International Edition 2004, 43 (44), 5980-5984. DOI: 10.1002/anie.200460598

[20] Zhu, Y.; Kaskel, S., Comparison of the *in vitro* bioactivity and drug release property of mesoporous bioactive glasses (MBGCs) and bioactive glasses (BGs) scaffolds. Microporous and Mesoporous Materials 2009, 118 (1), 176-182. DOI: 10.1016/j.micromeso.2008.08.046

[21] Li, Y.; Liu, Y.-Z.; Long, T.; Yu, X.-B.; Tang, T. T.; Dai, K.-R.; Tian, B.; Guo, Y.-P.; Zhu, Z.-A., Mesoporous bioactive glass as a drug delivery system: fabrication, bactericidal properties and biocompatibility. Journal of Materials Science: Materials in Medicine 2013, 24 (8), 1951-1961. DOI: 10.1007/s10856-013-4960-z

[22] Kaur, G.; Pandey, O. P.; Singh, K.; Homa, D.; Scott, B.; Pickrell, G., A review of bioactive glasses: Their structure, properties, fabrication and apatite formation. Journal of Biomedical Materials Research Part A 2014, 102 (1), 254-274. DOI: 10.1002/jbm.a.34690

[23] Li, W.; Nooeaid, P.; Roether, J. A.; Schubert, D. W.; Boccaccini, A. R., Preparation and characterization of vancomycin releasing PHBV coated 45S5 Bioglass®-based glass–ceramic scaffolds for bone tissue engineering. Journal of the European Ceramic Society 2014, 34 (2), 505-514. DOI: 10.1016/j.biomaterials.2011.01.004

[24] Hoppe, A.; Güldal, N. S.; Boccaccini, A. R., A review of the biological response to ionic dissolution products from bioactive glasses and glass-ceramics. Biomaterials 2011, 32 (11), 2757-2774. DOI:

[25] Wen, H.; Jung, H.; Li, X., Drug Delivery Approaches in Addressing Clinical Pharmacology-Related Issues: Opportunities and Challenges. AAPS J 2015, 17 (6), 1327-1340. DOI: 10.1208/s12248-015-9814-9

[26] Liu, W.; Chang, J., *In vitro* evaluation of gentamicin release from a bioactive tricalcium silicate bone cement. Materials Science and Engineering: C 2009, 29 (8), 2486-2492. DOI: 10.1016/j.msec.2009.07.015

[27] Dorati, R.; DeTrizio, A.; Genta, I.; Grisoli, P.; Merelli, A.; Tomasi, C.; Conti, B., An experimental design approach to the preparation of pegylated polylactide-co-glicolide gentamicin loaded microparticles for local antibiotic delivery. Materials Science and Engineering: C 2016, 58, 909-917. DOI: 10.1016/j.msec.2015.09.053

[28] Perni, S.; Martini-Gilching, K.; Prokopovich, P., Controlling release kinetics of gentamicin from silica nano-carriers. Colloids and Surfaces A: Physicochemical and Engineering Aspects 2018, 541, 212-221. DOI: 10.1016/j.colsurfa.2017.04.063

[29] Boonyang, U.; Li, F.; Stein, A., Hierarchical structures and shaped particles of bioactive glass and its *in vitro* bioactivity. Journal of Nanomaterials 2013, 2013 (Article ID 681391), 1-6. DOI: 10.1155/2013/681391

[30] Auniq, R. B.-Z.; Pakasri, N.; Boonyang, U., Synthesis and *in vitro* bioactivity of three-dimensionally ordered macroporous-mesoporous bioactive glasses; 45S5 and S53P4. Journal of the Korean Ceramic Society 2020. DOI: 10.1007/s43207-020-00050-z

[31] Kokubo, T.; Takadama, H., How useful is SBF in predicting *in vivo* bone bioactivity? Biomaterials 2006, 27 (15), 2907-2915. DOI: 10.1016/j.biomaterials.2006.01.017

[32] Zhang, X.; Wyss, U. P.; Pichora, D.; Goosen, M. F. A., An Investigation of poly (lactic acid) degradation. Journal of Bioactive and Compatible Polymers 1994, 9 (1), 80-100. DOI: 10.1177/088391159400900105

[33] Shen, S.; Wu, Y.; Liu, Y.; Wu, D., High drug-loading nanomedicines: progress, current status, and prospects. International Journal of Nanomedicine 2017, 12, 4085-4109. DOI: 10.2147/IJN.S132780

[34] Soundrapandian, C.; Datta, S.; Kundu, B.; Basu, D.; Sa, B., Porous bioactive glass scaffolds for local drug delivery in osteomyelitis: Development and *in vitro* characterization. AAPS PharmSciTech 2010, 11 (4), 1675-1683. DOI: 10.1208/s12249-010-9550-5

[35] Lucas-Girot, A.; Mezahi, F. Z.; Mami, M.; Oudadesse, H.; Harabi, A.; Le Floch, M., Sol–gel synthesis of a new composition of bioactive glass in the quaternary system SiO_2–CaO–Na_2O–P_2O_5: Comparison with melting method. Journal of Non-Crystalline Solids 2011, 357 (18), 3322-3327. DOI: 10.1016/j.jnoncrysol.2011.06.002

[36] Alothman, Z. A., A Review: Fundamental sspects of silicate mesoporous Materials. Materials 2012, 5 (12), 2874-2902. DOI: 10.3390/ma5122874

[37] Bell, D. C.; Zhang, K.; Yan, H.; Francis, L. F.; Stein, A., Three dimensionally ordered macroporous bioactive glasses. Microscopy and Microanalysis 2002, 8 (S02), 330-331. DOI: 10.1017/S1431927602100900

[38] Xichen, Z.; Wyss, U. P.; Pichora, D.; Goosen, M. F. A., A mechanistic study of antibiotic release from biodegradable poly(d,1-lactide) cylinders. Journal of Controlled Release 1994, 31 (2), 129-144. DOI: 10.1016/0168-3659(94)00011-5

Section 6

Challenges in Rietvild Refinement and Structure Visualization in Ceramic Materials

Chapter 12

Challenges in Rietveld Refinement and Structure Visualization in Ceramics

Touseef Ahmad Para and Shaibal Kanti Sarkar

Abstract

The most common and basic characterization in the field of material science is the almighty X-ray diffraction (XRD). In every institute, every research report and every manuscript, concerning material properties, the X-ray diffraction pattern is essentially found. Although the basis of these works relies on the fact that X-ray diffraction pattern was found to be matching with some structure in a database, the in depth significance of the various characteristic diffraction manifestations of various physical characters are rarely discussed. Most of the researchers (especially beginners) are either not aware of the prowess of X-ray based characterizations, or have not been introduced to it properly or may be sometimes they are not interested in its results at all. The decreased interest (later) in the results from such studies might be for not being productive enough for time spending or non-effectiveness in justifying the motivation of the work. The former two are more related to the availability and accessibility of study material for the development of core concepts. Most of the institutes always do not have access to the span-wide scientific literature and the researchers joining these institutions are partly affected. In this context the effective open-access and free availability of intech-open, it is prudent to at least attempt to accumulate, assimilated and aggregate the concepts related to X-ray diffraction in a single package. The chapter is an attempt in the path of this route.

Keywords: X-ray diffraction, space group, polyhedra, powder diffraction, Rietveld refinement, structure visualization

1. Introduction

Much has been written and learnt about powder diffraction in last two decades. The journey that began in 1910 with the Bragg father-son duo publishing their first paper on crystal structure determination using ionization spectrometer, a century later there are still perks and connives that have not been widely explored [1–3]. The meticulous solution to the single crystal NaCl structure by the Braggs was achieved by solving symmetry equations for thousands of positions within a unit cell of unknown symmetry, without the help of modern computational prowess [3–8]. As Mike Glazer put it in very powerful words, “It was the gifted mind of Lawrence Bragg seeing symmetries in space and numbers that enabled them to reach a solution much quickly than anticipated” [3, 5]. In addition, W L Bragg’s consideration of diffraction from crystals as merely reflections from crystal planes, simplified the theory around the structure determination considerably [9]. In just

IntechOpen

few months, Braggs determined structure of NaCl, KCl, KBr, CaF2, Cu2O, ZnS, NaNO3, some calcites and diamond from their respective single crystals [10].

The year 1914, Max von Laue was awarded Noble prize for his discovery of the diffraction of X-rays by crystals [11, 12] followed by 1915 prize for their services in the analysis of crystal structure by means of X-rays to W H and W L Bragg [6] itself concatenates the importance of crystal structure determination. In following years, Debye and Scherrer extended the theory from single crystal to powder diffraction, presenting the complete theory of powder diffraction patterns and crystal structures used today (*squared sums of hkl ordered triplets*) [13–16]. Although Scherrer, Debye and Hull solved structures of many materials, it was not until modern computational boom that new, more complex and low symmetry system could be solved via powder diffraction pattern [17–23]. In the quest of achieving a suitable pathway for attaining a solution of powder X-ray diffraction many niche-limited attempts like maximum likelihood method [24, 25], anomalous dispersion, maximum entropy method, line profile fitting [26] etc. were made abundantly in 1950s and 60s. Hugo Rietveld in 1960s came up with one such method, employing least square iteration principle to statistically estimate the weighted contribution of every point on a powder XRD pattern [27]. The method now known as Rietveld refinement was the first step towards full profile whole powder pattern fitting method for x-ray and neutron diffraction data.

2. Rietveld refinement

It was 1969, 27 copies of a 162 Kilobytes program were sent to different institutes all over the world. The program was accompaniment of paper published in Journal of Applied Crystallography titled "A Profile refinement Method for Nuclear and Magnetic Structures" by Hugo Rietveld. Within a span of a decade 200 structures were refined from powder diffraction data [26–28]. The method we all know as Reitveld refinement method, made possible to refine whole profile with parameters including half-width, zero shift, cell parameters [29], asymmetry correction [30, 31], preferred orientation correction [32, 33], overall scale factor, overall isotropic temperature factor, fractional coordinates of the atoms, atomic isotropic temperature, occupation numbers and the components of the magnetic vectors of each atom. The algorithm this program followed is summed up in **Figure 1**. In subsequent versions of the method, Rietveld introduced residual values (R values), allowing for a quantitative judgment of the refinement quality. Most of the findings and equations, which Rietveld published, are still used nowadays in their original form [14, 20, 21, 34–37].

In 1994, International Union of Crystallography (IUCr) constituted a commission on powder diffraction with the purpose of diving into the status of the world of scientific community in general and crystallographic community in particular and focus on the practical aspects of data collection, refinement software, data interpretations, future endeavors etc. [38]. The commission proposed certain protocols and few guidelines for data collection, background contribution, peak-shape function, refinement of profile parameters, Fourier analysis, refinement of structural parameters, geometric restraints, estimated standard deviation, interpretation of R values and some common problems with their possible solutions. Although the Hill and Cranswick [38–41] commission on powder diffraction formulated a set of general guidelines that encompassed the recommendations with some explanatory and cautionary notes regarding Rietveld refinement their application in the aspect for a newcomer are not totally encompassed [42, 43]. The reason is not the ineffectiveness, obsoleteness or incomprehension but rather the scattered nature of

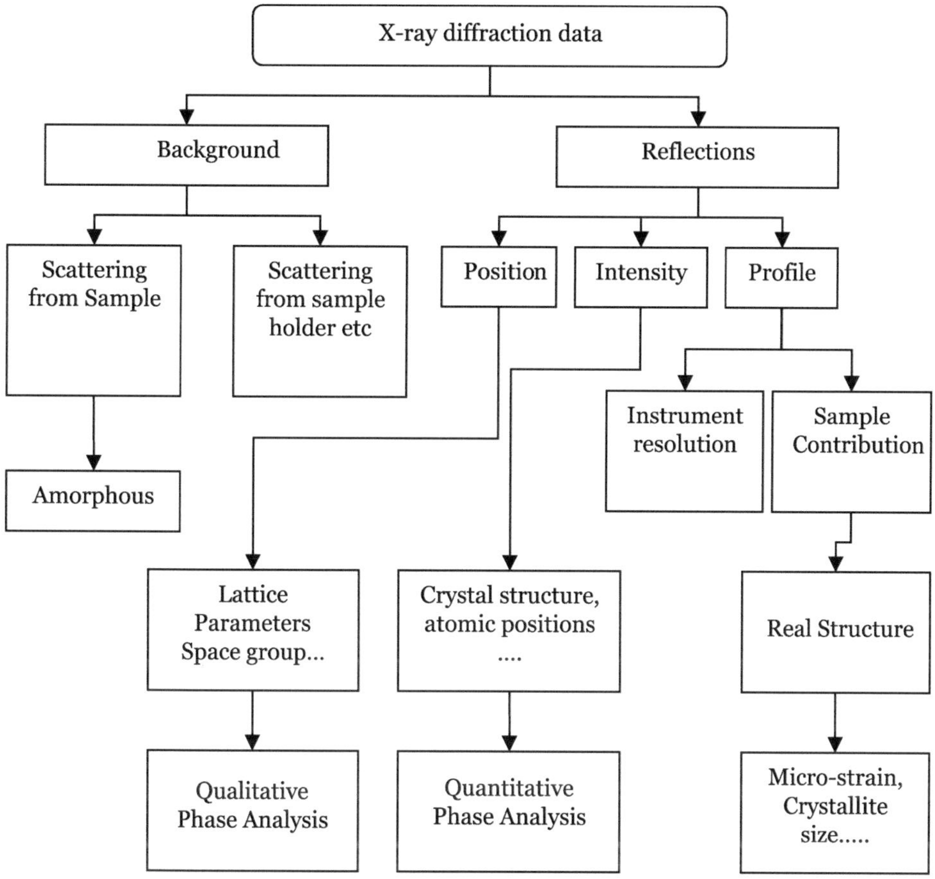

Figure 1.
The algorithm of whole profile refinement program developed by Hugo M Rietveld [reconstructed from the IUcr newsletter no. 26, Dec 2001].

current studies, antiquities and general information. This chapter attempts to accommodate most of these and present them in a more, newbie, newcomer friendly way. The chapter will follow a linear path from sample preparation, data collection to final results and conclusions accompanied by various current challenges, precautionary and explanatory notes.

3. Sample preparation

In order to understand different phases of sample preparation, we first need to define and understand the term "sample". The term "sample" encompasses a much broader meaning in scientific community with or without any restriction on size, quantity, quality etc. A sample may be a rather large portion of material, or a very tiny amount. A specimen on the other hand is the representative diminutive piece of a sample. Although there is a thin line of distinction between a sample and a specimen in X-ray diffraction, the term sample preparation generally means to prepare a specimen from a larger sample [43, 44].

The material, phase purity, homogeneity, density gradient etc. of a sample from which a specimen is taken are to be considered in advance. For a phase pure sample or mostly pure, a specimen is a good representative of the sample, so is the case with multiphase but homogenous samples. However a specimen from a multiphase and inhomogeneous sample may not be a good representative of the sample itself. The sample may consist of several phases, known or unknown, and may also include amorphous material. Depending on the technique and radiation, it may be small or large (neutron diffraction), it may be flat (Bragg–Brentano geometry), or cylindrical (Debye–Scherrer technique). In case of multiphase sample or amorphous contributions specimen should be taken with considerable representation of the sample such that during the refinement process quantitative contribution of each phase can be estimated more precisely.

In the length of this chapter the term "sample preparation" will be used to define collection of specimen, cleaning or remolding, mounting it on sample holder and all the processing necessary to prepare the diffracting material to its mounting on goniometer.

3.1 Precautionary/explanatory notes

Following few precautions are integral parts of sample preparation process

- Sample homogeneity/representative specimen
- Sample geometry
- Sample thickness
- Crystalline/Amorphous nature of sample
- Hygroscopic, gas absorbing nature and porosity of material
- Phase purity or at least the idea of chemical composition.

3.2 Current challenges

Despite the advances in current instrumentation and techniques we will not be able to obtain a 100% representative specimen from any sample, particularly powder samples. Grain size distribution, preferred orientation, inhomogeneous grain boundaries, defects and other microscopic differences will always act against it [45].

As world dives more and more into the nanoscale world, the sample thickness poses a problem with 1D and 2D materials.

Sample geometry can also not be obtained with certainty with nanoscale samples, especially with nano-morphologies and surface rough samples. A sample of 50–100 nm thickness and spiky morphology, with each spike of let us say 20 nm thickness and 50 nm length, will have so rough surface that there will be roughly 50% of thickness change while moving from one spike to another.

Another challenge will be the porosity of the samples. In nanomaterial samples the surface area to volume ratio increases leading to apparent amorphicity in actually crystalline samples.

4. Data collection

In order to perform a successful Rietveld refinement, it is essential that the powder diffraction data be collected appropriately. If relative intensities or the 2θ values (d-spacing) are recorded incorrectly, no amount of time spent on refinement will lead to any sensible results. The factors to be considered for effective and successful data collection are diffractometer geometry, instrument alignment, calibration, the radiation, the wavelength, slit size, necessary counting time and most importantly the alignment and positioning of incident beam [46].

It's important that the incident beam should always be kept on sample (specimen) such that the diffracting volume remains constant. In Bragg–Brentano configuration, the use of wide divergence slits must be accounted by a correction term. Introduction of this correction term is quite plainly geometry dependent, therefore, sample holder geometry has to be taken into consideration and an update to correction term should be applied. Most of the instruments correct this by using rotating circular sample, however, this does not always correct for low angle intensities. A more modern approach to this problem is the use of automated variable divergence slits which operate as a function of 2θ. At lower angles smaller slits are used, and at higher angles wider ones. A flat sample for Bragg–Brentano geometry is essential to ensure that focusing circle is always tangential to the sample surface. It is however practically more challenging to achieve surfaces with low roughness. At lower angles, the effect is negligible as the incident beam area is large, but at higher angles, as beam width decreases, the surface roughness can cause problems in collected data. A more common approach to this problem is to spend more time at collecting data at higher angles. Most of the modern diffractometers are equipped with such algorithms and generally adjust automatically as a function of θ. Next time you perform XRD measurements on your sample and fell the higher 2θ data collection are getting on your nerves. Remember, it is for the best [47].

Time is also an important factor to consider while data collection [48]. It is necessary to record suitable counts; therefore more time should be spent between each 2θ step. It is also necessary to record the data at suitable intervals (step size) to ensure recording of good profile and peak-broadening. As a rule of thumb, there should be at least 5 data points collected across a given peak. The maximum 2θ should always be kept to as low as you can go, however at least 50 2θ degrees should be measured to ensure statistical viability of data.

Sample transparency is yet another problem. The assumption for XRD in reflection geometry is satisfied only when the sample is infinitely thick. If the sample contains only light elements, this condition might not get fulfilled at all, therefore all the following assumptions will be invalid [38, 48–50].

i. The constant-volume assumption

ii. The intensities measured at higher angles

iii. The focusing circle adjustment etc. On the other hand, heavily absorbing samples can also be a problem, because the incident beam cannot penetrate the whole sample. The solution in the later case is much simpler than former one. Sample in later case may have to be diluted with a light-element material (e.g. diamond powder or glass beads).

Preferred-orientation effects can be very difficult to eliminate, especially for flat powder specimens. If the intensities show a strong hkl dependence (e.g. all hk0 reflections are strong and all h00 weak), preferred orientation of the crystallites should be suspected. Rietveld refinement can be done with many programs which are based on March model allowing a specific crystallographic vector based refinement of preferred-orientation parameter [32]. The elimination (or minimization) of the problem experimentally is to be preferred due to the crude nature of such models. Grain and particle morphology can also play a major role in preferential orientation. For large crystallite size the randomness of orientation of sample gets diminished i.e.

not all crystallite orientations are equally represented, creating a problem. In the underrepresented specimen, the preferred orientation parameter cannot be corrected at the refinement stage. Therefore the sample rotation method is strongly recommended in such cases. In smaller particle sizes, line-broadening effects due to crystallite size begin to become apparent which evidently decreases the intensity of peaks. The presence of large crystallites within such samples will cause the peaks from smaller particle size to be relatively very low or even reduced to background. In such cases also, the correction to preferred orientation parameter cannot be applied.

Another parameter to be considered in the diffractometer is to keep background to maximum peak ratio as low as possible.

Monochromatic radiation is to be preferred for all XRD measurements. Although longer data-acquisition times are required with monochromatic radiation, its use is particularly advantageous both in number of lines and the background observed.

Any temptation to smooth the diffraction data before doing a Rietveld refinement must be resisted. Smoothing introduces point-to-point correlations which will give falsely lowered estimated standard deviations in the refinement process.

The wavelength and zero offset should be calibrated with a reference material. The Si SRM 640b standard gives significantly broadened peaks, whereas the NIST LaB6 standard SRM 660 gives close to instrumental resolution and is probably a better choice.

The example of over, normal and under collection of data is shown in **Figure 2 (a)**, **(b)** and **(c)** respectively, while the presence of preferred orientation and normal XRD pattern of SnO_2 are shown in **Figure 3a** and **b** respectively.

4.1 Precautions and explanations

Specimen should be chosen in such a way that it represents the sample in every possible way (or at least nearly every way).

Uniform surface and thickness should be maintained across the sample.

In case of suspected preferred orientation, it should be a practice to repeat the experiment with newly prepared sample or specimen.

Many materials undergo phase transformation on exposure to humidity, Carbon Monoxide etc. In such cases, care should be taken to minimize the exposure.

Leveling of sample holder is essential to get an initial 2θ estimate.

In case of grazing incidence (GI) mode especially the background to peak height ratio is generally very low, therefore data collection is trickier. In order to minimize external errors thin film surfaces should be cleaned of any debris. Dust or other organic residues can sometimes reduce the quality of data by either hindering the path of beam or decreasing the intensity of peak recordings. This in some extreme cases can lead to inferences like oriented films, amorphous growth or preferred orientation errors.

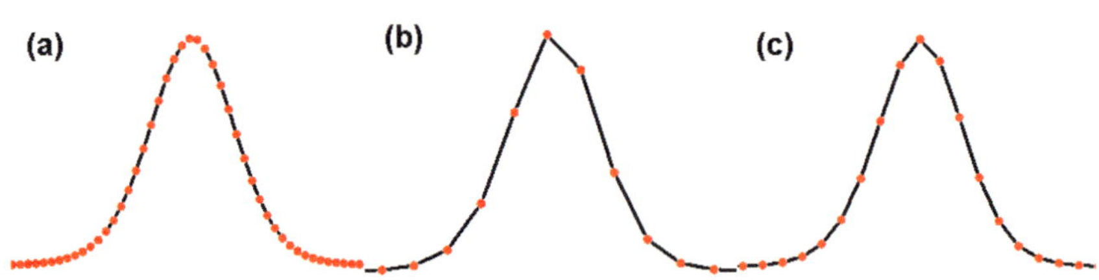

Figure 2.
The data (a) over collection (b) under collection (c) normal collection conditions for XRD data.

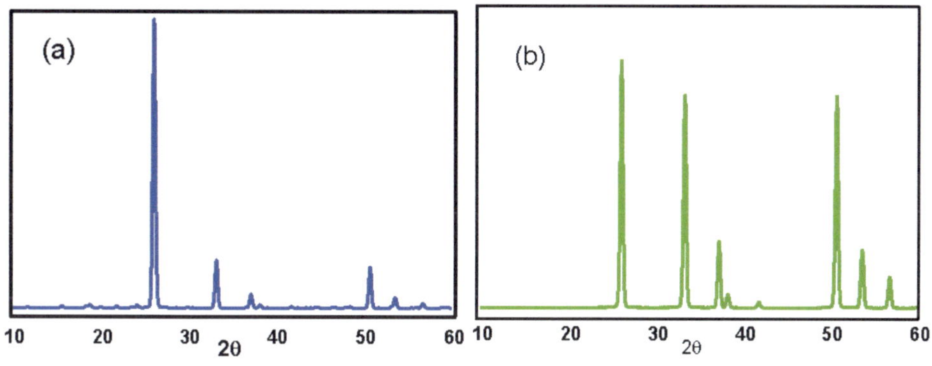

Figure 3.
SnO2: (a) the observation of preferred orientation due to poor particle distribution while sample preparation (b) data collected after 2 hr. grinding.

4.2 Challenges

The continuous motion of either or both detector and source arms of goniometer and the recording transit time of the cameras are one of the bigger challenges modern x-ray diffractometers face. Although the introduction of step size has essentially eliminated this problem, there are still concerns regarding too close and two far step sizes. Both can affect the peak geometry and background contribution in more effectual way. Too close step sizes, lower symmetry phases/peak splitting are bad combinations. Wide step sizes and nano-materials/GI mode/multiphase samples are also bad combinations. The time dependence of step size choice and effective counting times are the current limiting factors for diffractometers.

5. Background contribution

As discussed in previous section, the data collection should essentially be optimized to obtain least background. However, in practicality there are many possible unavoidable, yet necessary and characteristic reasons where background cannot be minimized after a certain degree without degradation of peak data quality. Although for pure phase materials, the background essentially remains negligible, till the particle or crystallite sizes are greater than 100 nm and grain boundaries are insignificant. For multiphase materials, the relative intensity difference between the peaks of different phases due to preferred orientation, crystallite size difference, peak broadening, quantitative presence, and sometimes amorphous phase do make background contributions a part of the X-ray reflection geometry [51].

Basically, the background contributions are dealt in two different ways in a powder diffraction pattern. Background can be modeled by an empirical/semi-empirical polynomial function with several refinable parameters or it can be estimated and at the end subtracted by a linearly interpolated set of points. Background subtraction although seems inelegant, is more sophisticated in circumstances where polynomial function cannot describe the background well. The normal procedure for background estimation should be an initial estimation using polynomial function, followed by (if required) linear interpolation and subtraction. This method is supposed to both preserve the estimates of standard deviations and correct for the background contribution optimally. It should also be noted that if a polynomial function does not describe the background well, no amount of refinement of its coefficients or increase in its order can fix the problem. In such cases for a complete and satisfactory refinement process, the estimation of background should be skipped and linear interpolation and subtraction procedure should be followed. While background is generally eliminated in refinement process, the peak base shapes are essentially a part of background and therefore at higher 2θ, more care should be taken in estimating the background. This is why background fitting using linear interpolation by cubic-splines should be generally avoided. The asymmetric peak shape especially at higher 2θ (where peak intensities are generally low) and non-careful background estimation or subtraction can affect the relative intensity of peaks and therefore degrade the overall refinement quality.

Figure 4(a), (b) and **(c)** respectively show contribution of amorphous, nanoscale and micrometer-scale phase towards background in $LaMnO_3$ samples.

5.1 Precautions/explanations

More time spent on measurement less significant background. This is somewhat misleading the background does not actually change with increased time spent per step. It is the increase in the number and intensity of counts per peak that increases which visibly smoothens the background. The precautions for background contribution during data collection have been discussed previously are almost entirely

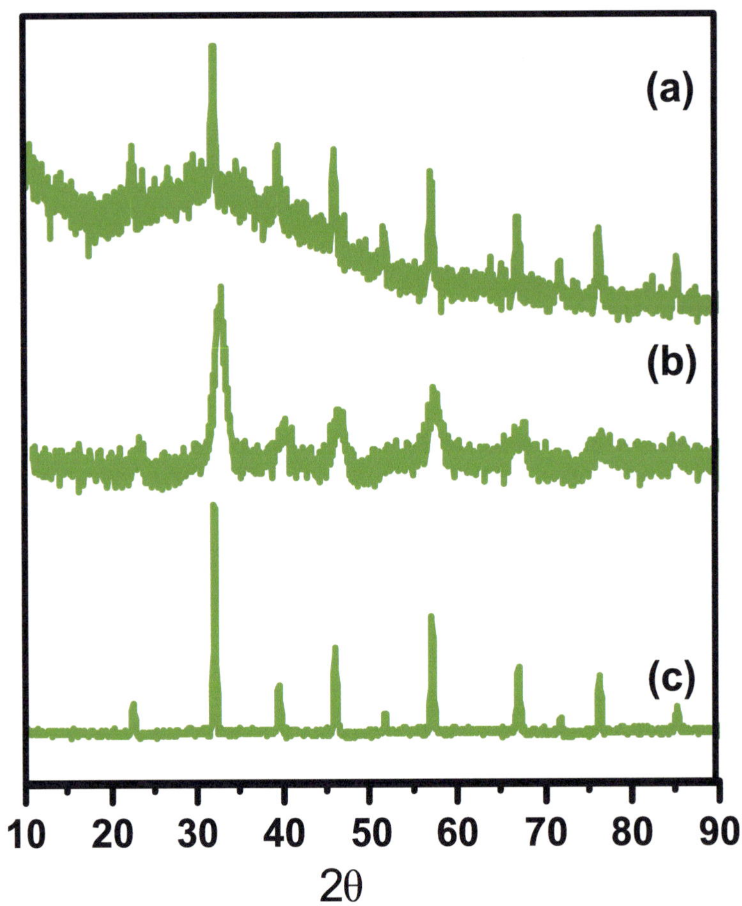

Figure 4.
XRD pattern for (a) mostly amorphous, (b) nanoscale and (c) micrometer-scale phase of $LaMnO_3$. The hump visible in (a) is a characteristic of amorphous phase, while the noisy background in (b) is characteristic of nanoscale phase due to low intensity counts.

complete set. During refinement and background estimation/subtraction, precautions need to be taken for segregating peak bases from background.

5.2 Current challenges

We are essentially in a nano-technological world right now and most of the materials applications around us have transitioned from bulk to micro to nanoscale. The complexities associated with the nanoscale XRD have also risen noticeably [52–55]. Nanoscale background contribution, irregular peak shapes, non-correctable preferred orientation/asymmetry parameters, sometimes odd combination of Lorentzian and Gaussian peak parameters. The porosity and reduced dimensionality (especially, 1D, 2D materials) are very difficult to characterize via normal XRD procedures.

6. Peak-shape function

The peak shape is one of the most important parameters in Rietveld refinement due to its dependence on crystallite/domain size, stress/strain, defects/vacancies, source/geometry, slit-size/detector resolution and 2θ/hkl indices [55]. An accurate description of the shapes of the peaks in a powder pattern is critical to the success of a Rietveld refinement. Poor description can lead to unsatisfactory refinement results, false minima and divergence. Peak shape analysis/function is the most complex parameter in Rietveld refinement, with dimensions into the space of unattainable and non-realistic. It is therefore essential for a working algorithm to make some assumptions/compromises on peak shape and sometimes neglect the otherwise essential aspect of peak shape. For x-ray and constant wavelength neutron data, the use of pseudo-Voigt approximated peak function is widely used. The pseudo-Voigt function is essentially a combination of Lorentzian and Gaussian peak function in a linear mode [30, 31, 56–61].

Voigt function is mathematically defined as

$$V(x,p,y) = \int_{-\infty}^{+\infty} G(x,p)L(x-x',y)dx' \qquad (1)$$

Where $G - \frac{e^{-x^2/2p^2}}{p\sqrt{2\pi}}$ is Gaussian function and $L = \frac{y}{\pi(x^2+y^2)^m}$ is Lorentzian function with m = 1 for symmetry.

The pseudo-Voigt function is described as

$$V_p(x,f) = \eta L(x,f) + (1-\eta)G(x,f) \qquad (2)$$

With

$$0 < \eta < 1$$

η is the full width half maximum parameter and the ratio of Gaussian and Lorentzian functions $\eta/(1-n)$ determines the mixing of these functions.

The graphical representation of the pseudo-Voigt function with variable η is shown in **Figure 5**.

Pearson VII peak-shape function (**Figure 6**) is used alternatively where the exponent m (Eq. 1) varies differently, but the same trends in line shape are observed. Although the Gaussian and Lorentzian components of Voigt function can be devolved into meaningful physical interpretations of stress/strain, microstructure and line broadening effects, no such interpretation can be drawn from Pearson

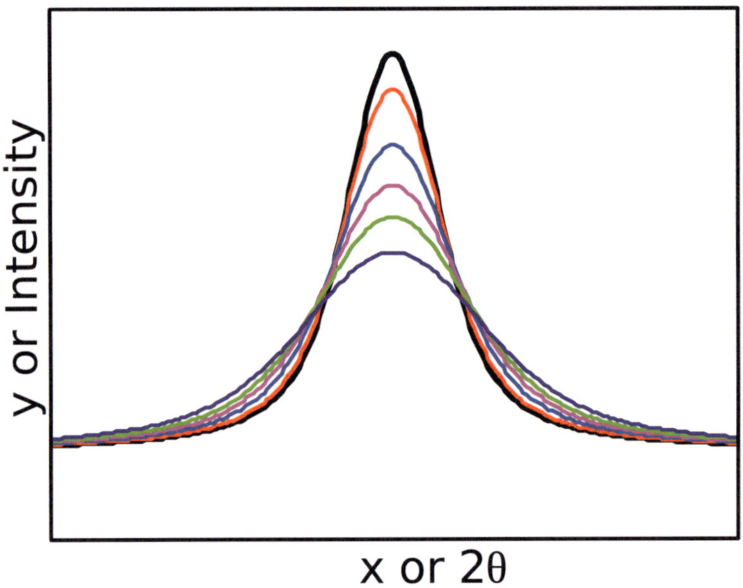

Figure 5.
Pseudo-Voigt peak function (black) and variation of peak shape (color) with η.

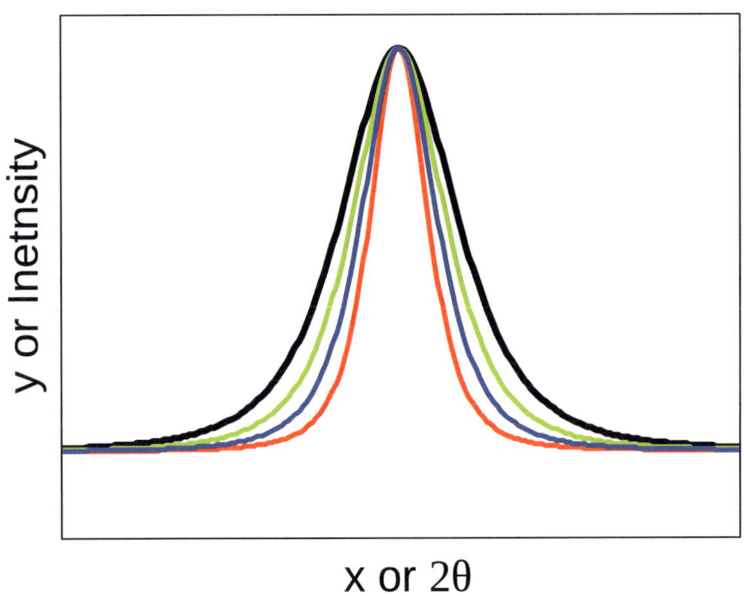

Figure 6.
Pearson VII peak function (black) and variation of peak shape (color) with $\mathbf{m}$.

VII function. Another advantage of pseudo-Voigt peak function over other functions is the separation of sample and instrument contributions.

$$I = I_o \frac{y^{2m}}{\left(y^2 + (x - x')^2\right)^m} \tag{3}$$

It is also imperative to point out that pseudo-Voigt peak fitting accounts for peak base asymmetry more rigorously while Pearson VII is more inclined towards the peak

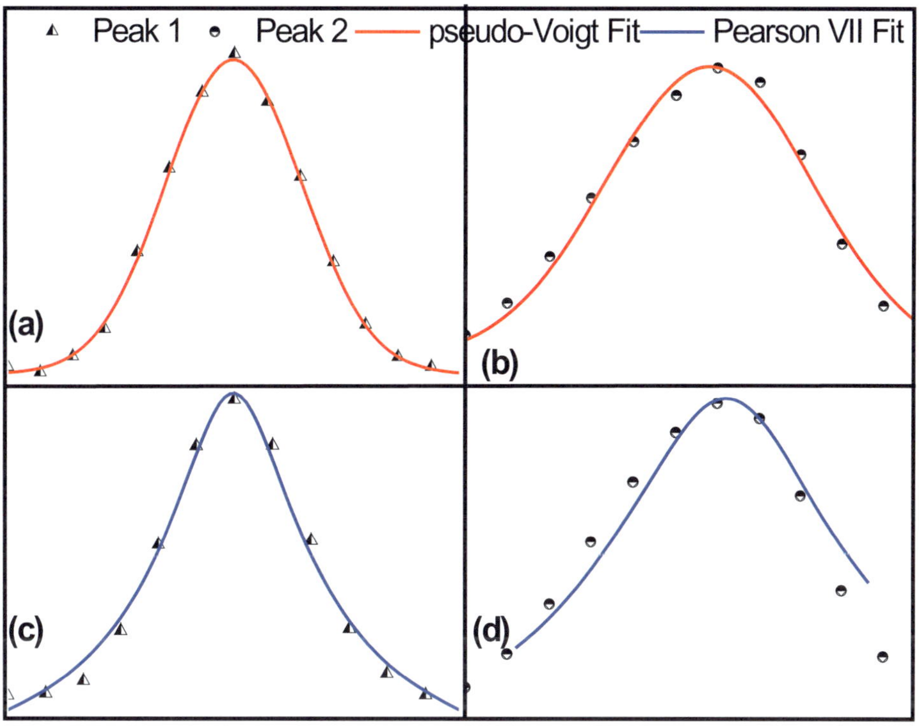

Figure 7.
Variation of fitting between two different shapes of peaks (triangles and circles) using (a), (b) pseudo-Voigt (red) and (c), (d) Pearson VII function (blue).

centre and Intensity offset. The property of these peak functions can be employed more efficiently by empirical evaluation of the peak shapes. Although both the functions provide similar results when the variation of peak shapes with 2θ is accounted.

The variation in chi-square fitting ($\chi^2 \sim 0.01$) of peak and base in both pseudo-Voigt and Pearson VII function can be visualized in **Figure 7(a)-(d)**.

7. Profile parameters

The profile parameters include every detail that a structural model packs in, except (background, peak shape and FWHM). Although FWHM is considered a part of profile in XRD, it is necessarily a variant under peak shape function. Therefore most of the available programs for Rietveld refinement list it under profile section. Practically, Clubbing of the asymmetry parameter, preferred orientation parameter and FWHM together due to their interdependence makes more sense. The structural model which is available should be complete otherwise the calculated profile will significantly deviate. The incorrect profile parameters during refinement process generally leads to refinement of FWHM, peak asymmetry, zero shift, etc. In such cases, it is more prudent to use methods that are structure independent. Le-Bail [23, 39, 62–65], Pawley [66, 67] etc. are suitable for obtaining initial values of profile parameters and extract a list of integrated intensities. The integrated intensities can then be used to calculate electron scattering densities and possible structure determination. In addition, the initial parameters can be refined to obtain more agreeable profile parameters. The information like crystallite size, defect concentration, microstrain etc. which can be extracted from XRD are derived from the profile

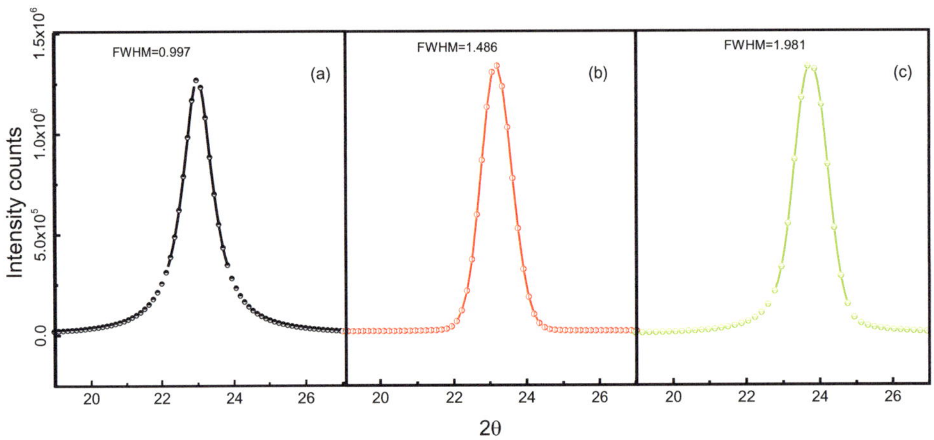

Figure 8.
The interplay of peak width, shape and its effect on FWHM with (a) symmetric profile (b) asymmetric profile without significant peak shift and (c) asymmetric profile resulting in significant peak shifting.

parameters. Although, the independence of profile parameters and peak shape is questionable due to their correlated nature, to make physical sense from the variation of either, the profile parameter needs a separate part in X-ray diffraction. The asymmetry in peak profile is another feature that arises from convolution of closely spaced multiple peaks. The two common and prevalent reasons for such conditions are; the evolution of microstrain and the defects. However the contributions from the instrument and the sample holder cannot be undermined. The interplay between FWHM, Peak position, shape and width is best visualized graphically (**Figure 8**).

7.1 Precautionary/explanatory notes

Although the basic idea about chemical composition, cell volume and density are needed to obtain a solution to an unknown phase, search using Le Bail, Pawley or ITO, DICVOL, TEROR, EXPO can always be widened to obtain initial profile parameters. However lower symmetry crystal systems like monoclinic and triclinic should not be included unnecessarily. These programs are likely to give multiple solutions to single set of reflection and it remains up to the user's judgment in these cases to choose a suitable solution. The multiple solutions are more prominent when lower symmetry systems are included, and sometimes the search criteria need to be adjusted to remove unrealistic solutions. In general, unrealistic solutions tend to possess either of the characteristics listed below or their combination:

i. Very large/small cell volume

ii. One or more of the cell parameters in extremely large/small[1]

iii. The fractional atomic coordinates are unrealistic

iv. The number of atoms per unit cell are either very high or low

v. Atomic overlapping

[1] The case of rhombohedral symmetry which is generally expressed in hexagonal axes format should be treated individually. It is normal in some materials which crystallize in rhombohedral symmetry to have large 'c/a' ratio when expressed in hexagonal axes. Few of the examples are Telurides, Selenides and lannonites.

8. Rietveld refinement: procedure and guidelines

After getting a complete structural model, suitable unit-cell parameters, the sufficient profile parameters and agreeable background, the Rietveld refinement of structural parameters can be started [68–71]. Refinement is usually done in sets of two to five cycles at a time but for effective refinements in simpler crystal systems hundreds of cycles will be required and thousands for complex systems. While the refinement is underway, we can monitor the progress either graphically or numerically [49]. While the numerical parameters can give us a statistical idea about the refinement, the most useful information about the profile fit is best seen graphically. However the parameter shifts are much more rigorously visualized numerically. Introduction of Reliability factors or R values by Rietveld enabled us to visualize profile fit between observed and calculated patterns more effective, although the graphical inspections still retain their superiority. The difference plot is also a good indication of the quality of profile fit, however the actual difference between the observed and calculated profiles and the origination of the deviation is not always quite evident from it. **Figure 9** shows the full profile Rietveld refinement of $LaMnO_3$ (couple more example of Rietveld refinement are given at the end of the chapter) while **Figure 10** shows the observed and calculated profile for a certain peak along with difference plots, while the corresponding R values for the whole profile are listed in the table. The increased R-values can be due to insufficient structural model, or inaccurate profile parameter. The atomic coordinates and positional parameters can cause changes in relative peak intensities at both high and low angles. The refinement of all the parameters has to be either done simultaneously or in a particular order to avoid numerical and statistical errors. The order and the explanation of the various refinable parameters and reliability factors will be done

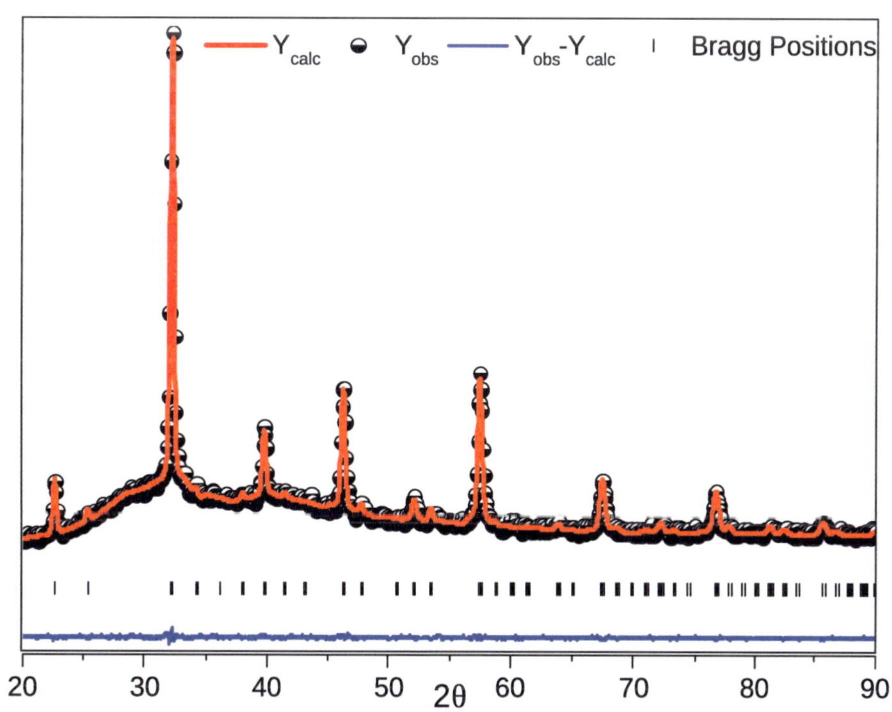

Figure 9.
Typical Rietveld refinement plot (LaMnO3) with observed (black circles), calculated (red), difference (blue) and Bragg positions (black bars).

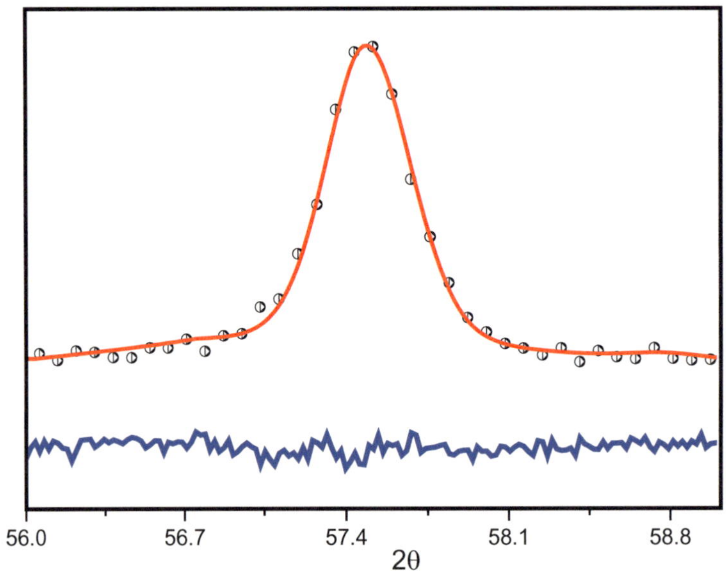

Figure 10.
*The zoomed in view of peak at around 57⁰ from **Figure 9** to visualize the goodness of fit.*

in later sections. It is important to know the source of errors in the refinement procedure for a effective and concise results. The most common error that occurs is due to the noisy data. The noisier the data the more refinement is needed for background parameter, this can sometimes lead to convolution of peak bases into background especially at higher angles. Zero shift and sometimes step size can also cause a range of errors to creep in. it is therefore a common procedure to first correct the data for zero shift and choose a more incredulous step size at the time of data collection. Apart from these, we need to look out for most of the other errors while the refinement process is underway. Sometimes lower estimated standard deviations can result from false minima observed due to unavailability of suitable structural model or unrealistic positional parameters (**Table 1**) [42, 43].

9. Refinement procedure

It is difficult to cover all the details of a full refinement, but an approximate strategy can be described. It is generally advised to begin the structural refinement first with the positions of the heavier atoms and then extend the refinement to positions of lighter atoms. It should however be always kept in mind that the statistical minima can sometimes attribute unrealistic positions to the atoms. All atomic positions, with constraints in place, can be refined simultaneously upon convergence. The scale, the thermal and the occupancy parameters are more

χ^2	R_{wp}	R_{exp}	R_F	R_{Bragg}
1.25	2.65	2.11	2.29	3.58

Table 1.
The chi (goodness of fit), and other Rietveld reliability factors (explanation of each factor in "R-factor" section ahead).

sensitive to the background correction due to their correlated nature. Positional parameters are somewhat independent of background. In order to reduce the number of thermal parameters to be refined in early stage, it is advisable to constrain the thermal parameters of similar atoms. Chemical constraints should be applied to maintain the physical sense of occupancy parameters. Refining a single structure using two independent data-sets e.g. x-rays and neutron diffraction the parameter correlation can be minimized. However, the experimental conditions for data collections such as pressure, temperature etc. in each case should be as similar as possible. Refinement of the profile parameters along with the structural parameters is also advisable. The structural model should be refined to convergence while care should be taken to retain the physical and chemical sense wherever applicable. Mere convergence with even a single parameter not making physical or chemical sense is all the efforts wasted. It is therefore necessary to always follow a certain procedure/pathway of refinement or at least at the earlier stages of refinement. The likely procedure of refinement pathway is given in **Figure 11**.

Because powder diffraction data are a one-dimensional projection of three-dimensional data, the inherent loss of information is always a problem. To partly compensate for this loss geometric information (bond distances and/or angles) taken from related structures is more appropriate method. The purpose of these constraints is to increase the number of observations by added geometric conditions. Another way to implement restraints is to follow rigid body model, this however results in decrease in the number of observations and complicating the structural model. The use of geometric restrains not only increases the number of observations but allows more parameters to be refined, while keeping the geometry of the structural model sensible. The set of geometric restraints can be treated as separate data set, with same rules of quantity minimization in the refinement. The geometric data set can be represented as:

$$S = S_y + c_w S_G \tag{4}$$

where S_y is the weighted difference between the observed *[y (obs)]* and calculated *[y(calc)]* diffraction patterns,

$$S_y = \sum_i w_i \left[y_i(obs) - y_i(calc)\right]^2 \tag{5}$$

S_G is the weighted difference between the prescribed *[G(obs)]* and calculated *[G(calc)]* geometric restraints,

$$S_G = \sum w[G(obs) - G(calc)]^2 \tag{6}$$

and c_w is a factor that allows a weighting of the geometric observations 'data-set' with respect to the diffraction data-set.

Geometric restraints can enhance a refinement considerably, allowing otherwise impossibly complex structures to be refined successfully. However care must to choose the bond distance and angles in order to accommodate the appropriate polyhedral geometry. It is imperative that the final structure model should fit both the geometric and the X-ray data satisfactorily.

9.1 Quantitative refinement

The methodology involved in qualitative and quantitative Rietveld refinements have been discussed at length by many authors [26, 27, 49, 72]. The theory behind

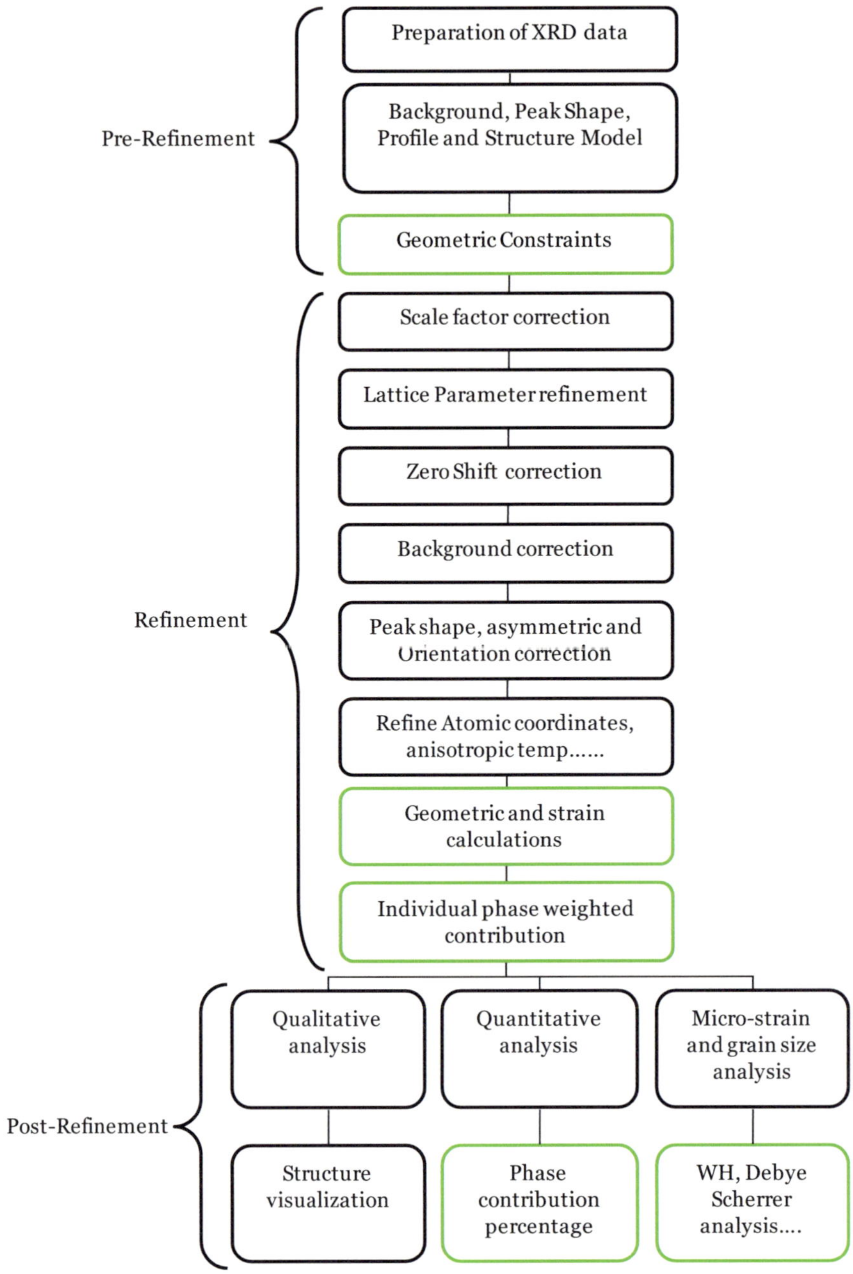

Figure 11.
The procedure typically followed during the refinement of XRD data via Rietveld method. The green boxes are optional calculations. The variation of procedure is necessary in cases with amorphous phases, anomalous reflections or sample induced asymmetry.

Rietveld quantitative analysis is identical to that implemented in most conventional quantitative analyses [34, 73–76]. The integrated intensity of X-rays diffracted by a randomly oriented infinitely thick [40, 76–81] polycrystalline sample in flat-plate geometry can be written for a particular reflection as:

$$I_{hkl} = K\left(\frac{1}{2\mu}\right)R_{hkl} \tag{7}$$

Where K and R_{hkl} are the hkl invariant and variant parameters.

The detailed discussion of the mathematical and physical interpretations of these quantities can be found abundantly in literature, particularly in the cited works [72, 82–84].

In a mixture, the intensity of *hkl* reflection originating from a particular phase (α) is written as

$$I_{\alpha,hkl} = C_\alpha K\left(\tfrac{1}{2\mu_m}\right)R_{\alpha,hkl} \tag{8}$$

Where C_α is the volume fraction of α phase with μ_m as linear absorption coefficient

In terms of weight fractions, which is statistically more convenient, the equation can be written as

$$I_{\alpha,hkl} = \frac{W_\alpha}{\rho_\alpha} K \frac{\rho_m}{2\mu_m} R_{\alpha,hkl} \tag{9}$$

Now the scale factor for alpha phase can be written as

$$S_\alpha = \frac{W_\alpha}{\rho_\alpha} K \frac{\rho_m}{2\mu_m} \tag{10}$$

For second phase (β), the weight fraction can be done similarly while the net contribution per phase can be sought from the equation below

$$W_\alpha = \frac{W_\alpha}{W_\alpha + W_\beta} \tag{11}$$

The equation can be solved by replacing weight fractions by equation above

$$W_\alpha = \frac{S_\alpha S_\beta}{S_\alpha \rho_\alpha + S_\beta \rho_\beta} \tag{12}$$

As scale parameters are refined we will get estimated weight fraction contribution of each phase.

9.2 R values

The numerical way of observing the quality/goodness of fit, although not as prudent as graphical visualization of difference plots, provides a good, intuitive numerical estimate. This is usually done in terms of agreement indices also called Residual values or Retiveld refinement indices or Rietveld discrepancy indices or R values [26, 27, 85–87] which are expressed as.

9.2.1 *The weighted-profile* R-*value*

The weighted profile *R* values (R_{wp}) is most straight forward an follows directly from the square root of minimized quantity, scaled using weighted intensities and is defined as:

$$R_{wp} = \left\{ \sum_i w_i \left[y_i(obs) - y_i(calc)\right]^2 / \sum_i w_i \left[y_i(obs)\right]^2 \right\}^{1/2} \tag{13}$$

where $y_i(obs)$ is the observed intensity, $y_i(calc)$ the calculated intensity, and w_i the weight at i^{th} step.

The numerator in Eq. (13) is the expression that is minimized during a Rietveld refinement procedure. Thus the inclusion or exclusion of background can have dramatic effect on the refinement. If the background has been excluded, and thus subtracted prior to refinement then, $y_i(obs)$ is the net intensity. However, the inclusion of background means the refinement of background pramaters. In such cases, $y_i(obs)$ includes both background and net intensity. Therefore, $y_i(obs)$ and $y_i(calc)$ both will likely include the background contribution. In the latter case when dealing with a high background to peak intensity ratio, most of intensity will be attributed to background, resulting in lowered value of R_{wp}. Therefore it is recommended to subtract background in such cases. R_{wp} for laboratory X-ray data are large ~10%. This is primarily due to the level of the background. In any publication, the type of agreement index used must be clearly specified. Ideally, the final R_{wp} should approach the statistically expected R value or R_{exp}.

9.2.2 The expected R-value

R_{exp} reflects both the quality of data and refinement and is expressed as

$$R_{exp} = \left\{ (N - P) / \sum_i w_i \left[y_i(obs) \right]^2 \right\}^{1/2} \quad (14)$$

where N is the number of observations and P the number of parameters.

However, the ratio between the R_{wp} and R_{exp}, called goodness of fit (χ^2), which is quoted quite often in the literature, should approach 1.

$$G^2 = \chi^2 = R_{wp} / R_{exp} \quad (15)$$

Most of the statistical errors in these R values can occur either due to under-collection or over –collection of data. The ratio will be less than one if data is under collected as R_{exp} will be much higher than R_{wp}. In case of over-collection the ratio will be greater than 1. It is always recommended to have over-collected rather than under-collected data. As estimated standard deviations [88] an also alter the ratio, there are other R values like R_F and R_{Bragg} which will improve the conclusivity of the data.

9.2.3 The structure factor R value

An R value based on structure factors, F_{hkl}, can also be calculated by distributing the intensities of the overlapping reflections according to the structural model.

$$R_F = \sum_{hkl} |F_{hkl}(obs) - F_{hkl}(calc)| / \sum_{hkl} |F_{hkl}(obs)| \quad (16)$$

R_F a derivative of structure factors is essentially biased towards the structural model. It can however give a clear indication of the reliability of structural refinement. Although not used actively while reporting the refinement of structure, it should necessarily decrease as the structural model improves in the course of the refinement.

9.2.4 The Bragg intensity R value

The Bragg-intensity R value (R_B) is essentially the structure factor R_F but in terms of Intensity I_{hkl}:

$$R_B = \sum_{hkl} |I_{hkl}(obs) - I_{hkl}(calc)| / \sum_{hkl} |I_{hkl}(obs)| \quad (17)$$

Where $I_{hkl} = mF_{hkl}^2$, m is multiplicity.

R values are useful indicators for the evaluation of a refinement, especially in the case of small improvements to the model which are not generally visible in difference plots. However, care should be taken while evaluating the *R* values as they are prone to over-interpretation. The most important questions that need to be asked for judging the quality of a Rietveld refinement are

i. Is the fit between observed data and calculated pattern good?

ii. Does the structural model make chemical sense?

iii. Are inter-atomic distances and angles realistic?

iv. Are the results from the refinement consistent with results from Raman, IR NMR etc. characterizations?

9.3 Common problems during refinement

Each structure refinement has its own idiosyncrasies and will present problems that require imaginative and selective solutions. However, some problems are of a more general nature and arise in many cases.

The most frequent source of difficulty in a Rietveld refinement is error in the input file. Most of these errors if occurring due to format or syntax can be corrected by conversion of files into suitable format using software like PowDLL from University of Ioannina.

The background does not seem to fit well

i. Try a different background function, increase the number co-efficient, change from linear to polynomial or vice versa [19]

ii. Try background subtraction

iii. Try combination of (i) and (ii)

The peak shapes are not suitably fitting

i. Check the difference plot and match with the **Figures 12(a)-(c)** to see if one of the characteristic difference profiles is shown. The respective profile parameter should be reset or further refined [20, 21]

ii. Use a different peak-shape function

iii. Perform asymmetry correction to the peak-shape function.

iv. Line broadening and shifting along with *2θ* dependence of FWHM can indicate microstructure contributions [89–91].

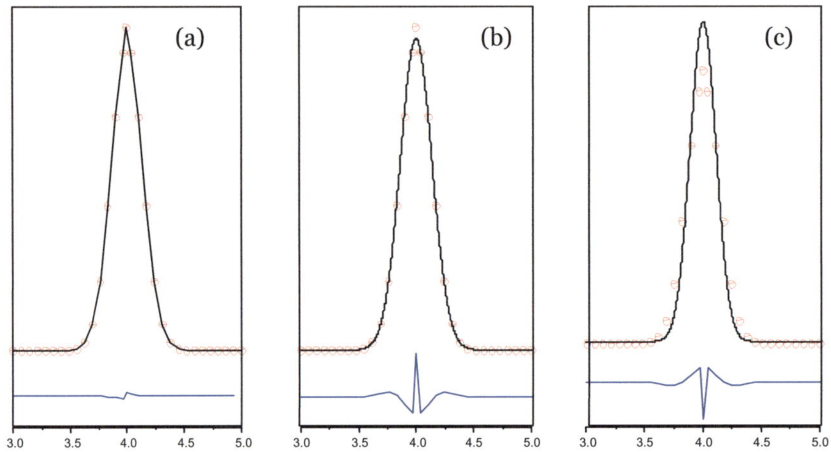

Figure 12.
(a) a good peak fit (b) Observed intensities are higher than calculated and (c) Observed intensities are lower than calculated (in both cases, possibly any of these might require to be reset or further refinement, (i) scale factor, (ii) preferred orientation, (iii) lattice parameters).

The peak positions in the calculated and observed patterns do not match

i. Check if unit cell parameters are correct

ii. Perform Zero shift refinement

iii. Determine the unit-cell parameters via independent indexing methods

The tails of the peaks in the calculated pattern are cut off prematurely

i. Increase the peak range used in the calculation

The relative intensities of a few reflections are high with very few low peaks

i. This is usually indicative of rock in dust problem concerned with poor particle statistics. The only solution is to recollect the data after proper sample preparation

There multiple un-indexed peaks in the diffraction pattern

i. Check for sample impurity

ii. Check whether the infinite sample thickness condition was fulfilled during data collection

iii. Check for peaks from sample holder

The refinement does not converge

i. Look at the observed/calculated profiles carefully and check these.

a. Are the observed peak shapes well defined by peak shape function?

b. Is there any mismatch between peak positions?

c. Is background refinement realistic and sensible?

d. Is the scale factor correct?

ii. Has structural model been completely described?

iii. Check for oscillations in the parameter shifts and apply damping factors as. Most modern refinement software perform this automatically

iv. Do not refine two parameters with high correlation together. Sometimes the high correlation is an indication of wrong space group

v. Refine fewer parameters initially

vi. Add geometric restraints

vii. If geometric restraints are already in use, are they correct?

viii. Fix thermal (atomic displacement) parameters at certain sensible values

ix. Use a different space group.

x. The number of parameters being refined is higher than what data can provide

The final structure is not chemically sensible (unrealistic inter-atomic distances)

i. Use restraints to keep inter-atomic distance sensible.

ii. Delete the offending atoms and try relocating them using Fourier maps [54, 59, 92, 93]

iii. Change restraints [94]

iv. Change the space group

v. Fix atomic thermal displacements and fractional coordinates in the beginning

Refinement converged but there are few peaks which are not fitted well

i. Check for Lorentz–polarization correction

ii. Apply absorption correction parameter if data permits [95]?

iii. Are atomic fractional co-ordinates correct?

iv. Is there preferred orientation in the sample?

10. Structural visualization

The most important aspect of a Rietveld analysis is the refinement of structure. The actual structure of the sample can be calculated taking into consideration the lattice parameter variations, the microstructure, stress strain contribution and other contributions. Effectively most of the currently available software for Rietveld refinement can easily generate the refined structure file. Visualization of structure at higher resolution has become easy with enhanced computational power. However the presentation of the structure is not standardized and most of the time the axial orientation is not mentioned. Although, it is not essentially a problem for the readers, the standard representation of the structure should be preferred. In cases where a non-standard representation is used, mention of plane, axial orientation, etc. should be clearly mentioned [90, 96]. The non standard representation of the structure can sometimes lead to wrong conclusions as shown in **Figure 13** for ZnO.

Apart from the problems discussed above, the tetrahedral and octahedral geometry should be visualized carefully (**Figure 14**). The actual polyhedral tilting, rotations or other geometric variations can be truly visualized only after symmetrised unit cell representation [97–99]. The **Figure 15(a)** and **(b)** show ZnO structure in symmetrised and non-symmetrised form. The difference in visualization is quite amazing [34, 100, 101].

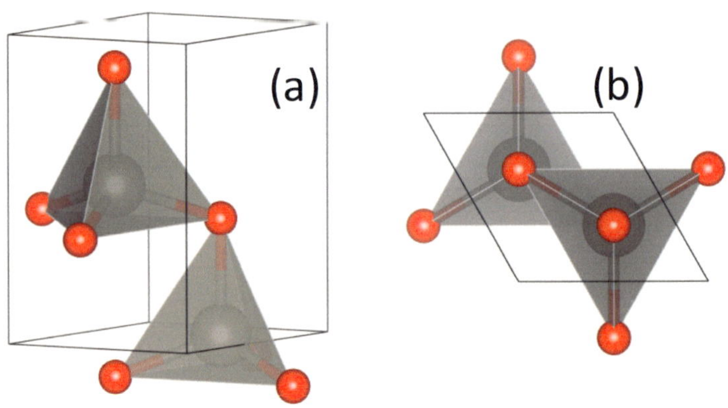

Figure 13.
(a) The standard view and (b) c axial view of hexagonal ZnO (●-Zn and ●-O) unit cell with non-standardized atomic positions.*

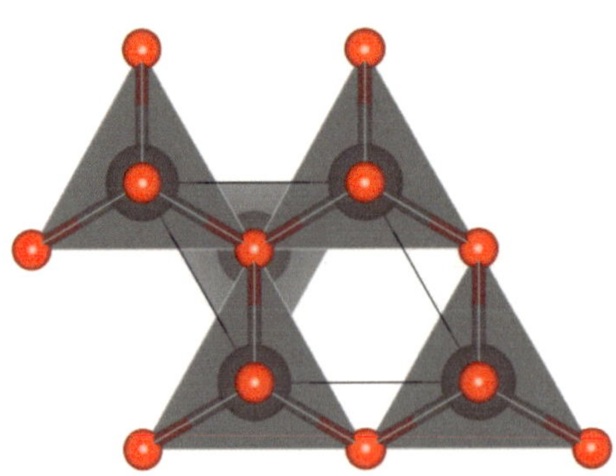

Figure 14.
The c axial view of hexagonal ZnO unit cell with standardized atomic positions. The transformation of structure to represent the hexagonal arrangement of ●-Zn and ●-O atoms is effectively visible within a single unit cell.*

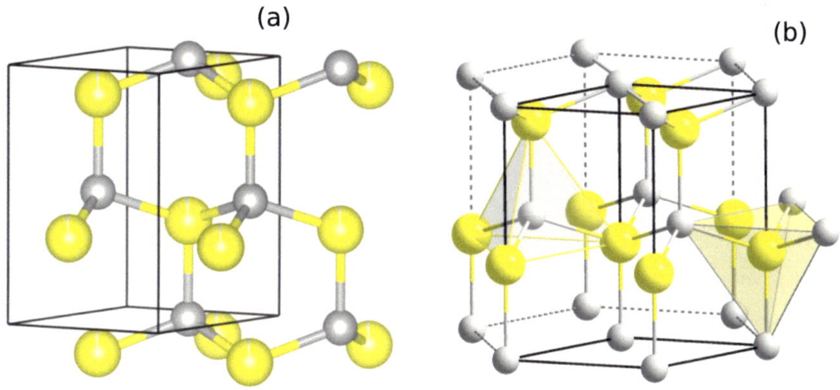

Figure 15.
*The non-standard viewing and primitive unit cell of wurtzite ZnO (●-Zn and ●-O). The atomic arrangement is not quite effectively legible and physically meaningful in non-standard viewing, while it is quite meaningful in case of primitive lattice (**Table 2**).*

11. Recommended software packages

1. X-ray diffraction data visualization:

    a. Winplotr [102]

    b. Panalytical X'Pert Highscore

    c. X-Powder

    d. Cyrstal Impact Match [103]

    e. PowderPlot

2. Inter-Conversion of XRD data between different formats

    a. X-powder

    b. PowDLL [104]

    c. Winplotr

3. Search and Match with database

    a. PCPDFWIN from ICDD

    b. Cyrstal Impact Match

    c. X-Powder

    d. Panalytical X'Pert Highscore [105]

4. Indexing

    a. EXPO2014

b. DICVOL

c. ITO

d. TREOR

5. Rietveld Refinement

a. EdPcr, fp2k from Fullprof suite [102, 106]

b. Crystal Impact Match

c. Profex [107]

d. QualX and Quanto

6. Fourier Map Visualization

a. GFourier

b. VESTA [96]

7. Structure Visualization

a. Crystal Impact Diamond

b. VESTA

12. Rietveld refinement examples

The case of Rietveld refinement of perovskite LaMnO3 along with various refinement parameters is given in “Retiveld Refinement ... .” section, while two

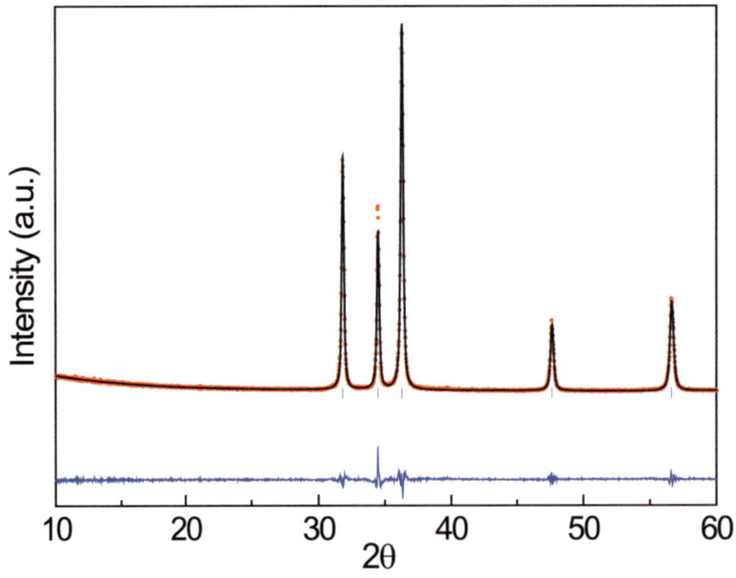

Figure 16.
Representative Rietveld refinement plot of wurtzite ZnO with observed (red circles), calculated (black), difference (blue) and Bragg positions (blue bars).

S. group	χ^2	R_p	R_{wp}	R_{exp}	R_F	R_B
P6₃mc	1.24	8.56	10.9	8.81	1.04	1.2
Lattice parameters			**Fractional coordinates**			
a (Å)	**c (Å)**		**Zn**		**O**	
3.254	5.212					
		x	**y**	**z**	**x**	**y**	**z**
		0.3333	0.6667	0.0	0.3333	0.6667	0.3820

Table 2.
Rietveld refined fractional co-ordinates, space group, lattice parameters, R- values (R_p: Un-weighted profile parameter, R_{wp}: Weighted profile parameter, R_{exp}: Expected profile parameter, R_F: Structure parameter, R_B: Intensity parameter), χ^2: Goodness of fit and other parameters of ZnO.

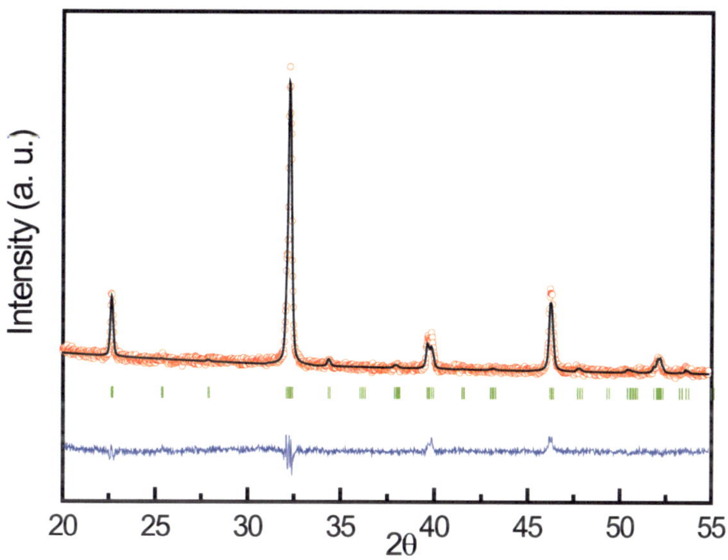

Figure 17.
Representative Rietveld refinement plot of double perovskite La_2FeCoO_6 with observed (red circles), calculated (black), difference (blue) and Bragg positions (green bars).

SG	χ^2	R_p	R_{wp}	R_{exp}	R_F	R_B		
P 21/n	1.24	4.87	6.34	4.73	5.51	6.05		
Lattice parameters			**Fractional coordinates**					
a (Å)	**b (Å)**	**c (Å)**		**La**			**O**	
5.566	5.501	9.666	**x**	**y**	**z**	**x**	**y**	**z**
α = γ = 90°			0.2434	0.0226	0.2486	0.1793	0.2280	0.9590
						0.23180	0.7221	0.9584
						0.3323	0.0122	0.7555
				Fe			**Co**	
β = 124.35°			**x**	**y**	**z**	**x**	**y**	**z**
			0.5	0.0	0.0	0.0	0.0	0.5

Table 3.
Rietveld refined fractional co-ordinates, space group (SG) lattice parameters, R- values (R_p: Un-weighted profile parameter, R_{wp}: Weighted profile parameter, R_{exp}: Expected profile parameter, R_F: Structure parameter, R_B: Intensity parameter), χ^2: Goodness of fit and other parameters of La_2FeCoO_6.

additional cases of Wurtzite ZnO and double-perovskite La_2FeCoO_6 are given here (**Figures 16 and 17**, **Tables 2** and **3**):

a. ZnO

b. La_2FeCoO_6

Acknowledgements

The chapter will remain incomplete without the mention of Dr. Vilas Shelke and the hour-long discussions regarding crystal structure and x-ray diffraction which sparked the interest and provided the motivation for detailed studies necessary for the completion of this work. Without availing the facilities and the healthy discussions with Dr. Mukhul Gupta, Dr. V Ganeshan, and Dr. D M Phase from UGC DAE CSR Indore and Prof. S P Sanyal from Barkatullah University Bhopal the work would have remained incomplete.

Author details

Touseef Ahmad Para* and Shaibal Kanti Sarkar
Devices and Interfaces Lab, Department of Energy Science and Engineering, Indian Institute of Technology Bombay, Mumbai, Maharashtra, India

*Address all correspondence to: drtouseefpara@yahoo.com

References

[1] Glazer AM. The first paper by W.L. Bragg-what and when? Crystallogr Rev. 2013;19(3):117–124. DOI: 10.1080/0889311X.2013.813494

[2] Wilkins SW. Celebrating 100 years of X-ray crystallography. Acta Crystallogr Sect A Found Crystallogr [Internet]. 2013;69(1):1–4. DOI: 10.1107/S0108767312048490

[3] Glazer AM. Celebrating the braggs—A personal account. Interdiscip Sci Rev. 2015;40(3):329–39. DOI: 10.1179/0308018815Z.000000000121

[4] Bragg WH, Bragg WL. The reflection of X-rays by crystals. Proc R Soc London Ser A, Contain Pap a Math Phys Character [Internet]. 1913 Jul;88(605):428–38. DOI: 10.1098/rspa.1913.0040

[5] Glazer AM. The Braggs. Ferroelectrics. 2002;267(1):35–41. DOI: 10.1080/00150190211007

[6] Liljas A. Background to the Nobel Prize to the Braggs. Acta Crystallogr Sect A Found Crystallogr [Internet]. 2012/12/20. 2013;69(1):10–5. DOI: 10.1107/S0108767312031133

[7] Thomson P. A tribute to W. L. Bragg by his younger daughter. Acta Crystallogr Sect A Found Crystallogr [Internet]. 2012/12/20. 2013;69(1):5–7. DOI: 10.1107/s0108767312047514

[8] BRAGG WL. A New Type of 'X-Ray Microscope.' Nature [Internet]. 1939 Apr;143(3625):678–678. DOI: 10.1038/143678a0

[9] Perutz MF. How W. L. Bragg invented X-ray analysis. Acta Crystallogr Sect A [Internet]. 1990;46(8):633–43. DOI: 10.1107/S010876739000410X

[10] Bragg WL. The structure of some crystals as indicated by their diffraction of X-rays. Proc R Soc London Ser A, Contain Pap a Math Phys Character. 1913;89(610):248–77. DOI: 10.1098/rspa.1913.0083

[11] Ewald PP. Max von Laue, 1879–1960. Acta Crystallogr [Internet]. 1960 Jul 10;13(7):513–5. DOI: 10.1107/S0365110X6000128X

[12] Laue M Von. Concerning the detection of X-ray interferences [Internet]. Nobel lecture. 1915.

[13] Scherrer P. Bestimmung der inneren Struktur und der Größe von Kolloidteilchen mittels Röntgenstrahlen. In: Kolloidchemie Ein Lehrbuch [Internet]. Berlin, Heidelberg: Springer Berlin Heidelberg; 1912. p. 387–409. DOI: 10.1007/978-3-662-33915-2_7

[14] Holzwarth U, Gibson N. The Scherrer equation versus the "Debye-Scherrer equation." Nat Nanotechnol. 2011;6(9):534. DOI: 10.1038/nnano.2011.145

[15] Patterson AL. The scherrer formula for X-ray particle size determination. Phys Rev. 1939;56(10):978–82. DOI: 10.1103/PhysRev.56.978

[16] Visser JW. Modern powder diffraction. Vol. 72, Sedimentary Geology. Walter de Gruyter GmbH & Co KG; 1991. 168–170 p. DOI: 10.1016/0037-0738(91)90134-y

[17] Chernyshev V V. Structure determination from powder diffraction. Russ Chem Bull [Internet]. 2001;50(12):2273–92. DOI: 10.1023/A:1015006807065

[18] Stephens PW, Bendele GM. X-Ray Powder Diffraction. In: Characterization of Materials [Internet]. Hoboken, NJ, USA: John Wiley & Sons, Inc.; 2002. DOI: 10.1002/0471266965.com071

[19] David WIF. Powder diffraction: Least-squares and beyond. J Res Natl Inst Stand Technol [Internet]. 2004/01/01. 2004 Jan;109(1):107. DOI: 10.6028/jres.109.008

[20] Bail A Le. Chapter 5. The Profile of a Bragg Reflection for Extracting Intensities. In: Powder Diffraction [Internet]. Cambridge: Royal Society of Chemistry; 2008. p. 134–65. DOI: 10.1039/9781847558237-00134

[21] Altomare A, Giacovazzo C, Moliterni A. Powder Diffraction. Powder Diffraction. 2008. DOI: 10.1039/9781847558237

[22] Dinnebier RE, Leineweber A, Evans JSO. Rietveld Refinement: Practical Powder Diffraction Pattern Analysis using TOPAS [Internet]. Vol. 52, Journal of Applied Crystallography. De Gruyter; 2019. 1238–1239 p. DOI: 10.1515/9783110461381

[23] Le Bail A, Duroy H, Fourquet JL. Ab-initio structure determination of LiSbWO6 by X-ray powder diffraction. Mater Res Bull [Internet]. 1988;23(3):447–52. DOI: 10.1016/0025-5408(88)90019-0

[24] Antoniadis A, Berruyer J, Filhol A. Maximum-likelihood methods in powder diffraction refinements. Acta Crystallogr Sect A [Internet]. 1990;46(8):692–711. DOI: 10.1107/S0108767390004500

[25] Hendrickson WA. Evolution of diffraction methods for solving crystal structures. Acta Crystallogr Sect A Found Crystallogr [Internet]. 2012/12/20. 2013;69(1):51–9. DOI: 10.1107/S0108767312050453

[26] Rietveld HM. Line profiles of neutron powder-diffraction peaks for structure refinement. Acta Crystallogr. 1967;22(1):151–2. DOI: 10.1107/s0365110x67000234

[27] Rietveld HM. A profile refinement method for nuclear and magnetic structures. J Appl Crystallogr. 1969;2(2):65–71. DOI: 10.1107/s0021889869006558

[28] Rietveld H. The Rietveld Method ? A Historical Perspective. Aust J Phys [Internet]. 1988;41(2):113. DOI: 10.1071/PH880113

[29] Sakata M, Cooper MJ. An analysis of the Rietveld refinement method. J Appl Crystallogr [Internet]. 1979 Dec 1;12(6):554–63. DOI: 10.1107/S002188987901325X

[30] Hester JR. Improved asymmetric peak parameter refinement. J Appl Crystallogr [Internet]. 2013;46(4):1219–20. DOI: 10.1107/S0021889813016233

[31] Finger LW, Cox DE, Jephcoat AP. Correction for powder diffraction peak asymmetry due to axial divergence. J Appl Crystallogr [Internet]. 1994;27(pt 6):892–900. DOI: 10.1107/S0021889894004218

[32] Dollase WA. Correction of intensities for preferred orientation in powder diffractometry: application of the March model. J Appl Crystallogr [Internet]. 1986 Aug 1;19(4):267–72. DOI: 10.1107/S0021889886089458

[33] Altomare A, Burla MC, Cascarano G, Giacovazzo C, Guagliardi A, Moliterni AGG, et al. Early Finding of Preferred Orientation: Applications to Direct Methods. J Appl Crystallogr [Internet]. 1996;29 PART 4 (4):341–5. DOI: 10.1107/s0021889896000271

[34] Para TA, Shelke V. Extreme blue-shifted photoluminescence from quantum confinement of core–shell ZnO. J Mater Sci Mater Electron [Internet]. 2017 Dec 1;28(24):18842–8. DOI: 10.1007/s10854-017-7835-0

[35] Balzar D, Popa NC. Analyzing microstructure by rietveld refinement *. J Rigaku. 2005;22(1):16–25.

[36] Audebrand N, Auffrédic JP, Louër D. An X-ray powder diffraction study of the microstructure and growth kinetics of nanoscale crystallites obtained from hydrated cerium oxides. Chem Mater. 2000;12(6):1791–9. DOI: 10.1021/cm001013e

[37] Von Dreele RB. Multipattern Rietveld refinement of protein powder data: An approach to higher resolution. J Appl Crystallogr [Internet]. 2007;40(1): 133–43. DOI: 10.1107/S0021889806045493

[38] Hill RJ, Cranswick LMD. International Union of Crystallography. Commission on Powder Diffraction. Rietveld refinement round robin. II. Analysis of monoclinic ZrO${\sb 2}$. J Appl Crystallogr [Internet]. 1994 Oct;27 (5):802–44. DOI: 10.1107/S0021889894000646

[39] Altomare A, Carrozzini B, Giacovazzo C, Guagliardi A, Moliterni AGG, Rizzi R. Solving Crystal Structures from Powder Data. I. The Role of the Prior Information in the Two-Stage Method. J Appl Crystallogr [Internet]. 1996;29 PART 6(6):667–73. DOI: 10.1107/s0021889896007467

[40] Madsen IC, Scarlett NVY, Cranswick LMD, Lwin T. Outcomes of the International Union of Crystallography Commission on Powder Diffraction Round Robin on Quantitative Phase Analysis: samples 1 a to 1 h. J Appl Crystallogr [Internet]. 2001 Aug 1;34(4):409–26. DOI: 10.1107/S0021889801007476

[41] Hill RJ. International union of crystallography commission on powder diffraction rietveld refinement round Robin. I. Analysis of standard x-ray and neutro n data for PbSO4. J Appl Crystallogr. 1992;25(pt 5):589–610. DOI: 10.1107/S0021889892003649

[42] Mccusker LB, Von Dreele RB, Cox DE, Louër D, Scardi P. Rietveld refinement guidelines. J Appl Crystallogr [Internet]. 1999;32(1):36–50. DOI: 10.1107/S0021889898009856

[43] Buhrke VE, Jenkins R, Smith DK, Kingsley D. Practical guide for the preparation of specimens for x-ray fluorescence and x-ray diffraction analysis. Wiley-VCH; 1998.

[44] Bish DL, Reynolds RC. SAMPLE PREPARATION FOR X-RAY DIFFRACTION. In: Bish DL, Post JE, editors. Modern Powder Diffraction [Internet]. Berlin, Boston: De Gruyter; 1989. p. 73–100. DOI: 10.1515/9781501509018-007

[45] Ewald PP. X-ray diffraction by finite and imperfect crystal lattices. Proc Phys Soc. 1940;52(1):167–74. DOI: 10.1088/0959-5309/52/1/323

[46] Hill RJ, Madsen IC. Data Collection Strategies for Constant Wavelength Rietveld Analysis. Powder Diffr. 1987;2 (3):146–62. DOI: 10.1017/S088571560001263X

[47] Von Dreele RB, Rodriguez-Carvajal J. Chapter 3. The Intensity of a Bragg Reflection. In: Powder Diffraction [Internet]. Cambridge: Royal Society of Chemistry; 2008. p. 58–88. DOI: 10.1039/9781847558237-00058

[48] Cooper MJ. The analysis of powder diffraction data. Acta Crystallogr Sect A [Internet]. 1982 Mar 1;38(2):264–9. DOI: 10.1107/S0567739482000564

[49] Young RA. The rietveld method. Vol. 6. Oxford university press Oxford; 1993.

[50] Mccusker LB, Von Dreele RB, Cox DE, Louër D, Scardi P. Rietveld refinement guidelines. J Appl Crystallogr. 1999;32(1):36–50. DOI: 10.1107/S0021889898009856

[51] David WIF, Sivia DS. Background estimation using a robust Bayesian

analysis. J Appl Crystallogr [Internet]. 2001 Jun 1;34(3):318–24. DOI: 10.1107/S0021889801004332

[52] Langford JI. Accuracy in powder diffraction. In: Natl Bur Stand Spec Publ. National Bureau of Standards, Gaithersburg, Maryland: US Dept. of Commerce, National Bureau of Standards: for sale by the Supt. of ... ; 1980. p. 255–69.

[53] Langford JI, Louër D, Scardi P. Effect of a crystallite size distribution on X-ray diffraction line profiles and whole-powder-pattern fitting. J Appl Crystallogr. 2000;33(3 II):964–74. DOI: 10.1107/S002188980000460X

[54] Langford JI, Delhez R, de Keijser TH, Mittemeijer EJ. Profile analysis for microcrystalline properties by the Fourier and other methods. Aust J Phys. 1988;41(2):173–87. DOI: 10.1071/PH880173

[55] Langford JI. Some applications of pattern fitting to powder diffraction data. Prog Cryst Growth Charact. 1987;14(C):185–211. DOI: 10.1016/0146-3535(87)90018-9

[56] Hepp A, Baerlocher C. Learned peak shape functions for powder diffraction data. Aust J Phys. 1988;41(2):229–36. DOI: 10.1071/PH880229

[57] Balzar D. Voigt-function model in diffraction line-broadening analysis. Int Union Crystallogr Monogr Crystallogr [Internet]. 1999;10:44.

[58] Prevey PS. The Use of Person VII Distribution Functions in X-Ray Diffraction Residual Stress Measurement. Adv X-ray Anal [Internet]. 1985 Mar 6;29:103–11. DOI: 10.1154/S037603080001017X

[59] Balzar D. Profile fitting of x-ray diffraction lines and fourier analysis of broadening. J Appl Crystallogr [Internet]. 1992;25(pt 5):559–70. DOI: 10.1107/S0021889892004084

[60] Howard SA, Preston KD. 8. Profile Fitting of Powder Diffraction Patterns. Mod Powder Diffr. 2018;20:217–76. DOI: 10.1515/9781501509018-011

[61] Langford JI. A rapid method for analysing the breadths of diffraction and spectral lines using the Voigt function. J Appl Crystallogr [Internet]. 1978 Feb 1;11(1):10–4. DOI: 10.1107/S0021889878012601

[62] Le Bail A. Monte Carlo indexing with McMaille. Powder Diffr. 2004;19(3):249–54. DOI: 10.1154/1.1763152

[63] Le Bail A, Louër D. Smoothing and validity of crystallite-size distributions from X-ray line-profile analysis. J Appl Crystallogr [Internet]. 1978;11(1):50–5. DOI: 10.1107/s0021889878012662

[64] Le Bail A. The Rietveld method using an experimental profile convoluted by adjustable analytical function. Acta Crystallogr Sect A Found Crystallogr [Internet]. 1984;40(a1):C369–C369. DOI: 10.1107/s0108767384089200

[65] Cranswick LMD, Le Bail A. Beyond classical Rietveld analysis using Le Bail fitting. Acta Crystallogr Sect A Found Crystallogr [Internet]. 2002 Aug 6;58(s1):c242–c242. DOI: 10.1107/S0108767302094709

[66] Pawley GS. Unit-cell refinement from powder diffraction scans. J Appl Crystallogr [Internet]. 1981;14(6):357–61. DOI: 10.1107/s0021889881009618

[67] Baharie E, Pawley GS. Counting statistics and powder diffraction scan refinements. J Appl Crystallogr [Internet]. 1983;16(4):404–6. DOI: 10.1107/s0021889883010699

[68] Hahn T, Shmueli U, Arthur JCW. International Tables for

Crystallography. Vol. 16, Journal of Applied Crystallography. Reidel Dordrecht; 1983. 284–284 p. DOI: 10.1107/s0021889883010444

[69] Prince E, Wilson AJC. International Tables for Crystallography. J Appl Crystallogr [Internet]. 1983 Apr 1;16(2): 284–284. DOI: 10.1107/ S0021889883010444

[70] Kopsky V, Litvin DB. International Tables for Crystallography. J Appl Crystallogr [Internet]. 1983 Apr 1;16(2): 284–284. DOI: 10.1107/ S0021889883010444

[71] Bryan RF. International Tables for Crystallography. J Appl Crystallogr [Internet]. 1983 Apr 1;16(2):284–284. DOI: 10.1107/S0021889883010444

[72] Cullity BD. Elements of X-ray Diffraction. Addison-Wesley Publishing; 1956.

[73] Von Dreele RB. Quantitative texture analysis by Rietveld refinement. J Appl Crystallogr. 1997;30(4):517–25. DOI: 10.1107/S0021889897005918

[74] Altomare A, Ciriaco F, Cuocci C, Falcicchio A, Fanelli F. Combined powder X-ray diffraction data and quantum chemical calculations in EXPO2014. Powder Diffr. 2017;32(S1): S123–8. DOI: 10.1017/ S088571561700015X

[75] Hill RJ. Expanded Use of the Rietveld Method in Studies of Phase Abundance in Multiphase Mixtures*. Powder Diffr. 1991;6(2):74–7. DOI: 10.1017/S0885715600017036

[76] Bish DL, Chipera SJ. Accuracy in Quantitative X-ray Powder Diffraction Analyses. Adv X-ray Anal. 1994;38:47–57. DOI: 10.1154/s0376030800017638

[77] Parrish W, Huang TC. Accuracy of the Profile Fitting Method for X-Ray Polycrystalline Diffractometry. Vol. 567, National Bureau of Standards, Special Publication. 1979. 95–110 p.

[78] Cline JP. Accuracy in powder diffraction III - Part 1 - Preface. J Res Natl Inst Stand Technol [Internet]. 2004 Jan;109(1):iii. DOI: 10.6028/jres.109.001

[79] Hill RJ, Flack HD. The use of the Durbin–Watson d statistic in Rietveld analysis. J Appl Crystallogr. 1987;20(5): 356–61. DOI: 10.1107/ S0021889887086485

[80] Block S, Hubbard CR. Accuracy in powder diffraction: proceedings of a Symposium on Accuracy in Powder Diffraction held at the National Bureau of Standards, Gaithersburg, Maryland, June, 11–15, 1979 [Internet]. Vol. 567. US Dept. of Commerce, National Bureau of Standards: for sale by the Supt. of~ ... ; 1980.

[81] Newsam JM, Deem MW, Freeman CM. Direct Space Methods of Structure Solution From Powder Diffraction Data. In: Accuracy in powder diffraction II: NIST Special Publication. 1992. p. 80–91.

[82] Woolfson MM. An Introduction to X-ray Crystallography [Internet]. Cambridge University Press; 1997. DOI: 10.1017/CBO9780511622557

[83] Smith F. Industrial Applications of X-Ray Diffraction [Internet]. Smith F, editor. CRC Press; 1999. DOI: 10.1201/ b16940

[84] Stanjek H, Häusler W. Basics of X-ray diffraction. Hyperfine Interact [Internet]. 2004;154(1–4):107–19. DOI: 10.1023/B:HYPE.0000032028.60546.38

[85] Toby BH. R factors in Rietveld analysis: How good is good enough? . Powder Diffr. 2006;21(1):67–70. DOI: 10.1154/1.2179804

[86] Post JE, Bish DL. Rietveld refinement of crystal structures using

powder x-ray diffraction data. In: Bish DL, Post JE, editors. Modern Powder Diffraction [Internet]. Berlin, Boston: De Gruyter; 1989. p. 277–308. DOI: 10.1515/9781501509018-012

[87] Cox DE, Papoular RJ. Structure Refinement with Synchrtron Data: R-Factors, Errors and Significance Tests. Mater Sci Forum [Internet]. 1996 Jul; 228–231(PART 1):233–8. DOI: 10.4028/www.scientific.net/MSF.228-231.233

[88] Berar JF, Lelann P. E.S.D.'s and estimated probable error obtained in rietveld refinements with local correlations. J Appl Crystallogr. 1991;24 (pt 1):1–5. DOI: 10.1107/S0021889890008391

[89] de Keijser T, Mittemeijer EJ, Rozendaal HCF. The determination of crystallite-size and lattice-strain parameters in conjunction with the profile-refinement method for the determination of crystal structures. J Appl Crystallogr [Internet]. 1983 Jun 1; 16(3):309–16. DOI: 10.1107/S0021889883010493

[90] Lutterotti L, Scardi P. Simultaneous structure and size-strain refinement by the rietveld method. J Appl Crystallogr. 1990;23(4):246–52. DOI: 10.1107/S0021889890002382

[91] Caglioti G, Paoletti A, Ricci FP. Choice of collimators for a crystal spectrometer for neutron diffraction. Nucl Instruments. 1958;3(4):223–8. DOI: 10.1016/0369-643X(58)90029-X

[92] Bienenstock A, Ewald PP. Symmetry of Fourier space. Acta Crystallogr. 1962;15(12):1253–61. DOI: 10.1107/s0365110x6200331x

[93] Bindzus N, Iversen BB. Maximum-entropy-method charge densities based on structure-factor extraction with the commonly used Rietveld refinement programs GSAS, FullProf and Jana2006. Acta Crystallogr Sect A Found Crystallogr [Internet]. 2012/10/19. 2012; 68(6):750–62. DOI: 10.1107/S0108767312037269

[94] Bushmarinov IS, Dmitrienko AO, Korlyukov AA, Antipin MY. Rietveld refinement and structure verification using Morse restraints. J Appl Crystallogr [Internet]. 2012;45(6):1187–97. DOI: 10.1107/S0021889812044147

[95] Alcock NW, Pawley GS, Rourke CP, Levine MR. An improvement in the algorithm for absorption correction by the analytical method. Acta Crystallogr Sect A [Internet]. 1972;28(5):440–4. DOI: 10.1107/S0567739472001159

[96] Momma K, Izumi F. VESTA: A three-dimensional visualization system for electronic and structural analysis. J Appl Crystallogr. 2008;41(3):653–8. DOI: 10.1107/S0021889808012016

[97] Glazer AM. The classification of tilted octahedra in perovskites. Acta Crystallogr Sect B Struct Crystallogr Cryst Chem. 1972;28(11):3384–92. DOI: 10.1107/s0567740872007976

[98] Glazer AM. Simple ways of determining perovskite structures. Acta Crystallogr Sect A. 1975;31(6):756–62. DOI: 10.1107/S0567739475001635

[99] Glazer AM, Mabud SA. Powder profile refinement of lead zirconate titanate at several temperatures. II. Pure PbTiO 3 . Acta Crystallogr Sect B Struct Crystallogr Cryst Chem. 1978;34(4): 1065–70. DOI: 10.1107/s0567740878004938

[100] Para TA, Reshi HA, Pillai S, Shelke V. Grain size disposed structural, optical and polarization tuning in ZnO. Appl Phys A [Internet]. 2016 Aug 12;122 (8):730. DOI: 10.1007/s00339-016-0256-8

[101] Para TA, Reshi HA, Shelke V. Synthesis of ZnSnO3 nanostructure by sol gel method. In: AIP Conference

Proceedings [Internet]. 2016. p. 050002. DOI: 10.1063/1.4947656

[102] Roisnel T, Rodríquez-Carvajal J. WinPLOTR: A Windows Tool for Powder Diffraction Pattern Analysis. Mater Sci Forum [Internet]. 2001 Oct; 378–381(I):118–23. DOI: 10.4028/www.scientific.net/MSF.378-381.118

[103] Putz H, Brandenburg K G. Match! – Phase identification from powder diffraction, crystal impact. Kreuzherrenstr. 102, 53227 Bonn, Germany [Internet]. Crystal Impact; 2018. p. 10.

[104] Kourkoumelis N. PowDLL, a reusable .NET component for interconverting powder diffraction data: Recent developments, ICDD Annual Spring Meetings [Internet]. O'Neill L, editor. Vol. 28, Powder Diffraction. 2013. p. 137–48. DOI: 10.1017/S0885715613000390

[105] Degen T, Sadki M, Bron E, König U, Nénert G. The HighScore suite. Powder Diffr [Internet]. 2014 Dec 30;29(S2):S13–8. DOI: 10.1017/S0885715614000840

[106] Rodríguez-Carvajal J. Recent advances in magnetic structure determination by neutron powder diffraction. Phys B Condens Matter [Internet]. 1993 Oct;192(1–2):55–69. DOI: 10.1016/0921-4526(93)90108-I

[107] Doebelin N, Kleeberg R. Profex: A graphical user interface for the Rietveld refinement program BGMN. J Appl Crystallogr [Internet]. 2015/10/27. 2015; 48(5):1573–80. DOI: 10.1107/S1600576715014685